C. A. Lee

OPERATIONAL AMPLIFIERS

OPERATIONAL AMPLIFIERS
The Devices and Their Applications

Charles F. Wojslaw
San Jose City College
San Jose State University

Evangelos A. Moustakas
San Jose State University

John Wiley & Sons
New York   Chichester   Brisbane   Toronto   Singapore

Copyright © 1986, by John Wiley & Sons, Inc.

All rights reserved. Published simultaneously in Canada.

Reproduction or translation of any part of
this work beyond that permitted by Sections
107 and 108 of the 1976 United States Copyright
Act without the permission of the copyright
owner is unlawful. Requests for permission
or further information should be addressed to
the Permissions Department, John Wiley & Sons.

Library of Congress Cataloging in Publication Data:

Wojslaw, Charles F.
  Operational amplifiers.

  Includes index.
  1. Operational amplifiers.   I. Moustakas, Evangelos A.
II. Title.
TK7871.58.O6W65   1985   621.381'735   85-6345
ISBN 0-471-80646-3

Printed in the United States of America

10 9 8 7 6 5 4 3 2 1

To our students,
past, present, and future

Preface

The operational amplifier is the key active device used in low and moderate frequency analog circuits. This heavily used and widely applied device has spawned a discipline that is now fundamental to the understanding of electronics in the analog domain. The primary reasons for the present role of the op amp are twofold: price and performance. The integrated circuit versions of the operational amplifier cost less than most transistors, and their behavior in many applications approaches the ideal. Designing with operational amplifiers, especially when compared to bipolar transistors, is much simpler and the cost of the end products and their development time have been dramatically reduced. The importance of the operational amplifier is also reflected in the electrical engineering and technology programs that now use the device throughout their curricula to illustrate basic circuit and electronics principles.

The philosophy of this book reflects the current status of the op amp by providing in-depth coverages of the device, the circuits in which it is used, and the practical means to minimize the device's nonidealities. The authors, both strong proponents of teaching excellence and practical engineering, have attempted to blend fundamental concepts and principles with reality to provide the reader with the ability to translate the knowledge of operational amplifiers into a working skill.

The book is divided into five parts. Part I provides an overview of the device and the way it is fundamentally used. Part II covers the numerous applications of the device and is organized into eight broad areas. Initially, the operational amplifier is treated as an ideal device to allow the reader to concentrate on its various applications. Fundamental principles are employed to develop an understanding of the operation of the op amp circuits and the input-output relationships. Part III discusses the practical limitations of the nonideal operational amplifier. The effects of these primary and secondary parameters on circuit performance are described, qualitatively and quantitatively, using two basic circuits as a vehicle. The practical design examples in Part IV apply the principles described in the first three parts and translate this information into a functional circuit. The operational amplifier is revisited in Part V where an in-depth analysis of the two most important integrated circuit op amps, the 741 and 101A, is provided.

The book proceeds from the elementary to the more complex. Selected topics from Chapters 1 to 8 are suitable for lower division engineering and electronic technology courses. Chapters 1 to 13 and selected material in Chapters 14 and 15 may be used in an upper division electrical or mechanical engineering course. The entire book could form the basis for a graduate level course or any course whose content embodies circuit theory, electronics, and control theory concepts.

The book is the product of a great number of years of both teaching and practical experience. Most of the material has been classroom tested at San Jose State University and the University of Santa Clara and was selected because of its relevance to industry. Most of the students in the graduate classes at these universities are employed in high-technology industrial firms in Silicon Valley, which makes imperative the down-to-earth treatment of the various topics in the book.

We would like to acknowledge the love, patience, support, and understanding of our families, and our students to whom this book is dedicated. A special thank you also goes to Dr. Shu-Park Chan at the University of Santa Clara.

Charles F. Wojslaw
Evangelos A. Moustakas
San Jose, California

Symbol Convention

With few exceptions, the following symbol convention is used in this book. DC quantities, such as the operational amplifier's bias current I_B, are represented by uppercase symbols with uppercase subscripts. Small-signal, instantaneous, or incremental quantities, such as small-signal transistor collector current i_c, are represented by lowercase symbols with lowercase subscripts. Uppercase symbols with lowercase subscripts are used to represent the RMS values of small-signal quantities, voltages, currents, and impedances that are a function of the complex variable s, and quantities that have dc and ac components.

Contents

PART I
OPERATIONAL AMPLIFIERS   1

Chapter 1
Introduction   3

Chapter 2
The Device   12

PART II
APPLICATIONS   23

Chapter 3
Voltage Amplifiers   25

Chapter 4
Mathematical Operators   46

Chapter 5
Convertors   75

Chapter 6
Oscillators   96

xii Contents

Chapter 7
Special-Purpose Circuits — 123

Chapter 8
Filters — 148

Chapter 9
Active Filters — 176

Chapter 10
Analog ICs and Subsystems — 210

PART III
PRACTICAL LIMITATIONS — 241

Chapter 11
Voltage Gain and Bandwidth — 243

Chapter 12
Input and Output Resistance — 259

Chapter 13
Secondary Parameters — 277

PART IV
PRACTICAL IMPLEMENTATIONS — 309

Chapter 14
Design Examples — 311

PART V
INTEGRATED CIRCUIT AMPLIFIERS — 335

Chapter 15
Circuit Descriptions — 337

Appendix A
Data Sheets: 741, 101A, 108A, Op-07, 111, 140, 555 — 363

Appendix B
Butterworth, Chebyshev, and Bessel Charts — 378

Appendix C
Answers to Selected Problems — 381

REFERENCES — 389

INDEX — 393

OPERATIONAL AMPLIFIERS

PART I

OPERATIONAL AMPLIFIERS

Chapter 1

Introduction

1.1 WHAT IS AN OPERATIONAL AMPLIFIER?

In general, an *operational amplifier* implies a voltage amplifier with the characteristics of high voltage gain, high input resistance, low output resistance, and a moderate bandwidth. An *ideal* operational amplifier is an amplifier with infinite voltage gain, infinite input resistance, zero output resistance, and infinite bandwidth, among other things.

1.2 HISTORY

The most popular and widely used linear IC (integrated circuit) in the world is a low-frequency amplifier, the monolithic operational amplifier or op amp. It is sold by the tens of millions of devices per year. The op amp can be found in diverse applications and derives its name from amplifier circuits that were used to perform mathematical operations in analog computers. It is the most important analog component after the transistor.

The amplifier, as a functional block, was first used in the 1940s and was then implemented with vacuum tubes. Analog computers, with vacuum-tube amplifier circuits, were used to simulate systems whose behavior was mathematically modeled

4 Introduction

using differential equations. Amplifier circuits were configured to perform the addition, subtraction, scaling, and integration operations for the equations involved.

In the early 1960s, small, transistorized, plastic-encapsulated op amp modules were introduced. These plug-in discrete amplifier units were quickly followed by integrated versions and the op amp's rise to dominance in the design of low-frequency analog systems followed.

Presently, the small, lightweight, and inexpensive IC operational amplifier can be found in diverse applications such as filters, oscillators, convertors, regulators, signal processors, and control circuits, just to mention a few of them. The device comes in many sizes, packages, and configurations and is used as a key building block in most analog systems.

1.3 CHARACTERISTICS

The op amp is a two-port device; that is, it has two input terminals that sense a difference of potential and one output terminal that is referenced to the circuit ground. The amplifier, the schematic symbols for which are shown in Fig. 1.1, requires a minimum of five terminals. Two terminals are used for input purposes and are designated as inverting ($-$) and noninverting ($+$) input terminals. Two other terminals are used for the dc supplies while the fifth one is for the output. The amplifier is typically powered by a positive and a negative dc voltage source connected to the pins designated V^+ and V^-. The reference for the output of the amplifier is the common connection of these supplies. Additional pins, usually three, may be provided to allow the user to compensate for internal amplifier deficiencies, such as offset voltage (V_{OS}), as well as to control its frequency response.

As previously mentioned, the op amp is a "high-gain" voltage amplifier with "high" input resistance, "low" output resistance, and "moderate" bandwidth. Figure 1.2 lists the numbers associated with these characteristics for several generations of monolithic amplifiers.

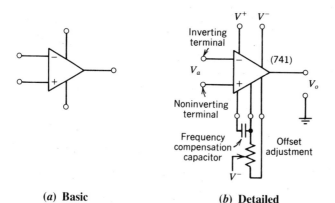

(a) Basic   (b) Detailed

Figure 1.1 IC operational amplifier schematic symbols.

| | Amplifier | | | | State of |
Characteristic	709A	741A	101A	108A	the art
Voltage gain	50,000	200,000	200,000	300,000	10^7
Input resistance	700 kΩ	6 MΩ	6 MΩ	70 MΩ	10^{12} Ω
Output resistance	150 Ω	75 Ω	75 Ω	75 Ω	1 Ω
Unity-gain bandwidth	5 MHz	1 MHz	1 MHz	1 MHz	1 GHz
		⸺ Monolithic ⸺			

Figure 1.2 Typical specifications of several generations of operational amplifiers.

There are an extremely large number of different types of op amps. The monolithic devices listed are high-volume, low-cost, general-purpose op amps. The majority, however, have been designed to optimize the performance of one or more of the amplifier's characteristics, sometimes at the expense of other characteristics or cost. Microelectronic (hybrid or monolithic) amplifiers are available whose voltage gain is 10^7, input resistance is 10^{12} Ω, output resistance is 1 Ω, and unity-gain bandwidth is 1 GHz. These high performance specifications describe a group of devices, not any particular one.

An equivalent circuit that reflects the characteristics of the op amp, while in the linear region of its operation, is shown in Fig. 1.3. In the equivalent circuit, the source AV_a is an ideal, voltage-dependent voltage source, where A is the amplifier's (open-loop) voltage gain. The output of this source depends on the value of V_a, where V_a is the differential input voltage of the op amp. The equivalent differential, small-signal input resistance is R_i, and the resistance in series with the output is the equivalent output resistance R_o.

If the input voltage to the amplifier is such that the inverting input ($-$) is positive with respect to the noninverting input ($+$), the output voltage will be negative with respect to ground. If the noninverting ($+$) input is positive with respect to the inverting input ($-$), the output voltage will be positive with respect to ground; thus, their designation as inverting and noninverting input terminals. In the op amp's linear region, the magnitude of the output voltage is A times the difference of potential between its

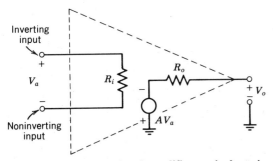

Figure 1.3 IC operational amplifier equivalent circuit.

6 Introduction

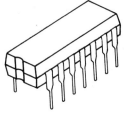
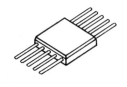

(a) TO-5   (b) Dual-in-line   (c) Flat pack

Figure 1.4 Common IC op amp packages.

input terminals. The amplifier output can go positive or negative with respect to ground because the device is typically powered with a pair of symmetric (V^+, V^-) dc supplies.

The op amp comes in a variety of configurations, sizes, shapes, and ratings. The device can be purchased as a single amplifier, a dual, or a quad. The cost for most types is relatively low and does not economically hinder the use of several amplifiers in multiple-amplifier applications. There are a large number of packages used for the op amps, but the three most popular packages (Fig. 1.4) are the 8 pin TO-5, the 8 and 14 pin DIP (dual-in-line package), and the 10 pin flat pack. The specific package used is, generally speaking, a function of whether the application is industrial, commercial, military, or aerospace. There are an extremely large number of different op amps available. Some of them are general purpose, but most of them are specialized amplifiers, designed to optimize one or more of their characteristics.

1.4 INPUT-OUTPUT RELATIONSHIP

The input-output relationship of an op amp may be expressed in two ways: graphically, by a transfer characteristic showing the variation of the output voltage as a function of its differential input voltage and, mathematically, through an equivalent circuit and its associated equations.

The transfer characteristic of Fig. 1.5 is a plot of V_o (to ground) versus V_a. The polarity of the output, as mentioned in Section 1.3, is a function of the polarity of the voltage applied between the amplifier's input terminals. The slope of the linear characteristic in the active region of its operation is $-A$ and illustrates the linear relationship of the input and output. This linear relationship

$$V_o = -AV_a \qquad (1.1)$$

is maintained until the amplifier output becomes voltage-limited. The limit in the positive direction is called positive saturation and in the negative direction is, correspondingly, called negative saturation. These limits are a function of the supply voltages and the operating characteristics of the transistors in the amplifier's output stage.

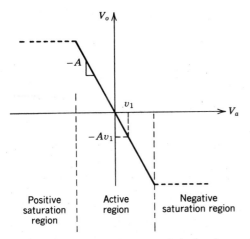

Figure 1.5 Transfer characteristic for the op amp.

1.5 HOW IS THE AMPLIFIER USED?

The op amp has a high voltage gain, where "high," typically, translates to approximately 200,000 for operation near dc. The maximum output voltage from the amplifier is typically ±15 V. Dividing the maximum output voltage by the gain, we get the typical maximum input voltage required, which is ±150 μV. Signals at this level are extremely difficult to process because of noise and interference. In addition, gains of 200,000 are seldom required. Why, then, is the op amp so popular and how is it really used?

1.5.1 Stability Through Feedback

Op amps are used primarily in circuits that contain feedback, usually negative feedback. Feedback, generally speaking, implies a situation in which output quantities in a system are allowed to affect the input quantities that caused them in the first place. Negative feedback is applicable when the output signal is purposely allowed to reduce the effect of the input signal. This, of course, results in the reduction of the net (closed-loop) gain of the circuit. Negative feedback is typically introduced in an actual op amp circuit by connecting a component, usually a resistor, from the amplifier output back to its inverting (−) input terminal.

What are the tradeoffs in amplifier feedback circuits? We lose the maximum obtainable gain that would be the open-loop gain of the amplifier itself. In return, we get stability, primarily gain stability. Most applications do not require high gain, but they do require the gain to be precise and stable. On an op amp data sheet, the amplifier gain is guaranteed to be a certain minimum amount, but the maximum is not specified, and for a given amplifier type, it will vary at least by 10 to 20%. In addition to this relatively large variation, the gain of any specific amplifier varies with temperature, age, and operating conditions. In fact, temperature is a crucial factor in all integrated circuits. Bipolar ICs use bipolar transistors whose base-to-emitter voltage is temperature dependent (-2.2 mV/°C) and whose current gain (beta) is also temperature dependent. These two factors directly contribute to amplifier gain variations. A negative feedback configuration reduces these effects by several orders of magnitude. Besides gain stability, negative feedback tends to improve linearity, reduce distortion, decrease output resistance and, in some configurations, to increase input resistance. The price paid for these advantages is, in most instances, indeed modest. Another advantage of amplifiers employing negative feedback is ease of design. In negative-feedback circuits, the control of the performance characteristics is shifted to external (relative to the amplifier) components whose price, size, shape, accuracy, and stability are easier to control. Today, it is a much easier task to design (and make work!) circuits using operational amplifiers than those employing discrete transistors.

Most voltage-amplification circuit requirements vary from about 0.1 to 100. The number 200,000 associated with the operational amplifier is referred to as the open-loop gain. If we configure the amplifier in a circuit employing negative feedback so that the circuit's gain is 20, then 20 is called the closed-loop gain. Similarly, we can talk about open- and closed-loop bandwidths, input resistance, and output resistance.

The application of negative feedback to an operational amplifier yields a circuit with the closed-loop characteristics determined primarily by the values of discrete feedback components whose accuracy and stability are much easier to control.

1.5.2 The Feedback Concept

A control system is a system that regulates an output variable with the objective of producing a given relationship between it and the input variable or of maintaining the output at a fixed value. In a feedback control system, at least part of the information used to change the output variable is derived from measurements performed on the output variable itself.

A feedback control system, in block diagram form, is illustrated in Fig. 1.6. The input variable is V_s and the output variable is V_o. The system contains an amplifier with an open-loop voltage gain A and a feedback network whose transfer function is defined as F, oftentimes referred to as the feedback factor. The feedback network is usually implemented with discrete, passive components. The circles are summation

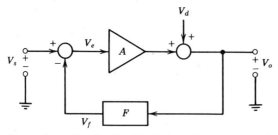

Figure 1.6 Negative feedback control system model.

points. Negative feedback systems produce many positive characteristics and to illustrate the stability characteristic of the system, a voltage V_d is introduced in series with the amplifier's output. V_d can represent an externally induced voltage or a change in the output voltage due to changing circuit (internal) characteristics. We seek to find an expression relating the input and output variables, V_d, and circuit constants.

V_e, the "error" voltage, is a function of V_s and the feedback voltage V_f. It is the difference between the two,

$$V_e = V_s - V_f \tag{1.2}$$

The output voltage V_o is a function of V_e, the gain of the amplifier A, and the disturbance voltage V_d,

$$V_o = AV_e + V_d \tag{1.3}$$

The signal path for the feedback network is from the amplifier's output back to the input summation point; the input signal of the feedback network is V_o, and its output is V_f. These are related by

$$V_f = FV_o \tag{1.4}$$

Eliminating V_e and V_f between (1.1), (1.3), and (1.4) and then solving for V_o, we obtain

$$V_o = \frac{A}{1 + AF} V_s + \frac{1}{1 + AF} V_d \tag{1.5}$$

In the above expression, the product AF is called the loop gain, or loop transmission, and is typically a large number; A, at low frequencies, is typically in the hundreds of thousands and F varies from 0.01 to 1. Thus in general, AF has a magnitude that is much larger than unity.

Now, V_o is equal to some constant $A/(1 + AF)$ times V_s, and summed with this term is another constant $1/(1 + AF)$ times V_d. The disturbance voltage V_d will be attenuated, at the output, by $1/(1 + AF)$ as a result of the negative feedback contributed by F. Hence, the overall input-output characteristic will only change by an extremely small percentage ($<<1\%$) of the change(s) occurring within the system itself. The specific amount will vary depending on the loop gain and where the change occurs.

Neglecting V_d, we obtain the output voltage of the configuration

$$V_o = \frac{A}{1 + AF} V_s \tag{1.6}$$

If we assume a large magnitude for the loop gain AF,

$$1 + AF \cong AF$$

10 Introduction

and

$$V_0 \cong \frac{A}{AF} V_s$$

or

$$V_o \cong \frac{1}{F} V_s$$

F is a dimensionless number and is usually a function of discrete components. The ratio

$$\frac{V_o}{V_s} \cong \frac{1}{F} \tag{1.7}$$

defines the overall closed-loop gain of the configuration of Fig. 1.6.

The key to making a circuit's gain be, primarily, a function of external components is to have a high loop gain, which in turn, depends on the amplifier having a high open-loop gain.

In summary, using amplifiers in negative feedback configurations desensitizes the resulting circuit from various changes it may be subjected to and, simultaneously, makes the design process easier to deal with.

Negative feedback circuits also possess other advantages, but we will identify them in the appropriate contexts. Operational amplifier applications abound and most of them use amplifiers in feedback configurations. Regulators, oscillators, amplifier circuits, and filters employ op amps in the presence of feedback. The feedback circuit theory, relative to these applications, will be discussed under the individual topics.

1.6 APPLICATIONS

There was a time when it was very difficult to find published information on op amps and their applications. Today, countless books are available that try to cover a myriad of op amp applications. The op amp is so widely and diversely used that any book can only summarize its usage. The focus of this book is to extend and update the op amp's applications and to illustrate how to accommodate the amplifier's limitations.

As a general statement, we can say that the operational amplifier is used in process and control circuits and systems. More specifically, we find the op amp designed into test and measurement equipment, biomedical instrumentation, control systems of various sorts and types, analog computers, commercial products, and communication equipment. In this text book, the applications of the op amp have been broken down into the functions of the circuits in which the amplifier is used. These are (1) voltage amplifiers, (2) mathematical operators, (3) convertor circuits, (4) signal sources or oscillators, (5) special-purpose circuits, (6) active filters, and (7) analog ICs and subsystems. The circuits presented in the chapters that follow are limited in number and serve as a basis for illustrating circuit principles and demonstrating the use of the op amp in a specific applications area. They are by no means all-inclusive, nor are they unique.

We will look at a representative number of circuits in each of the above areas, once we have taken an initial, brief look at the operational amplifier itself.

Chapter 2

The Device

2.1 PROLOGUE

Before we get into the details of the various circuits in which the op amp is used, let us briefly look at the device itself. The purpose of our simplified coverage in this chapter is to provide the reader with a basic understanding of the characteristics of the amplifier and how they are achieved, and to introduce some of the amplifier's limitations and their origin. A detailed description of the design of an integrated operational amplifier is provided in Chapter 15.

2.2 THE EVOLUTION OF COMMERCIAL DEVICES

The first generally accepted IC op amp, the μA702, was introduced by Fairchild in 1963. For a number of application-related reasons, the 702 was not universally used. However, it did lay the groundwork for future IC amplifier designs and is historically important because it initiated the IC design philosophy of using matched components, a precedent that still prevails today. The μA709, introduced in 1965 by Fairchild, improved the performance of the 702 by increasing the voltage gain, input and output voltage range, and output current, reducing the input current, and using symmetrical 15 V supplies. The 709 was universally accepted and applied and, even though newer designs have eclipsed many of the 709's individual performance characteristics, it is still used today in large quantities.

Just as the 709 was an improved version of the 702, National Semiconductor's LM101A and Fairchild's μA741, introduced in 1968, were improved versions of the 709. These devices reduced the complexity of the frequency-compensation network and the power consumption, improved the short-circuit protection circuitry, and increased the differential-input voltage and supply-voltage ranges. These second-generation, general-purpose devices today sell in the millions of devices per year, are universally used, and represent an optimum compromise between performance, flexibility, and economy.

The 108A is a precision IC op amp and is the first device to use "super-β" transistors in the input stage. It has excellent input characteristics, a low power-consumption rating, and will operate over a wide power-supply range. It is a frequent choice in precision, high-impedance circuits used in battery-operated applications. Most of what will be said in this book is directed at the characteristics, performance, and applications of the 709, 101A, 741, and 108A op amps, with an emphasis on the 741 and 101A.

Usually three versions or grades of each of the above amplifiers are sold by IC manufacturers; military or aerospace (-55 to 125 °C), industrial (-25 to 85 °C), and commercial (0 to 70 °C). The performance specifications, and the range of temperatures over which the specifications are guaranteed, are tightest for the aerospace-grade devices and loosest for the commercial-grade, with industrial-grade specifications in the middle of the range. Of course, the price of the device varies accordingly. The commercial versions of the aerospace-grade devices 709A, 101A, 741A, and 108A are the 709C, 301A, 741C, and the 308A, respectively.

The block diagrams for these four amplifiers are similar. The detailed circuits for each amplifier vary, and newer IC circuit techniques are employed in the more recently designed devices. However, the functional blocks and design philosophy are relatively consistent from one device to another. The 741 and 101A are not only two of the more popular general-purpose devices, but their designs are also relatively close. To provide us with a background of the operational amplifier itself, we will describe the characteristics and performance of the general-purpose bipolar op amp by looking at a simplified but functional block diagram of the 741 and 101A operational amplifiers.

2.3 BLOCK DIAGRAM

The general purpose bipolar op amp contains three basic circuits: the differential amplifier, the common-emitter amplifier, and the emitter-follower. A basic three stage integrated-circuit op amp and its simplified discrete-circuit equivalent version are shown in Figs. 2.1a and 2.1b, respectively.

An operational amplifier has high voltage gain, high input resistance, low output resistance, and moderate bandwidth. The high voltage gain is developed by the first two stages of the device; the high input resistance is a characteristic of the first stage; and the low output resistance is a characteristic of the last stage. The op amp's bandwidth

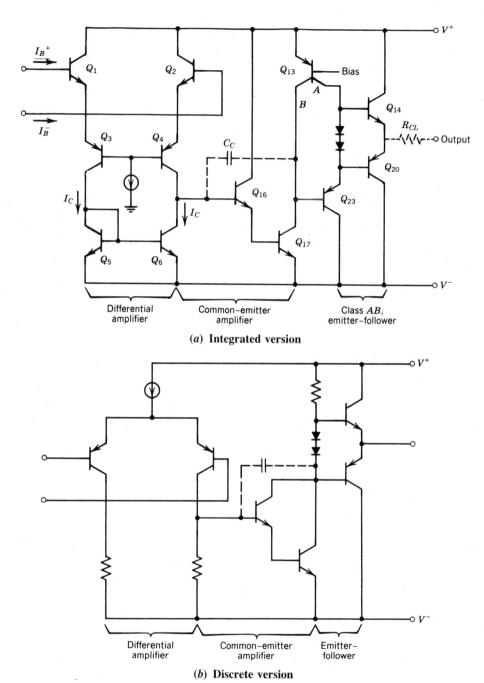

(a) Integrated version

(b) Discrete version

Figure 2.1 Simplified, conceptual schematics of the 741/101A type amplifier.

is limited by the active devices in the first two stages and is controlled in the second stage.

2.4 DIFFERENTIAL AMPLIFIER CIRCUIT

The discrete version of the differential amplifier circuit shown in Fig. 2.2a, is characterized by a pair of matched transistors with a common emitter. It is a circuit whose input V_{id} is a differential voltage applied to the transistors' bases and whose output V_{od} is a differential voltage taken from the collectors. Typically, only one of the collector outputs is used to drive the next circuit and the circuit is referred to as single ended.

The voltage gain for this circuit is the product of the equivalent collector resistance R_C and transistor transconductance. In order to achieve a high voltage gain, R_C must be very large, a prohibitive situation in the integrated circuit because of the resulting large die area requirement. In the IC version in Fig. 2.2b, the current sources, implemented by Q_5 and Q_6, are used as active loads to increase the effective collector resistance. This minimizes the die area and maximizes the voltage gain. The effective collector resistance is the parallel combination of the output resistance of the first stage and the input resistance of the second. For the first stage of the IC op amp, the voltage gain is

$$|A_{v1}| \cong 500 \qquad (2.1)$$

A_{v1} is the gain from the differential input to its single-ended output.

The operational amplifier's input (bias) currents I_B are the base currents of the transistors in the differential amplifier stage of Fig. 2.1b. To make these base currents

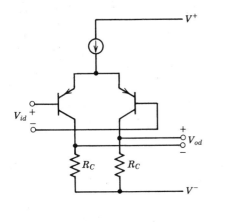
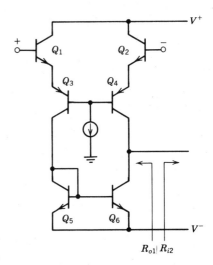

(a) Discrete   (b) Integrated

Figure 2.2 Differential amplifier.

16 The Device

as small as possible, the stage is designed for a low value of I_C (10 μA) and a high value of β, near 250. The end result is a bias current in the 40 nA region

$$I_B = \frac{I_C}{\beta} = \frac{10 \, \mu A}{250} = 40 \, nA \tag{2.2}$$

Newer IC designs, like the 108A, use super-β transistors that reduce the bias current to the 4 nA level. Operational amplifiers, which use FET transistors in the input stage, have bias current ratings in the picoamp range.

The differential input resistance of the op amp is

$$R_i \cong 1.5 \, M\Omega \tag{2.3}$$

The input resistance of the amplifier is the sum of the input resistances of transistors Q_1 through Q_4. It is four times the equivalent input resistance of one of the transistors because, as we go from one differential input to the other, we see the input resistances of the four transistors effectively in series.

Ideally, if the differential input voltage is zero, the differential output voltage is zero. However, because of the slight mismatch in the base-to-emitter voltages of the differential stage transistors, the output voltage will not necessarily be zero. To make the output go to zero volts, we apply an "offset" voltage to the amplifier's input terminals. This input offset voltage is of the order of a few millivolts and can be compensated for by using an external nulling circuit. The input offset voltage V_{OS} is statistically determined and is given as

$$V_{OS} = 2.6 \, mV \tag{2.4}$$

Op amps, whose offset voltage is factory trimmed, have V_{OS} ratings in the microvolt range.

2.5 COMMON-EMITTER CIRCUIT

The common-emitter circuit, Fig. 2.3a or its equivalent Fig. 2.3b, is driven from one of the collectors of the differential amplifier. This technique converts the differential input of the op amp to a single-ended signal required at the output. The gain for this stage is also given by the product of the equivalent collector resistance and the stage transconductance. For the second stage of the IC,

$$|A_{v2}| \cong 500 \tag{2.5}$$

Since the small-signal voltage gain of the last stage is near one, the overall voltage gain of the operational amplifier is

$$A = A_v \cong A_{v1} A_{v2} = 250,000 \tag{2.6}$$

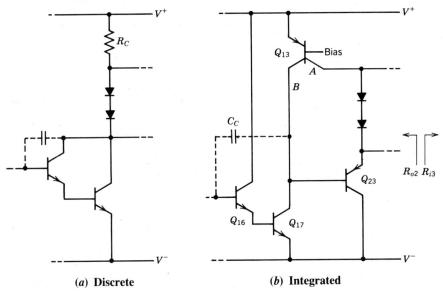

(a) Discrete (b) Integrated

Figure 2.3 Common-emitter amplifier.

The frequency response of the op amp is controlled by connecting a capacitor C_c between the collector and base of the Darlington pair, Q_{16}–Q_{17} in Fig. 2.3b, in the second stage. This capacitor is internal and fixed (about 30 pF) for the 741 and is externally connected for the 101A. The cutoff frequency associated with this capacitor is about 5 Hz.

For small-signal conditions, the cutoff frequency f_c is established by the equivalent of a simple, first-order RC low-pass filter whose f_c is given by

$$f_c \cong 5 \text{ Hz} \tag{2.7}$$

Although the cutoff frequency for the op amp is very low, the excess gain at dc and above f_c makes the device useful at frequencies up to 1 MHz. The gain-frequency performance of the op amp is not measured in terms of the cutoff frequency but in terms of a figure of merit called the gain-bandwidth product. This figure of merit, GB, is equal to the frequency where the open-loop voltage gain of the op amp is unity; it is typically in the order of 1 MHz.

The compensation capacitor not only affects the small-signal performance of the amplifier through the gain-bandwidth product, but also limits the large-signal frequency response through a parameter called the slew rate (S_R). Slew rate describes the maximum rate at which the output voltage can change. Under large-signal conditions, the second stage acts like an integrator. The maximum rate of change of the output voltage is a function of the maximum available current from the input stage and the value of the

18 The Device

compensation capacitor. The typical value for S_R for the 741 op amp is

$$S_R = \left.\frac{dv_o}{dt}\right|_{max} \cong 0.6 \text{ V}/\mu\text{sec} \tag{2.8}$$

The parameters GB and S_R are discussed in detail in Chapter 11 and in Section 13.5.1, respectively.

2.6 EMITTER-FOLLOWER CIRCUIT

The emitter-follower, Fig. 2.4, has the characteristics of low output resistance, high input resistance, wide bandwidth, a voltage gain near unity, and moderate power and current gain. This class AB, push-pull stage provides the low output resistance characteristic of the operational amplifier.

The small-signal, output resistance R_o of the operational amplifier consists of two parts; the current-limiting resistance and the output resistance of the emitter-follower circuit.

The output of the operational amplifier is current-limited at about 20 to 30 mA. A resistor R_{CL} (26 Ω), in series with the output, monitors the output current and triggers the current-limiting circuitry (not shown) to protect the output stage when excessive demands are made by the load.

The output resistance is somewhat load and frequency dependent, and

$$R_o \cong 75 \text{ }\Omega \tag{2.9}$$

is used as a nominal number for the 741 operational amplifier.

A feature of the push-pull output stage is that it can provide both polarities of output voltage and thus sink or source output current.

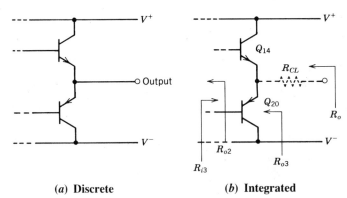

(a) Discrete   (b) Integrated

Figure 2.4 Emitter follower.

2.7 AMPLIFIER LIMITATIONS

The limitations of an amplifier fall into two categories:

1. The finite and nonzero values associated with the *primary characteristics* of gain and bandwidth, and input and output resistance.
2. *Secondary* or *data sheet parameters*.

Secondary parameters are nonidealities that are specified by the manufacturer and are a result of the circuit design and the manufacturing process. In many cases, the errors incurred because of the data sheet parameters eclipse those incurred by the primary characteristics. The key data sheet parameters are input offset voltage V_{OS}, input current I_B, and slew rate S_R.

2.8 IDEAL OPERATIONAL AMPLIFIER

An ideal operational amplifier is an amplifier with infinite voltage gain, infinite bandwidth, infinite input resistance, and zero output resistance, among other things.

The use of an amplifier devoid of all its nonidealities has a practical value in analyzing and synthesizing the circuits in which the device is used. The numbers associated with the recently designed amplifiers are obviously not ideal, but they come very close to the ideal in many respects.

The next eight chapters deal with the applications of the operational amplifier. To focus on the applications areas, we assume ideal op amps. Later, in the chapters on the practical limitations of the amplifier, we will examine how, and to what extent, the finite and nonzero values of the characteristics affect the performance of the circuit in which it is used.

If the operational amplifier is idealized, two constraints may be applied to the input. These fundamental input constraints or axioms will be formally stated and then justified.

2.8.1 Input Constraints

Voltage Constraint *In an ideal operational amplifier used in an effectively negative feedback configuration, the potential difference between its inverting and noninverting input terminals is zero.*

This statement can be verified using the following argument. From (1.1) and Fig. 1.3,

$$V_a = V_i - V_{ni} = \frac{V_o}{-A}$$

where V_i and V_{ni} are the voltages at the inverting and noninverting inputs, respectively.

20 The Device

As A approaches infinity, that is,

$$A \longrightarrow \infty$$

then V_a will approach zero,

$$V_a \longrightarrow 0$$

As the input voltage approaches zero, the voltages at the inverting and noninverting input terminals will approach each other, that is,

$$V_i \longrightarrow V_{ni}$$

and, as a limit in the ideal case,

$$V_a = 0$$

and

$$V_i = V_{ni}$$

This verifies the voltage constraint.

Current Constraint *The current drawn by either of the input terminals of an ideal operational amplifier is zero.*

The voltage and the current constraints are simple to remember and can immensely simplify the amount of work necessary in analyzing or designing circuits that use operational amplifiers. The voltage constraint is a direct consequence of the small and limited input voltage range that allows the amplifier to remain in its active region; the higher the magnitude of the voltage gain, the smaller this range becomes. If the circuit-level signals are in the millivolt region or greater, then indeed, when compared to the microvolt region of the amplifier's input voltage, we can say that, for all practical purposes, the potential difference between the input terminals of the real amplifier is *zero*.

If the circuit signal currents are in the microamp region or greater, then again, when compared to the nanoamp region of the amplifier's input currents, we can say that, for all practical purposes, the amplifier's input currents are *zero*.

2.9 SUMMARY

The high voltage gain and input resistance, and low output resistance, of the integrated operational amplifier are accomplished using three basic stages. These are (1) the differential amplifier, (2) the common-emitter circuit, and (3) the emitter-follower.

The voltage gain is developed in the first two stages, and its bandwidth is controlled in the second. The high input resistance and low output resistance are characteristics of the first and third stages, respectively.

The key parameters of an "ideal" operational amplifier are either infinite or zero. As a result of idealizing these quantities, we can employ two constraints related to the input terminals: (1) the differential input voltage is zero, and (2) the input currents are zero.

PART II

APPLICATIONS

Chapter 3

Voltage Amplifiers

3.1 PROLOGUE

A *voltage amplifier* is a circuit that *creates,* at its output, *a scaled replica of its input voltage.* Typically, the magnitude of the output signal is greater than that of the input signal and, thus, we say that the circuit amplifies or has gain. The gain is constant over a certain range of frequencies and is possible only because the circuit contains active devices within it. Basic amplifier circuits, employing operational amplifiers as active devices, not only are used to perform the analog function of amplification, but they are also used as building blocks in complex analog systems.

3.2 AMPLIFIER CHARACTERISTICS

Amplifiers are circuits, IC or discrete, that create at their output replicas of their input signal. In the case of voltage amplifiers, they typically increase the magnitude of the input voltage.

The most important *characteristics* of a voltage amplifier are:

(A) Voltage gain.
(B) Upper cutoff frequency.
(C) Lower cutoff frequency.
(D) Input impedance.
(E) Output impedance.

26 Voltage Amplifiers

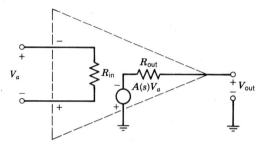

Figure 3.1 Equivalent circuit of an amplifier.

The *voltage gain* is essentially constant over a certain frequency range (passband). The value of the gain in the passband is usually established by circuit constants and is only possible because of the active device(s) in the circuit. For our case, the active device is the operational amplifier.

The *upper frequency,* above which the gain begins to decrease appreciably, is called the upper cutoff frequency. The rate of decrease of the amplitude (typically, in decibels per octave or decade of frequency) beyond the cutoff frequency is a function of the number of high-frequency poles in the amplifier's amplitude response.

The *lower frequency* below which the gain begins to decrease appreciably, is called the lower cutoff frequency. The cutoff frequency (either upper or lower) is normally defined as the -3 dB point or the frequency where the gain is down to 0.707 of its mid-frequency value. The difference between the upper and lower cutoff frequency is defined as the amplifier's bandwidth.

The *input impedance* of an amplifier is defined as the equivalent impedance seen looking into its input terminals. It is the impedance that the driving source or previous circuit sees under steady-state ac conditions. The *output impedance* is defined as the equivalent impedance seen looking into the output terminals of an amplifier circuit. These two impedance characteristics are important when interfacing one circuit with another because, in the cascaded circuits that process voltage signals, it is desirable to have a low output impedance drive a high input impedance. The voltage gain of a system is maximized when circuits, which have a high input impedance and low output impedance, are cascaded. The terms input and output resistance are often interchanged with input and output impedance. For most cases, the input and output impedance of an amplifier are primarily resistive in the amplifier's passband. The circuit of Fig. 3.1 reflects the voltage amplifier's key characteristics, and its similarity to the operational amplifier's equivalent circuit should be noted. This is because the operational amplifier is a voltage amplifier with some very special characteristics.

3.3 DC AMPLIFIERS

There are two fundamental op amp configurations: the *noninverting circuit* and the *inverting circuit*. These two basic circuits are frequently used and serve as a basis for more complex configurations.

3.3.1 Noninverting Amplifier

The circuit of Fig. 3.2a represents a noninverting amplifier. The operational amplifier is a 741 and the pin numbers are identified for a TO5 package. The analysis of the circuit is done using a step-by-step procedure.

Step 1: As a result of the voltage constraint and Kirchhoff's Voltage Law (KVL),

$$V_{R1} = V_s$$

Step 2: Because of the current constraint and voltage division,

$$V_{R1} = V_{out} \frac{R_1}{R_1 + R_2}$$

Step 3: If we equate the expressions from steps 1 and 2 and solve for the closed-loop gain G, we obtain

$$G = \frac{V_{out}}{V_s} = 1 + \frac{R_2}{R_1} \qquad (3.1)$$

The minimum gain for the noninverting amplifier of Fig. 3.2a is unity. This gain is obtained when $R_2 = 0\ \Omega$, independently of R_1. The magnitude of the ideal closed-loop gain depends only on the ratio of R_2 and R_1 or, in the general case, the ratio of the impedances $Z_1(s)$ and $Z_2(s)$ of the corresponding elements. The actual magnitudes of R_1 and R_2 can be important since they set, in conjunction with V_{out} and the load resistor R_L, the current I_o that the operational amplifier will have to source or sink. This current is

$$I_o = \frac{V_{out}}{R_1 + R_2} + \frac{V_{out}}{R_L}$$

Typically, the output current I_o should be less than 20 mA.

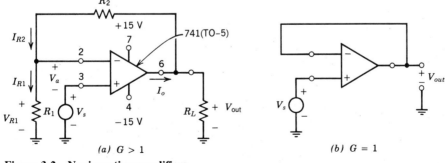

Figure 3.2 Noninverting amplifiers.

28 Voltage Amplifiers

Large values of R_1 and R_2 can also present problems; they may cause excessive noise at the output, or inaccuracy in the closed-loop gain if the current through R_1 and R_2 approaches the order of magnitude of the input (bias) currents of the operational amplifier. Resistors in the range from 1 to 100 kΩ are recommended.

In summary, the gain of the noninverting circuit is established by R_1 and R_2. The ideal bandwidth is infinite, because there are no external capacitors to limit it and the op amp is ideal. The circuit input impedance is infinite because the source sees the differential input impedance of the ideal amplifier, which is infinite. The output impedance of this circuit is zero because the ideal amplifier's output impedance is zero.

Voltage Follower A special case for the noninverting configuration of Fig. 3.2a occurs when R_1 is infinite and R_2 is zero. The resultant circuit, shown in Fig. 3.2b, is called a voltage follower or buffer. The voltage follower is a noninverting amplifier with unity gain, that is,

$$G = \frac{V_{out}}{V_s} = 1 \tag{3.2}$$

It is frequently used to "buffer" or separate a high output resistance circuit from the circuit coupled to it. The follower, in addition to its unity gain, has an extremely high input resistance, a very low output resistance, and a relatively high bandwidth.

Example

Assuming in the circuit of Fig. 3.2a, $R_1 = 2$ kΩ, $R_2 = 10$ kΩ, and $V_s = +1.5$ V, calculate the ideal circuit gain, V_{out}, V_{R1}, V_{R2}, I_{R1}, and I_{R2}.

Solution

$$G = \frac{V_{out}}{V_s} = 1 + \frac{R_2}{R_1} = 6$$

If $V_s = +1.5$ V,

$$V_{out} = GV_s = 9.0 \text{ V}$$

Since $V_a = 0$,

$$V_{R1} = V_s = 1.5 \text{ V}$$

The current through R_1 will be

$$I_{R1} = \frac{V_{R1}}{R_1} = 0.75 \text{ mA}$$

Because of the current constraint,

$$I_{R1} = I_{R2} = 0.75 \text{ mA}$$

The voltage drop across R_2 will be

$$V_{R2} = I_{R2} R_2 = +7.5 \text{ V}$$

Applying KVL to the R_1–R_2 circuit, we obtain

$$V_{out} = V_{R1} + V_{R2} = +9.0 \text{ V}$$

which confirms the value of V_{out} found by using (3.2) ∎

3.3.2 Inverting Amplifier

Figure 3.3 represents possibly the most common operational amplifier configuration. The active device is a 101A operational amplifier and the pin numbers are identified for a TO5 package. This particular configuration provides a 180° phase difference between its input and output voltages. The closed-loop gain G of the inverting amplifier may be found as follows:

Step 1: As a result of the voltage constraint, KVL, and Ohm's law.

$$I_{R1} = \frac{V_s}{R_1}$$

and

$$I_{R2} = -\frac{V_{out}}{R_2}$$

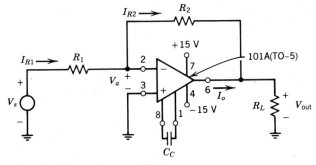

Figure 3.3 Inverting amplifier.

Step 2: Because of the current constraint,

$$I_{R1} = I_{R2}$$

Step 3: If we equate the expressions for I_{R1} and I_{R2} in step 1 and then solve for G, we get

$$G = \frac{V_{out}}{V_s} = -\frac{R_2}{R_1} \tag{3.3}$$

The ideal closed-loop gain for the inverting amplifier and the noninverting amplifier is a function of the ratio of R_2 and R_1. The amplifier output current I_o for the noninverting circuit is

$$I_o = -I_{R2} + \frac{V_{out}}{R_L} = V_{out}\left(\frac{1}{R_2} + \frac{1}{R_L}\right) \tag{3.4}$$

Because the inverting terminal of the operational amplifier is at zero volts or virtual ground, R_2 and R_L are effectively in parallel.

In the general case, resistors R_1 and R_2 are replaced by impedances $Z_1(s)$ and $Z_2(s)$, respectively. The voltage-gain transfer function, in this case, is

$$G(s) = \frac{V_{out}(s)}{V_s(s)} = \frac{-Z_2(s)}{Z_1(s)} \tag{3.5}$$

This expression applies to the modified inverting amplifier of Fig. 3.4 where

$$Z_1(s) = R_1$$

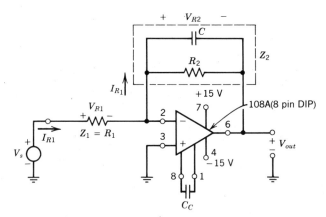

Figure 3.4 Modified inverting amplifier.

and

$$Z_2(s) = \frac{1}{1/R_2 + sC} = \frac{1}{C(s + 1/R_2C)}$$

The voltage-gain transfer function will then be

$$G(s) = \frac{V_{out}}{V_s} = \frac{(-R_2/R_1)(1/R_2C)}{s + 1/R_2C} \tag{3.6}$$

This transfer function has the form of a low-pass amplifier (or filter),

$$G(s) = \frac{G_o\omega_c}{s + \omega_c}$$

where

$$\omega_c = \frac{1}{R_2C}$$

and

$$G_o = -\frac{R_2}{R_1}$$

The frequency ω_c is the upper cutoff frequency of the amplifier, and G_o is the passband or low-frequency voltage gain. The linearized magnitude Bode plot is shown in Fig. 3.5; it represents a single-pole response.

For frequencies much lower than ω_c, the capacitor C may be treated as an open circuit and the circuit gain is $-R_2/R_1$. At frequencies much higher than the cutoff frequency, the magnitude of the closed-loop gain becomes inversely proportional to the frequency. This corresponds to a constant rolloff of -6 dB/octave (-20 dB/decade).

The technique of adding a small capacitor in shunt with R_2 provides the circuit designer with the ability to control the frequency response of the circuit without de-

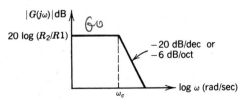

Figure 3.5 Linearized amplitude response of a modified inverting amplifier.

pending on the frequency characteristics of the operational amplifier. Of course, the bandwidth of the operational amplifier must be at least 20 to 100 times greater than the bandwidth of the circuit.

The input impedance of the inverting amplifier circuit of Fig. 3.3 is R_1. In general, inverting circuits have a lower input impedance than their noninverting counterparts. The output impedance of both configurations is ideally zero.

Example

For the circuit in Fig. 3.4, $R_1 = 2\ \text{k}\Omega$, $R_2 = 10\ \text{k}\Omega$, $C = 0.0022\ \mu\text{F}$, and $V_s = +1.5$ V. The active device is a 108A operational amplifier, and the pin numbers are identified for an 8 pin dual-in-line package. Calculate the ideal circuit gain at dc, V_{out}, V_{R1}, V_{R2}, I_{R1}, I_{R2}, and the upper cutoff frequency f_c.

Solution

Using (3.3), we get

$$G = -\frac{R_2}{R_1} = -5 \text{ (dc gain)}$$

If $V_s = 1.5$ V,

$$V_{out} = GV_s = -7.5 \text{ V}$$

Because of the voltage constraint, $V_a = 0$ V and

$$V_{R1} = V_s = +1.5 \text{ V}$$

The current through R_1 will be

$$I_{R1} = \frac{V_{R1}}{R_1} = 0.75\ \text{mA}$$

Because of the current constraint, $I_{in} = 0$ A and

$$I_{R1} = I_{R2} = 0.75\ \text{mA}$$

Hence,

$$V_{R2} = I_{R2}R_2 = 7.5 \text{ V}$$

The voltage-gain transfer function of the circuit is

$$G(s) = \frac{-1}{4.4 \times 10^{-6}(s + 2\pi f_c)}$$

where, using (3.6), we obtain

$$f_c = \frac{\omega_c}{2\pi} = \frac{1}{2\pi R_2 C}$$
$$= 7.2 \text{ kHz} \qquad \blacksquare$$

3.4 AC AMPLIFIERS

3.4.1 Noninverting

The basic noninverting amplifier of Fig. 3.2 may be converted to an ac-coupled amplifier by adding capacitors in series with the input and the output. The circuit is shown in Fig. 3.6. The input and output coupling capacitors, C_1 and C_2, introduce two transmission zeros at dc in the gain expression and are used to control the lower cutoff frequency. The pole associated with C_2 is usually made nondominant by placing it near the origin. The pole associated with C_1 is usually made dominant, which means that C_1 and R_3 will establish the lower cutoff frequency. The upper cutoff frequency, on the other hand, is a function of the frequency characteristic of the amplifier and the closed-loop gain of the circuit. This topic is covered in detail in Chapter 11.

The impedances of C_1 and C_2 are small in the amplifier passband and, thus, the circuit's gain is still established by R_1 and R_2, that is,

$$G = 1 + \frac{R_2}{R_1} \qquad (3.7)$$

With the impedance of C_1 small in the passband, the input impedance of the circuit is essentially determined by R_3.

3.4.2 Single-Supply Amplifier Circuit

Most operational amplifier circuits are biased with a pair of symmetric dc supply voltages. The most common values are $+$ and -15 V. The common point of these

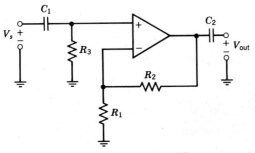

Figure 3.6 AC noninverting amplifier.

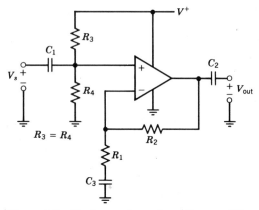

Figure 3.7 Single supply noninverting amplifier.

supplies is the reference (ground) for the op amp's internal circuitry as well as the components connected externally. The maximum output voltage of the amplifier reaches within a few volts of the supplies and is load sensitive.

There are, however, a number of applications that use a single supply. Battery-operated portable equipment and automobile electronics are examples. Since the op amp output, in a single supply circuit, cannot swing to ground without going into its nonlinear region, its reference must be shifted from zero volts. Typically, the reference for the circuit is set to one-half the supply value.

The voltage divider formed by R_3 and R_4 in Fig. 3.7 establishes the circuit's reference. If these resistors are equal, the reference is one-half the value of the single supply and the amplifier, to dc, is a voltage follower. Its output is biased at the value established by R_3 and R_4 in conjunction with the supply voltage. In the amplifier's passband, ac signals are amplified and the output moves above and below the dc reference value. The signals are ac-coupled in and out of the circuit by capacitors C_1 and C_2. Capacitor C_3 is large in value and appears as a short circuit in the passband. Thus, the ac circuit gain is established by R_1 and R_2.

The lower cutoff frequency is, typically, established by the parallel combination of R_3 and R_4, and C_1, but the upper cutoff frequency is determined primarily by the op amp. The circuit's input resistance, at midband, is equal to the parallel combination of R_3 and R_4.

Example

Design a single supply ($+12$ V) noninverting amplifier, using the circuit of Fig. 3.7, to meet the following passband specifications:

$$G = 10$$
$$R_{in} \geqslant 20 \text{ k}\Omega$$
$$V_o(p\text{--}p) = \text{maximum}$$

Assume that C_1, C_2, and C_3 are large, that is, $C_1 = C_2 = 10$ μF and $C_3 = 100$ μF.

Solution

Resistors R_1 and R_2 are used to satisfy the midband gain requirement. Let $R_1 = 2\ \text{k}\Omega$. From

$$G = 10 = 1 + \frac{R_2}{R_1}$$

$$R_2 = 18\ \text{k}\Omega$$

For a maximum peak-to-peak output signal, the amplifier output must be dc biased to one-half the supply value or $+6$ V. This means that R_3 and R_4 must be made equal. To meet the input resistance requirement ($R_{in} \geq 20\ \text{k}\Omega$),

$$(R_3 \parallel R_4) \geq 20\ \text{k}\Omega$$

Let $R_3 = 47\ \text{k}\Omega$. Thus,

$$R_4 = R_3 = 47\ \text{k}\Omega$$

and

$$R_{in} = 47\ \text{k}\Omega \parallel 47\ \text{k}\Omega = 23.5\ \text{k}\Omega > 20\ \text{k}\Omega$$

The passband impedances of the capacitors are small relative to the resistances they are associated with. ∎

3.4.3 Inverting

The addition of coupling capacitors convert the basic inverting configuration of Fig. 3.3 to an ac-coupled amplifier as shown in Fig. 3.8a. Similar to the case of the noninverting ac-coupled amplifier, coupling capacitors C_1 and C_2 introduce two transmission zeros at dc. The pole caused by C_1 is usually made dominant allowing C_1,

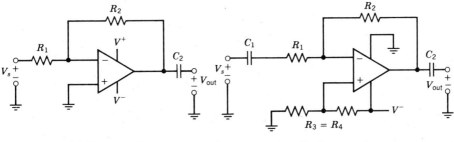

(a) Two supplies   (b) One supply

Figure 3.8 AC inverting amplifiers.

along with R_1, to establish the lower cutoff frequency. The ac inverting amplifier is, in essence, a first-order high-pass filter, neglecting the effects of the output coupling capacitor. In the passband, the impedances of the coupling capacitors are small, and the gain is established by R_1 and R_2.

The circuit in Fig. 3.8a uses two supplies, while the circuit shown in Fig. 3.8b uses only one supply. The addition of R_3 and R_4 in Fig. 3.8b establishes the dc level at the output of the op amp. Making $R_3 = R_4$ will bias the noninverting input of the amplifier to one-half the negative supply. Notice that the op amp with respect to this dc voltage is a noninverting amplifier.

The midband gain is still established by R_1 and R_2, that is,

$$G = -\frac{R_2}{R_1}$$

3.5 INSTRUMENTATION AMPLIFIERS

Instrumentation amplifiers are high performance voltage amplifiers. These circuits are differential amplifiers exhibiting high gain, high input impedance, and low output impedance. Instrumentation amplifiers are the circuit-level equivalent of op amps; however, they have more stable circuit characteristics because of the negative feedback. The stability is a result of the excess gain in the system and the negative feedback configuration. These amplifiers are used as the first stages to amplify signals from strain gauges, temperature sensors, and pressure transducers. Oftentimes, the input signal will be the difference of two large voltages.

3.5.1 High Input Resistance Version

The two op amp instrumentation amplifier in Fig. 3.9 has a differential voltage gain of 101. V_{s1} and V_{s2} are the two input sources, the difference of which is to be amplified. The circuit's output voltage, in terms of V_{s1} and V_{s2}, may be found using superposition.

$$V_{out1} = \left(1 + \frac{R_2}{R_1}\right)\left(\frac{-R_4}{R_3}\right)V_{s1} = -101V_{s1}$$

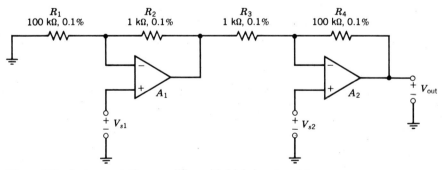

Figure 3.9 Instrumentation amplifier with high input resistance.

for $V_{s2} = 0$, and

$$V_{out2} = \left(1 + \frac{R_4}{R_3}\right)V_{s2} = 101V_{s2}$$

for $V_{s1} = 0$.
The output V_{out} is the sum of the individual responses from each input. Hence,

$$V_{out} = 101(V_{s2} - V_{s1}) \tag{3.8}$$

If precision resistors are used, the circuit will respond primarily to the differential input signal and not to the individual values of V_{s1} and V_{s2}. The circuit has a very high input resistance because each source sees a resistance greater than the differential input resistance of the op amps. Part III will cover the practical limitations of the amplifier and will show how negative feedback causes the circuit's input resistance (R_{in}) to be larger than the device's differential input resistance (R_i) in noninverting configurations. As a preview, R_{in} is approximately the loop gain AF times the device's R_i.

3.5.2 High Input Voltage Version

The instrumentation amplifier in Fig. 3.10 uses two op amps in the inverting mode. This circuit establishes at its output a small difference in potential between two very large input voltages, V_{s1} and V_{s2}. The voltage gain is only one, but the main attribute of this circuit is its ability to amplify the small difference between the large (typically up to 100 V) input voltages. The gain may be found by using an approach similar to that employed previously in the case of the high input resistance instrumentation amplifier. Using superposition, we obtain

$$V_{out1} = \left(\frac{-R_3}{R_1}\right)\left(\frac{-R_6}{R_4}\right)V_{s1} = V_{s1}$$

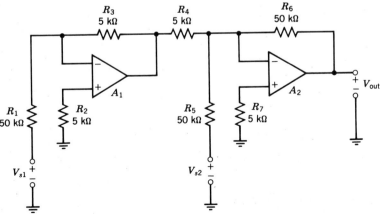

Figure 3.10 Instrumentation amplifier with high input voltage capability.

38 Voltage Amplifiers

for $V_{s2} = 0$ and

$$V_{out2} = -\frac{R_6}{R_5} V_{s2} = -V_{s2}$$

for $V_{s1} = 0$. Summing the two responses produces the output voltage,

$$V_{out} = V_{s1} - V_{s2} \qquad (3.9)$$

The inverting configuration allows the circuit to handle large input voltages, but the tradeoff is the lower input resistance which, for the values shown, is a modest 100 kΩ. Resistors R_2 and R_7 are added to minimize the effect of the amplifier's input or bias currents and do not affect the first-order input-output relationship.

3.5.3 High Common-Mode Rejection Version

A third version of an instrumentation amplifier is shown in Fig. 3.11. This circuit has a high common-mode rejection, which means that the error contributed by the average value of the input voltages V_{s1} and V_{s2} is minimal at the output. Common-mode rejection is a parameter of the nonideal amplifier as will be seen later.

The circuit is somewhat complex, but like most multiple-input op amp applications, is best handled using superposition.

For $V_{s2} = 0$ (Fig. 3.12a), amplifier A_1 is a noninverting amplifier. Its output is

$$V_{31} = V_{s1}\left(1 + \frac{R_1}{R_3}\right)$$

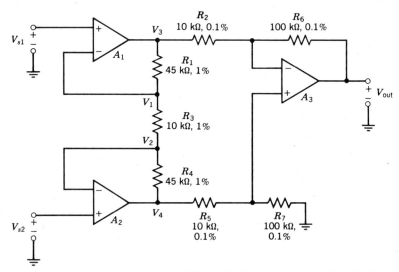

Figure 3.11 Instrumentation amplifier with high common-mode rejection.

Instrumentation Amplifiers

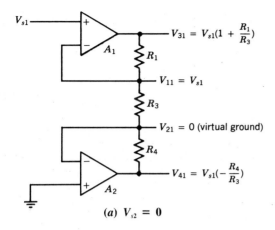

(a) $V_{s2} = 0$

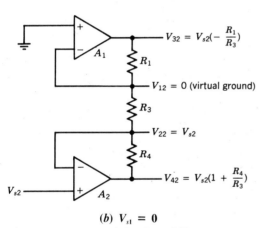

(b) $V_{s1} = 0$

Figure 3.12 Determining V_3 and V_4.

Amplifier A_2, with respect to V_{11} (V_{s1}), is an inverting amplifier. With $V_{s2} = 0$, the noninverting input terminal of A_2 is at ground. Its output is

$$V_{41} = V_{s1}\left(-\frac{R_4}{R_3}\right)$$

For $V_{s1} = 0$ (Fig. 3.12b), amplifier A_2 is a noninverting amplifier, and its output is

$$V_{42} = V_{s2}\left(1 + \frac{R_4}{R_3}\right)$$

Amplifier A_1, with respect to V_{22} (V_{s2}), is an inverting amplifier, and with $V_{s1} = 0$, the noninverting input terminal of A_1 is at ground. Its output is

$$V_{32} = V_{s2}\left(-\frac{R_1}{R_3}\right)$$

The total response at the outputs of A_1 and A_2 is

$$V_3 = V_{31} + V_{32}$$
$$= \left(1 + \frac{R_1}{R_3}\right)V_{s1} + \left(-\frac{R_1}{R_3}\right)V_{s2}$$
$$= 5.5V_{s1} - 4.5V_{s2}$$

and

$$V_4 = V_{41} + V_{42}$$
$$= -\frac{R_4}{R_3}V_{s1} + \left(1 + \frac{R_4}{R_3}\right)V_{s2}$$
$$= -4.5V_{s1} + 5.5V_{s2}$$

The circuit composed of A_3, R_2, R_5, R_6, and R_7 is a difference amplifier. This particular circuit is explained in detail in Chapter 4, but here we can treat it like a functional block whose output is equal to the difference between its two input voltages V_4 and V_3. For the circuit values shown,

$$V_{out} = V_o = 10(V_4 - V_3)$$

and finally,

$$V_{out} = V_o = 100(V_{s2} - V_{s1}) \quad (3.10)$$

The circuit output voltage V_{out} is designated V_o in subsequent chapters.

3.6 OTHER AMPLIFIERS

There are a large number of amplifiers, especially their IC versions, which are designed for a specific applications area or to maximize some characteristic. The reader can be convinced of this fact by looking at the product catalogs of the major semiconductor manufacturers. Integrated circuit amplifiers are available that have been designed for high speed, high output voltage, high output current, and low drift, low distortion, low power, low noise, and low input current. Amplifiers that have been designed for a specific applications area, like audio equipment, include preamplifiers and audio power amplifiers. The Norton, or current mode, amplifier is the dual of the voltage-operated op amp. It was specifically designed to operate in a single supply system, and many of the circuits we will discuss in this book have Norton amplifier equivalents.

Instrumentation Amplifiers

3.7 SUMMARY

Amplifiers are circuits that create at their output, typically, a larger replica of the input signal. An amplifier circuit and an operational amplifier possess similar characteristics; both have gain, bandwidth, and an input and output resistance. Amplifier circuits that use the operational amplifier as the active device have two basic forms: noninverting and inverting. Both types can be used to amplify either dc or ac input signals and serve as a foundation for more complex amplifier applications. Amplifiers, both integrated and discrete, are often designed to maximize the performance of a particular characteristic; instrumentation amplifiers are such examples.

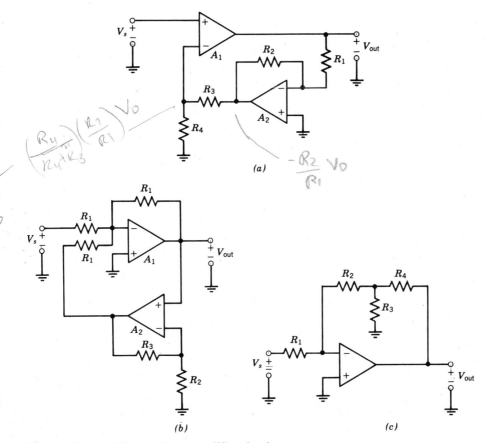

Figure 3.13 Additional voltage amplifier circuits.

42 Voltage Amplifiers

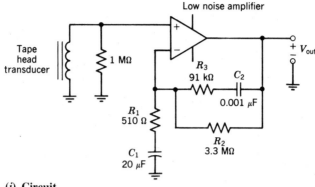

(i) Circuit

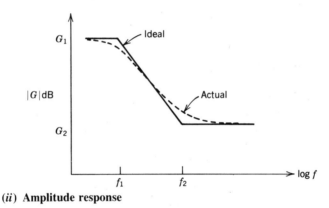

(ii) Amplitude response

(d) NAB tape head preamplifier

Fig. 3.13 Additional voltage amplifier circuits. (*Continued*)

PROBLEMS

3.1 What is the required ratio of R_2 and R_1 for a voltage gain V_{out}/V_s of 10 in the noninverting amplifier in Fig. 3.2a? What are the smallest values of R_1 and R_2 that can be used in this circuit if the load resistor R_L is 2 kΩ, the maximum output current of the op amp ± 10 mA, and the maximum output voltage of the circuit is ± 10 V?

3.2 Calculate all of the currents and voltages in the noninverting amplifier circuit of Fig. 3.2a if $R_1 = 100$ kΩ, $R_2 = R_L = 50$ kΩ and $V_s = -5$ V.

3.3 In the inverting amplifier circuit of Fig. 3.3, $R_1 = 2$ kΩ and $R_2 = 12$ kΩ: (a) determine the voltage gain V_{out}/V_s as a ratio and in decibels. (b) If the range of the output voltage is from 0.1 to 12 V, and the input current of the nonideal op amp is 100 nA, what is the maximum percentage of error of the output voltage contributed by the input current?

3.4 Calculate all of the voltages and currents in the inverting amplifier circuit of Fig. 3.3 if $R_1 = 100$ kΩ, $R_2 = R_L = 50$ kΩ, and $V_s = -5$ V.

3.5 Sketch the amplitude Bode plot for the inverting circuit in Fig. 3.4 if $R_1 = 5$ kΩ, $R_2 = 10$ kΩ, and $C = 0.001$ μF.

3.6 Sketch the phase Bode plot for the inverting circuit in Fig. 3.4 if $R_1 = 5$ kΩ, $R_2 = 10$ kΩ, and $C = 0.001$ μF.

3.7 Design a dc-coupled, inverting amplifier circuit (Fig. 3.4) that meets the following specifications:
(a) $V_{out}/V_s = -5$ at dc
(b) $R_{in} = 10$ kΩ
(c) $\omega_c = 2 \cdot 10^4$ rad/sec

3.8 Design a dc-coupled, inverting amplifier circuit (Fig. 3.4) that meets the following specifications:
(a) $V_{out}/V_s = -1$ at dc
(b) $R_{in} = 100$ kΩ
(c) $f_c = 100$ kHz

3.9 Find R_1, R_2, and R_3 in the noninverting circuit of Fig. 3.6 to meet the following specifications:
(a) $V_{out}/V_s = 15$ in the passband
(b) $R_{in} = 10$ kΩ in the passband
(c) $R_1 + R_2 = 15$ kΩ

3.10 Design a noninverting circuit (Fig. 3.6) that meets the following specifications:
(a) $V_{out}/V_s = 2$ in the passband
(b) $R_{in} = 100$ kΩ in the passband
(c) $R_1 + R_2 = 100$ kΩ
(d) $f_c = 50$ Hz

3.11 For the circuit in Fig. 3.6, show that

$$\frac{V_{out}(s)}{V_s(s)} = \frac{s(1 + R_2/R_1)}{s + \omega_c}$$

where

$$\omega_c = \frac{1}{R_3 C_1}$$

3.12 Discuss what will establish the upper cutoff frequency in the circuit of Fig. 3.6 if the amplifier is nonideal.

44 Voltage Amplifiers

3.13 For the circuit in Fig. 3.7, show that

$$\frac{V_{out}(s)}{V_s(s)} = \frac{s(1 + R_2/R_1)}{s + \omega_c}$$

where

$$\omega_c = \frac{1}{C_1(R_3 \parallel R_4)}$$

Assume that C_3 is large.

3.14 Discuss what will establish the upper cutoff frequency in the circuit of Fig. 3.8a if the amplifier is nonideal.

3.15 For the single supply amplifier circuit in Fig. 3.7, $R_1 = R_2 = R_3 = R_4 = 5$ kΩ, $V^+ = 12$ V, $C_1 = 2$ μF, and C_2 and C_3 are large. Calculate V_{out}/V_s and R_{in} in the passband, ω_c, and $V_{out}(dc)$.

3.16 What change(s) in the circuit of Fig. 3.7 is (are) required to have the circuit operate from a single -12 V supply?

3.17 The gain expression for the ac-inverting amplifier in Fig. 3.8a may be found using

$$\frac{V_{out}(s)}{V_s(s)} = -\frac{Z_2(s)}{Z_1(s)}$$

where

$$Z_2(s) = R_2 \quad \text{and} \quad Z_1(s) = R_1 + \frac{1}{sC_1}$$

Show that

$$\frac{V_{out}(s)}{V_s(s)} = \frac{s(-R_2/R_1)}{s + \omega_c}$$

where

$$\omega_c = \frac{1}{R_1 C_1}$$

3.18 What change(s) in the circuit of Fig. 3.8a is (are) required to have the circuit operate from a single $+24$ V supply?

3.19 Design a single supply, ac-coupled inverting amplifier (Fig. 3.8b) that meets the following specifications:
(a) $V^- = -15$ V
(b) $V_{out}/V_s = -5$ in the passband
(c) $\omega_c = 500$ rad/sec
(d) $R_{in} = 10$ kΩ in the passband
(e) $V_o(p\text{–}p)$ = maximum

3.20 Draw the dc and ac (midband) equivalent circuits for the circuit in Fig. 3.8b.

3.21 Calculate all of the voltages and the currents in the circuit of Fig. 3.9 if $V_{s1} = +0.1$ V and $V_{s2} = +0.2$ V.

3.22 Calculate all of the voltages and the currents in the circuit of Fig. 3.10 if $V_{s1} = +100$ V and $V_{s2} = +90$ V.

3.23 Calculate V_1, V_2, V_3, V_4, and V_{out} in the circuit of Fig. 3.11 if $V_{s1} = +10.0$ V and $V_{s2} = +9.90$ V.

3.24 Find V_{out}/V_s for the circuit in Fig. 3.13a.

3.25 Find V_{out}/V_s for the circuit in Fig. 3.13b.

3.26 Find V_{out}/V_s for the circuit in Fig. 3.13c.

3.27 The circuit in Fig. 3.13d(i) is a tape head preamplifier or active filter equalizer. Equalizer circuits, basically, provide the inverse response to that of the signal out of the transducer, so that the frequency response to the next amplifier stage is essentially flat. The amplitude Bode and actual responses (called NAB or National Association of Broadcasters) of the circuit are shown in Fig. 3.13d(ii). Estimate the corner frequencies f_1 and f_2 and the low- and high-frequency gains G_1 and G_2. The difference in the gains, because of the application, should be about 30 dB. What causes the gain to decrease for frequencies below f_1?

A companion to this circuit is the RIAA phono preamplifier, which is presented, as a problem, in Chapter 8.

Chapter 4

Mathematical Operators

4.1 PROLOGUE

Mathematical operators are operational amplifier circuits that *perform mathematical operations*. These circuits were initially used as functional blocks that simulated mathematical functions in analog computers; however today, they serve as general-purpose building blocks in signal-conditioning analog systems.

The inverting amplifier, when viewed as a mathematical operator, functions as a scaler. It scales (and inverts) the input voltage by a constant and is the most elementary circuit of the inverting class. The noninverting amplifier, when viewed as a mathematical operator, also scales the input voltage by a constant. Many other mathematical operators are derived from these basic configurations that add and subtract, integrate and differentiate, generate log and antilog functions of their input, multiply and divide, and find the absolute value.

4.2 SUMMING AMPLIFIERS

4.2.1 Inverting

The summing amplifier in Fig. 4.1a provides an output voltage equal to the weighted algebraic sum (inverted) of the input voltages. This circuit can have n inputs and is a member of the inverting amplifier class.

Summing Amplifiers 47

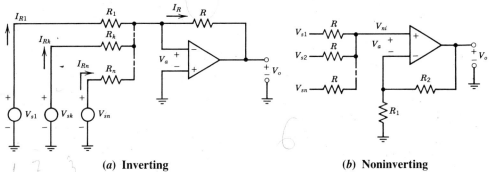

(a) Inverting (b) Noninverting
Figure 4.1 Summing amplifiers.

Each of the input voltage sources contributes a current

$$I_{Rk} = \frac{V_{sk}}{R_k}$$

Because of the current constraint, the sum of these currents is equal to the current through the feedback resistor R,

$$I_R = \sum_{k=1}^{k=n} I_{Rk}$$

or

$$I_R = \sum_{k=1}^{k=n} \frac{V_{sk}}{R_k}$$

The inverting input terminal of the operational amplifier of this circuit is connected to the node where these currents sum and is often referred to as the "summing junction." In the ideal case, the potential difference between the amplifier inputs is zero and, since the noninverting input terminal is grounded, then the inverting terminal is also at zero volts. The inverting terminal in inverting amplifier configurations is also referred to as "virtual ground." Thus, the output voltage is the same as the potential difference across the feedback resistor,

$$V_o = -I_R R$$

or in terms of the input voltages, equals

$$V_o = -R \sum_{k=1}^{k=n} \frac{V_{sk}}{R_k} \tag{4.1a}$$

48 Mathematical Operators

Example

For the two-input version of the circuit in Fig. 4.1a, $R_1 = 2\ \text{k}\Omega$, $R_2 = 4\ \text{k}\Omega$, and $R = 8\ \text{k}\Omega$. Express I_R and V_o in terms of V_{s1}, V_{s2}, and circuit constants.

Solution

From (4.1a),

$$I_R = \frac{V_{s1}}{R_1} + \frac{V_{s2}}{R_2} = 0.5V_{s1} + 0.25V_{s2}\quad (\text{mA})$$

$$V_o = -(4V_{s1} + 2V_{s2}) = -4V_{s1} - 2V_{s2}\qquad\blacksquare$$

4.2.2 Noninverting

The noninverting summer of Fig. 4.1b, like the inverting version, offers the possibility of the simultaneous application of n sources. The sources are summed through the input resistor divider to establish V_{ni}, which is then amplified by the gain of the noninverting circuit.

The voltage V_{ni} is determined using superposition. For a circuit with n sources, voltage V_{ni} may be expressed as

$$V_{ni} = \frac{1/R}{1/R + (n-1)R}(V_{s1} + V_{s2} + \cdots + V_{sn})$$

or

$$V_{ni} = \frac{V_{s1} + V_{s2} + \cdots + V_{sn}}{n}$$

Thus,

$$V_o = \left(1 + \frac{R_2}{R_1}\right)V_{ni}$$

or

$$V_o = \left(1 + \frac{R_2}{R_1}\right)\frac{V_{s1} + V_{s2} + \cdots + V_{sn}}{n} \qquad (4.1b)$$

Example

Design a summer to implement the function

$$V_o = V_{s1} + V_{s2} + V_{s3}$$

Solution

Using (4.1b), we find that the circuit of Fig. 4.1b will provide the desired output if

$$1 + \frac{R_2}{R_1} = n = 3$$

or

$$R_2 = 2R_1$$

Somewhat arbitrarily, we let $R = R_1 = 10$ kΩ. Hence,

$$R_2 = 20 \text{ k}\Omega$$ ∎

4.3 DIFFERENCE AMPLIFIERS

A difference, or differential voltage, amplifier generates an output voltage proportional to the weighted difference of the two voltages applied to its input terminals. This circuit is a combination of the inverting and noninverting amplifiers.

The elementary difference amplifier in Fig. 4.2 is implemented using an inverting amplifier A_1 and an inverting summing amplifier A_2. The magnitude of the gain of each of the circuits is one; hence the output voltage is equal to the difference of the input voltages,

$$V_o = -(-V_{s1} + V_{s2}) = V_{s1} - V_{s2} \quad (4.2)$$

This configuration was used before the advent of the IC operational amplifier since, at that time, it was not possible to connect an op amp with its noninverting terminal electrically above or below ground.

The difference amplifier in Fig. 4.3 uses a single operational amplifier to generate an output voltage proportional to the difference of the weighted input voltages V_{s1} and

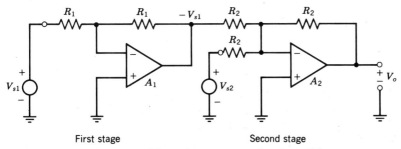

First stage   Second stage

Figure 4.2 Difference amplifier using two operational amplifiers.

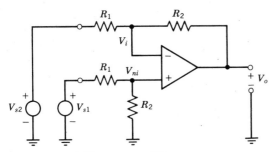

Figure 4.3 Difference amplifier.

V_{s2}. The expression for the output voltage V_o may be obtained using superposition and Figs. 4.4 and 4.5;

$$V_o = V_{o1} + V_{o2}$$

where

$$V_{o1} = \left(V_{s1} \frac{R_2}{R_2 + R_1}\right)\left(1 + \frac{R_2}{R_1}\right)$$

$$= V_{s1} \frac{R_2}{R_1}$$

and

$$V_{o2} = -V_{s2} \frac{R_2}{R_1}$$

The circuit of Fig. 4.5 acts like a basic inverting amplifier because the currents and voltages associated with R_1 and R_2 are zero.

For the difference amplifier,

$$V_o = (V_{s1} - V_{s2}) \frac{R_2}{R_1} \tag{4.3}$$

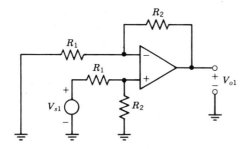

Figure 4.4 Difference amplifier modified for finding V_{o1}.

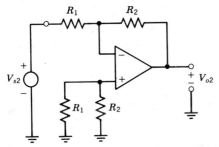

Figure 4.5 Difference amplifier modified for finding V_{o2}.

Example

For the difference amplifier in Fig. 4.3, $R_1 = 10\ \text{k}\Omega$, $R_2 = 20\ \text{k}\Omega$, $V_{s1} = -2\ \text{V}$, and $V_{s2} = +3\ \text{V}$. Calculate V_o and the voltage at the amplifier's input terminals (V_i, V_{ni}) with respect to ground.

Solution

From (4.3),

$$V_o = 2(-2\ \text{V} - 3\ \text{V})$$
$$= -10\ \text{V}$$

The ideal difference amplifier is a linear application that has several characteristics: (1) the op amp is operating in its linear region, (2) V_a is zero, and (3) each amplifier input is at the same potential. This potential, with respect to ground, is

$$V_i = V_{ni} = \frac{R_2}{R_1 + R_2} V_{s1} = 2/3\ (-2\ \text{V})$$
$$= -1.33\ \text{V}$$

Two practical limitations of the differential amplifier, $CMRR$ and A_c, are discussed in detail in Section 13.2.6. ∎

4.4 INTEGRATORS

4.4.1 Ideal and Practical

An integrator, as the term implies, provides an output voltage proportional to the integral of its input.

In the ideal integrator circuit of Fig. 4.6a, which is similar to a charge amplifier, the input current i is provided by the combination of the voltage source v_s in series

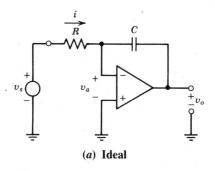

(a) Ideal

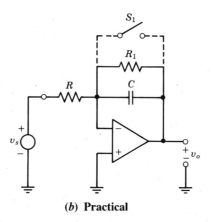

(b) Practical

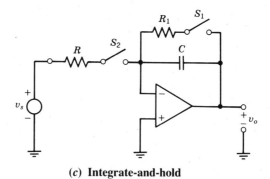

(c) Integrate-and-hold

Figure 4.6 Integrators.

with resistor R. If we realize that V_a is zero as a result of the voltage constraint, the current i is

$$i = \frac{v_s}{R}$$

The output voltage is proportional to the integral of this current,

$$v_o = -\frac{1}{C}\int_0^t i\, dt + v_c(0)$$

or

$$v_c = -\frac{1}{RC}\int_0^t v_s\, dt + v_c(0) \quad (4.4)$$

If the input is a dc voltage, the output will be a voltage whose magnitude increases linearly with time. The slope of this voltage "ramp" is established by the magnitudes of R, C, and V_s. Assuming $v_c(0) = 0$, we have

$$v_o = -\frac{V_s}{RC}t \quad (4.5)$$

The input-output relationship in the s domain is found by treating the ideal integrator of Fig. 4.6a as an inverting amplifier with a feedback impedance $Z_2(s)$ and an input impedance $Z_1(s)$.

$$V_o(s) = -\frac{Z_2(s)}{Z_1(s)}V_s(s) = -\frac{1}{sRC}V_s(s)$$

or

$$V_o(s) = -\frac{1}{RC}\frac{V_s(s)}{s} \quad (4.6)$$

The ideal integrator in Fig. 4.6a is seldom used in that form. The small dc, voltage and current, errors associated with the input of the real op amp force the operational amplifier into saturation because it effectively operates open loop at dc. The modified (practical) integrator of Fig. 4.6b uses resistor R_1 in the feedback loop to limit the low-frequency gain of the circuit to $-(R_2/R_1)$. R_1 is selected so that the magnitude of worst-case dc voltage across it is below the magnitude of the saturation voltage of the op amp. If, for example, the saturation is 12 V and the maximum expected dc current in the feedback loop is 100 nA, then R_1 should be 120 MΩ or less.

Another circuit technique used to overcome the input-current error is to use a switch across C in the circuit of Fig. 4.6b. The switch is closed when the circuit is not integrating the input signal. During integration, the switch is opened. However when the switch is open, the bias current will still sum with v_s/R and cause an error voltage at the output of the integrator. To minimize this error, the signal current should be made large compared to the amplifier's input current. The switch may be solid state or mechanical and can be manually operated or computer controlled. The switch also ensures that the initial condition for the capacitor is zero.

4.4.2 Integrate and Hold

A derivative of the practical integrator is shown in Fig. 4.6c. It has three modes of operation: reset, integrate, and hold. The reset and hold modes extend the capability of the basic integrator by providing initialization and storage features.

In the reset mode, switch S_1 is closed and the capacitor is discharged to zero. Resistor R_1 is used to limit the current and control the discharge rate when S_1 is closed. In the integrate mode, S_1 is opened and S_2 is closed, and the output is proportional to the integral of v_s. In the hold mode, S_1 and S_2 are both open, and the circuit will maintain the voltage across the capacitor. During the hold mode, the capacitor will discharge at a rate determined, essentially, by the input current of the op amp. This particular integrator is similar to a sample-and-hold circuit except that this circuit samples the *integral* of the input voltage rather than the input itself.

Integrators are frequently used in timing and sweep circuit applications. These circuits are readily adaptable to microprocessor-controlled systems because the switches can be controlled by logic-level signals, and a digital-to-analog convertor can be used for v_s. The integrator is also used as a basic building block in many filter and oscillator circuits.

Example

Assuming $V_s = -1$ V and zero initial conditions ($t = 0$), specify a set of values for R and C in the integrator of Fig. 4.6a so that $V_o = +5$ V at $t = 100$ μsec.

Solution

From (4.5),

$$5 = -\frac{(-1 \text{ V})(10^{-4})}{RC}$$

or

$$RC = 2 \cdot 10^{-5} \text{ sec}$$

We have the option of arbitrarily specifying either R or C to provide us with the necessary value for the RC product. Since we usually have a greater selection of resistor values, we specify the capacitor value

$$C = 0.001 \ \mu F$$

For the resistor,

$$R = 20 \ k\Omega$$

If a large value of C is chosen, the amplifier may oscillate because of the additional phase shift caused by the capacitor. If C is too small, the distributed capacitance may add an apreciable error to its nominal value. Generally, capacitor values from 0.001 to 0.1 µF are preferred; capacitors in this range are generally small and low in cost. ∎

4.5 DIFFERENTIATORS

4.5.1 Ideal

A differentiator, as the term implies, provides an output voltage proportional to the derivative of its input. The ideal differentiating amplifier in Fig. 4.7a is implemented by interchanging the positions of resistor R and capacitor C in the integrating amplifier of Fig. 4.6a. In the time domain, the input v_s and the output v_o are related by

$$v_o = -RC \frac{dv_s}{dt} \qquad (4.7)$$

The relationship in the s domain is found in the same manner employed for that of the integrator and is given by

$$V_o(s) = -RCsV_s(s) \qquad (4.8)$$

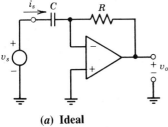
(a) Ideal

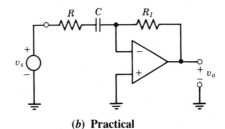
(b) Practical

Figure 4.7 Differentiators.

4.5.2 Practical

The output of the ideal differentiator of Fig. 4.7a is proportional to the rate of change of its input voltage v_s. In a practical situation, noise or any fast-rising portion of v_s will cause the output voltage to distort because of saturation or the slew-rate limitations of the op amp.

This problem can be overcome by limiting the high-frequency gain of the ideal differentiator as shown in Fig. 4.7b. As seen from the linearized Bode plot of Fig. 4.8, the closed-loop gain of the practical circuit cannot exceed $-(R_2/R_1)$ at high frequencies. Thus, the circuit will no longer operate as a differentiator for input frequencies above ω_c, the -3 dB point of the amplitude response. This cutoff frequency may be found from the voltage-gain transfer function $G(s)$ of the practical circuit of Fig. 4.7b:

$$G(s) = \frac{V_o(s)}{V_s(s)} = -\frac{R_1}{R + 1/sC}$$

$$= -\frac{(R_1/R)s}{s + 1/RC} = -\frac{(R_1/R)s}{s + \omega_c} \tag{4.9}$$

where ω_c is the magnitude of the pole of $G(s)$, that is,

$$\omega_c = \frac{1}{RC}$$

The transfer function for this circuit, and others like it, is found by taking the negative of the ratio of the feedback and input network impedances. Here, as in the case of the practical integrator, the frequency response of the design is purposely limited to reduce the problems associated with the nonideal operational amplifier.

Example

Calculate the current provided by source v_s in the ideal differentiator of Fig. 4.7a if (1) v_s increases linearly from 0 to 1 V at a rate of 10 V/μsec, (2) $C = 0.1$ μF, and (3) $R = 10$ kΩ.

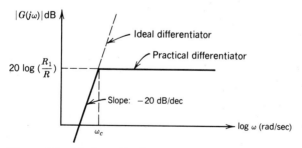

Figure 4.8 Ideal amplitude response of a practical differentiator.

Solution

Since the source current is the same as the capacitor current,

$$i_s = C \frac{dv_s}{dt}$$
$$= (0.1 \ \mu F) (10 \ V/\mu sec) = 1 \ A$$

Obviously, the op amp cannot sink such a high a value of current. This example brings out one of several limitations of the ideal differentiator. Assuming that the maximum value of the output current in the op amp is ± 10 mA, we see that the corresponding value of dv_s/dt that this circuit can handle is ± 0.1 V/μsec. ∎

4.6 ABSOLUTE VALUE AMPLIFIER

The absolute value amplifier in Fig. 4.9 consists of two basic op amp circuits: a half-wave rectifier and a summing amplifier. It is a circuit that, additionally, can be used as a full-wave rectifier or an ac-to-dc convertor. The specific name that is used is a function of its application.

The half-wave rectifier, Fig. 4.10a, is an inverting amplifier with a limiting element (diode) in its feedback loop. Since this circuit is nonlinear, the input and output are more commonly related using a graph called a transfer characteristic rather than a mathematical model. The transfer characteristic for the half-wave rectifier is shown in Fig. 4.10b.

For positive values of the half-wave rectifier's input voltage, the output will go negative. Negative values of the output voltage will reverse-bias diode D_1 in the feedback loop, and the circuit will behave like an inverting amplifier with a gain of -1. For negative values of the input voltage, the output will go positive and forward-bias the diode. If we assume an ideal diode with a forward voltage drop of zero volts, then the output will be held to zero volts because of the connection of the diode from the output to the virtual ground at the inverting input terminal of the operational

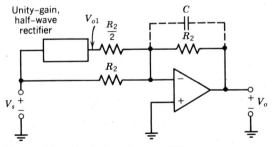

Figure 4.9 Absolute value amplifier.

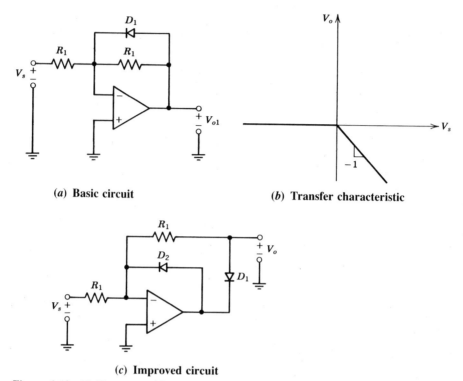

(a) Basic circuit

(b) Transfer characteristic

(c) Improved circuit

Figure 4.10 Half-wave rectifier.

amplifier. The input and output for the circuit of Fig. 4.10a are mathematically related as follows:

$$V_{o1} = -V_s \quad \text{for} \quad V_s > 0$$

and

$$V_{o1} = 0 \quad \text{for} \quad V_s < 0$$

The name "half-wave rectifier" comes from dc power supply considerations where the input is sinusoidal. For this case, the output is a half-wave rectified waveform.

When the input voltage to the basic half-wave rectifier is negative, the output will be offset from zero by the forward voltage drop of diode D_1. The circuit in Fig. 4.10c eliminates this error. For positive values of V_s, D_2 is reverse-biased, D_1 is forward-biased, and the output is an exact replica of the input. For negative values of V_s, D_2 is forward-biased and provides a negative feedback path, while reverse-biased D_1 disconnects the amplifier output from the circuit output. This will force the circuit output to zero volts because of the virtual ground at the inverting terminal of the op amp.

Absolute Value Amplifier

The full-wave rectifier of Fig. 4.9 is obtained when a summing amplifier is used to add the input signal and the output of the half-wave rectifier of Fig. 4.10a. The output of the summing amplifier is given by

$$V_o = \frac{-V_{o1} R_2}{R_2/2} + \frac{-V_s R_2}{R_2} = -2V_{o1} - V_s$$

When V_s is negative, the ideal diode D_1 in the circuit of Fig. 4.10a will be conducting and V_{o1}, the output of the half-wave rectifier, is zero volts. The output of the summing amplifier is then a function of the input signal V_s only,

$$V_o = -V_s \quad \text{for} \quad V_s < 0$$

For the situation where V_s is positive, the output of the half-wave rectifier is $-V_s$. This time the output of the summing amplifier becomes

$$V_o = -2(-V_s) - V_s = V_s \quad \text{for} \quad V_s > 0$$

Thus

$$V_o = |V_s| \tag{4.10}$$

If the input signal is sinusoidal, the output is a full-wave rectified waveform.

4.6.1 AC-to-DC Convertor

If a capacitor is added in parallel with the resistor in the summing amplifier's feedback loop in Fig. 4.9, the circuit becomes a pseudo ac-to-dc convertor. The parallel combination of the resistor and capacitor averages the full-wave rectified sinusoidal signal. A requirement for the capacitor is that

$$R_2 C \geq 5T$$

where T is the period of the lowest frequency input sine wave. The summing amplifier, in this case, behaves like a low-pass filter with a very low cutoff frequency $f_c = 1/(2\pi R_2 C)$. It can be shown for the ac-to-dc convertor that

$$V_o = \frac{2}{\pi} V_s(\text{peak}) = 0.637 V_s(\text{peak})$$

The magnitude of the feedback resistor R_2 in Fig. 4.9 can be scaled to change the 0.637 factor to 0.707 so that this configuration provides an output equal to the root-mean-square (RMS) value of its input sinusoidal signal. This last circuit is a pseudo RMS convertor applicable to sinusoids only since it does not give, in the general case, an output proportional to the square root of the average of the input waveform squared (the true RMS value of the input periodic waveform).

4.7 LOG AND ANTILOG AMPLIFIERS

A logarithmic amplifier generates an output voltage proportional to the logarithm of its input voltage. The elementary log amplifier of Fig. 4.11 uses a forward-biased semiconductor diode to generate the nonlinear, logarithmic function. The I–V relationship of a forward-biased semiconductor diode is approximately equal to

$$I_d \cong I_S \exp \frac{V_d}{V_T} \qquad (V_d > 0, I_d \gg I_s)$$

I_S is the reverse saturation current of the diode, a function of temperature and the physical characteristics of the device, and V_T, also temperature dependent, is approximately 26 mV at room temperature.

For the elementary logarithmic amplifier of Fig. 4.11,

$$I_d = \frac{V_s}{R} \cong I_S \exp \frac{V_d}{V_T}$$

for $V_s > 0$. Since $V_o = -V_d$,

$$\frac{V_s}{R} \cong I_S \exp \frac{-V_o}{V_T}$$

If we solve the last expression for V_o, we get

$$V_o \cong -V_T \ln V_s + V_T \ln (I_S R) \qquad (V_s > 0) \qquad (4.11)$$

The circuit of Fig. 4.11 produces an output voltage that is linearly related to the logarithm (to any base) of its positive input voltage.

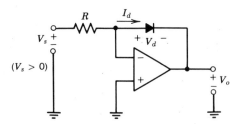

Figure 4.11 Elementary logarithmic amplifier.

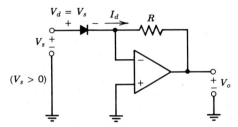

Figure 4.12 Elementary antilogarithmic amplifier.

The corresponding antilogarithmic amplifier is shown in Fig. 4.12. For this circuit,

$$V_o = -I_d R$$

and

$$I_d = -\frac{V_o}{R} = I_S \exp \frac{V_s}{V_T} \quad (V_s > 0)$$

or, solving for V_o,

$$V_o = -I_S R \exp \frac{V_s}{V_T} \quad (V_s > 0) \quad (4.12)$$

Since by definition, if $y = \ln x$, then $x = \ln^{-1} y = \exp y$, the output of the circuit of Fig. 4.12 is indeed linearly related to the inverse logarithm (to any base) of its positive input voltage.

A possible application of log and antilog amplifiers is in analog multipliers. A multiplier, shown in block diagram form in Fig. 4.13, uses two log amplifiers, an antilog amplifier, and a summer. The output voltage of this circuit is

$$V_o = \log^{-1} (\log V_{s1} + \log V_{s2})$$
$$= \log^{-1} (\log V_{s1} V_{s2})$$

or

$$V_s = V_{s1} V_{s2}$$

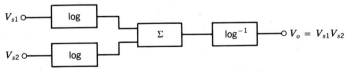

Figure 4.13 Block diagram of an analog multiplier.

62 Mathematical Operators

The key to multiplying in the analog domain is the logarithmic relationship of the voltage and current in a semiconductor diode. This fundamental relationship is the basis of multiplier designs. The conversion of current to voltage to form an input voltage-output voltage relationship is straightforward.

The practical problems with the models presented are the temperature dependence of V_T and I_S, and the low level voltages and currents in the circuit. In discrete circuits, a transistor is used instead of a semiconductor diode to increase the current level, and additional circuitry is used to compensate for the temperature dependence of V_T and I_S.

4.8 ANALOG MULTIPLIERS

Analog multipliers are frequently used in control and communication systems; they function as phase detectors, modulators, demodulators, and analog controllers. In discrete form, multipliers can be implemented using log and antilog amplifiers. In integrated form, they are usually implemented using a Gilbert multiplier cell.

The block diagram of the four-quadrant, analog multiplier in Fig. 4.14 consists of a Gilbert multiplier cell, a pair of differential voltage-to-current convertors, and a differential current-to-single ended voltage convertor. Like the elementary op amp multiplier, the integrated multiplier also depends on the exponential transfer function of the bipolar transistor. The detailed analysis of the multiplier using a Gilbert cell is somewhat complex and more appropriately discussed in an integrated circuits text.

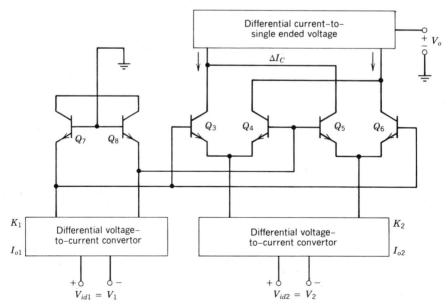

Figure 4.14 Analog multiplier using a Gilbert multiplier cell.

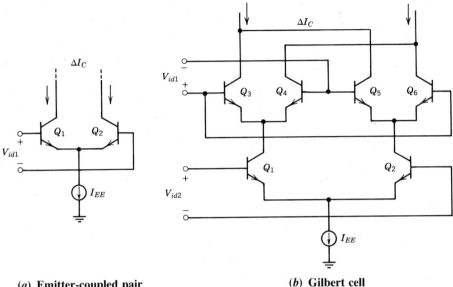

(a) Emitter-coupled pair   **(b)** Gilbert cell

Figure 4.15 Gilbert multiplier circuit.

However, this type of multiplier is frequently used in conjunction with other amplifier circuits. Thus, we present only a qualitative description of its operation.

The Gilbert multiplier cell is a modification of an emitter-coupled transistor pair. For a single emitter-coupled transistor pair, shown in Fig. 4.15a, it can be shown that the difference between the collector currents, ΔI_c, is related to the differential input voltage v_{id1} and the bias current I_{EE}.

The Gilbert cell shown in Fig. 4.15b has an emitter-coupled pair (Q_1, Q_2) in series with a cross-coupled, emitter-coupled pair (Q_3,–Q_6). Simply, the bias current I_{EE} for the latter transistor pair is made proportional to the second input voltage V_{id2} relating ΔI_C to the two input voltages v_{id1} and v_{id2}. The differential voltage-to-current convertors, and Q_7 and Q_8, shown in the complete multiplier in Fig. 4.14 introduce a nonlinearity to compensate for the transfer characteristic of the basic cell. It can be shown that the overall transfer characteristic for the multiplier is

$$\Delta I_c = I_{EE} \frac{K_1 V_1}{I_{o1}} \frac{K_2 V_2}{I_{o2}} \tag{4.13}$$

I_{EE} is the emitter bias current for Q_1 and Q_2, where K_1 and K_2 are the transconductances of the voltage-to-current convertors, and I_{o1} and I_{o2} are the dc currents that flow in the output of the convertors when the input voltage is zero. The differential output current is then converted to a ground-referenced voltage, and the constants are usually chosen such that

$$V_o = 0.1 V_1 V_2 \tag{4.14}$$

4.8.1 Analog Arithmetic Using Multipliers

The symbolic representation of an analog multiplier is shown in Fig. 4.16. The multiplier, as a functional block, can be used in op amp circuits to generate the divide, square, cube, and square root functions. The block diagrams of these functions are shown in Fig. 4.17.

In the analog divider of Fig. 4.17a, the multiplier is connected in the feedback loop of a modified summing amplifier. For practical multipliers, the inputs and output of the multiplier are commonly related by

$$V'_o = \frac{V_x V_y}{10} \tag{4.15}$$

Because of the virtual ground at the inverting input terminal of the op amp,

$$V'_o = -V_1 = \frac{V_o V_2}{10}$$

or

$$V_o = \frac{-10 V_1}{V_2} \tag{4.16}$$

To insure negative feedback around A_1, V_o and V'_o must have the same polarity and V_2 must be positive.

The multiplication of a signal by itself, Fig. 4.17b, produces the square function,

$$V_o = \frac{V_1^2}{10} \tag{4.17}$$

If the output of the squaring circuit, along with the original signal, is applied to a second multiplier, the result is the cube function

$$V_o = \frac{V_1^3}{100} \tag{4.18}$$

The circuit is shown in Fig. 4.17c. This concept can be extended to obtain higher-order powers of input functions.

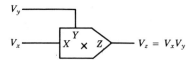

Figure 4.16 Symbolic representation of the analog multiplier.

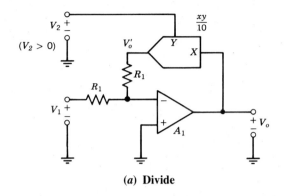

(a) Divide

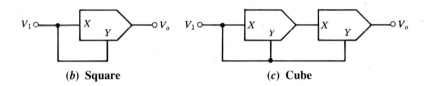

(b) Square (c) Cube

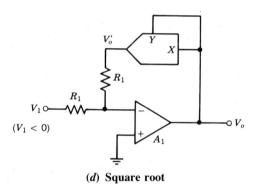

(d) Square root

Figure 4.17 Analog arithmetic functions.

The square root function is implemented by connecting a multiplier, configured as a squarer, in the feedback loop of an amplifier. The circuit is shown in Fig. 4.17d. If we define the output of this circuit as V_o, the output of the multiplier is

$$V'_o = \frac{V_o^2}{10}$$

V'_o and V_1 are related by the R_1 resistor divider,

$$V_1 = -V'_o$$

Hence

$$V_o = \sqrt{-10V_1} \qquad (4.19)$$

Input V_1 must be negative to insure negative feedback around the op amp.

The analog functions discussed in this section are not used for computational purposes; obviously, digital computers do a better job. Instead, they are primarily used in low-frequency control applications to produce an analog signal related arithmetically to other system signals.

4.9 ANALOG COMPUTATION

In an analog computer, the numbers representing the variables in a problem are represented continuously by voltages. The mathematical operations involved in the solution of a problem are performed in the computer as analogous operations on these voltages. The analog computer comprises many of the basic mathematical operators discussed in this chapter, and a simple example will help illustrate how these fundamental circuits are employed.

The second-order linear differential equation

$$m \frac{d^2x}{dt^2} + b \frac{dx}{dt} + kx = 0 \qquad (4.20)$$

describes the mechanical vibrations of a body of mass m on the end of a spring of force constant k in the presence of viscous damping described by the damping constant b as shown in Fig. 4.18a. The displacement of the body from its equilibrium position is represented by the variable x.

The current i in the electrical analog of this mechanical system, Fig. 4.18b, is described by a differential equation of the same form. The corresponding equation is

$$L \frac{d^2i}{dt^2} + R \frac{di}{dt} + \frac{i}{C} = 0 \qquad (4.21)$$

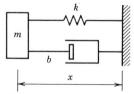

(a) Mechanical vibrating system   (b) Electrical analog

Figure 4.18 Mechanical and electrical systems.

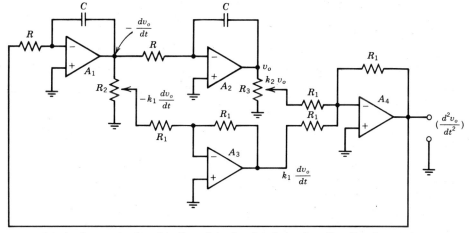

(a) Analog computer circuit used to simulate behavior of a vibrating system

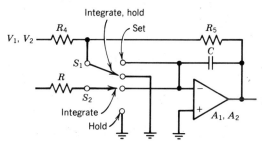

(b) Integrate, hold, and set modes of an integrator used in an analog computer

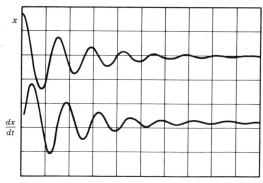

(c) Solutions for x and dx/dt

Figure 4.19 Application of an analog computer circuit.

The initial conditions are given as

$$x(0) = X_o \quad \text{and} \quad i(0) = I_o$$

Solving (4.20) for d^2x/dt^2 in the mechanical system, we get

$$\frac{d^2x}{dt^2} = \left(\frac{-b}{m}\right)\frac{dx}{dt} + \left(\frac{-k}{m}\right)x$$

Assuming a voltage is available that represents d^2x/dt^2, we can see that we need two integrators to give us voltages proportional to the dx/dt and x terms. Figure 4.19a shows one arrangement of four operational amplifiers that will solve (4.21). The configuration consists of two integrators, A_1 and A_2, and the equivalent of a summing amplifier, A_3 and A_4. Assuming $RC = 1$ sec and an output v_o for A_2, the outputs of A_1 and A_4 will be $-dv_o/dt$ and d^2v_o/dt^2, respectively. The output of A_4 is

$$v_o(A_4) = -k_1\frac{dv_o}{dt} - k_2 v_o \quad (k_1 \geq 0,\ k_2 \geq 0)$$

where k_1 and k_2 depend on the settings of potentiometers R_2 and R_3, respectively. Thus,

$$\frac{d^2v_o}{dt^2} = -k_1\frac{dv_o}{dt} - k_2 v_o$$

The differential equation is identical to (4.20) if $k_1 = b/m$, $k_2 = k/m$, and v_o is assumed to represent the unknown x.

In general, the analog computer circuit will force the voltages, which are used to represent the variables, to vary in a manner described by the equation to be solved. Voltages V_1 and V_2 in Fig. 4.19b, representing the initial values of x and dx/dt, may be used to set the outputs of the integrator circuits of A_1 and A_2. The computation is initiated by simultaneously switching S_1 and S_2 to the "integrate" mode. The configuration of the integrate, hold, and set modes of an integrator is shown in Fig. 4.19b. Possible solutions for x and dx/dt are shown in Fig. 4.19c and may be displayed using an oscilloscope or chart recorder.

4.10 SUMMARY

Mathematical operators are operational amplifier circuits that perform mathematical operations. The amplifier circuits that simulate such functions include scalers, summing and differencing amplifiers, integrators and differentiators, log and antilog amplifiers, absolute value amplifiers, and analog multipliers.

A summing amplifier is implemented using either of the basic amplifier circuits. The multiple input sources, in each case, are summed using a multiple resistor input network. The difference amplifier algebraically subtracts two voltages and is an extension of the inverting amplifier. The integrator is a frequently applied circuit and, when its input is a dc voltage, its output is a voltage ramp that serves as a time base in sweeplike circuits. The ideal differentiator circuit is beset with practical related problems and, when applied, is used in a modified form.

Log and antilog amplifiers utilize the logarithmic relationship between the current and the voltage in a diode to develop a logarithmic relationship between the circuit's input and output voltages. These circuits serve as models in analog multipliers where the output voltage is the product of the two input voltages. The analog multiplier, as a functional block, is diversely applied in circuits whose functions vary from mathematical operations to phase detection.

The analog computer is a system level application of most of the mathematical operators discussed in this chapter. The mathematical operations involved in the solution of a problem are performed in the computer as analogous operations.

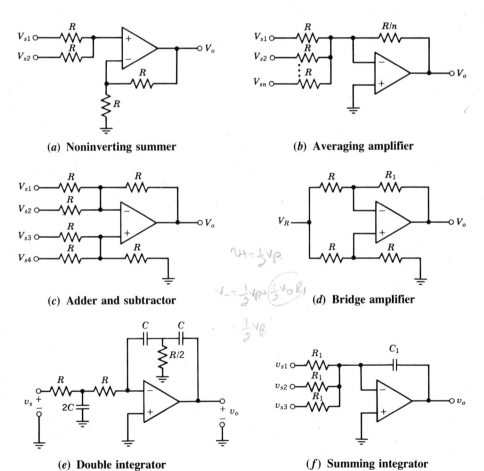

Figure 4.20 Additional mathematical operator circuits.

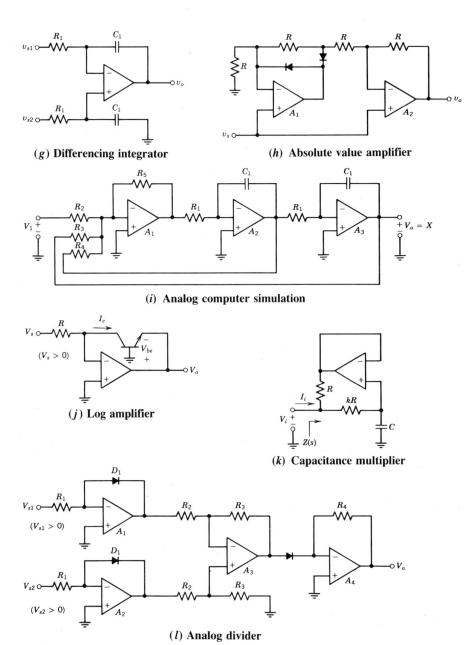

Fig. 4.20 Additional mathematical operator circuits. (*Continued*)

PROBLEMS

4.1 Design a circuit to implement the function

$$V_o = V_{s1} - 2V_{s2} + 3V_{s3}$$

where V_{s1}, V_{s2}, and V_{s3} are available source voltages.

4.2 Calculate the voltages and currents in the summing amplifier of Fig. 4.1a if $R_4 = 2R_3 = 4R_2 = 8R_1 = 0.5R = 16\ \text{k}\Omega$, $V_{s1} = V_{s2} = V_{s3} = V_{s4} = 10\ \text{V}$. What is special about the relationship between the currents in the input resistors?

4.3 Design a circuit to implement the function

$$V_o = \frac{V_{s1}}{2} - \frac{V_{s2}}{4}$$

4.4 Design a circuit to implement the function

$$V_o = \tfrac{1}{4}(V_{s1} + V_{s2} + V_{s3} + V_{s4})$$

The load for each source must be 50 kΩ or greater.

4.5 Find V_o as a function of V_{s1} and V_{s2} for the difference amplifier in Fig. 4.21.

4.6 Specify a set of values for the resistors in the difference amplifier of Fig. 4.21 to implement the function

$$V_o = 2V_{s1} - 3V_{s2}$$

4.7 Specify a set of values for the integrate-and-hold circuit in Fig. 4.6c, and the proper timing signals for S_1 and S_2, so that the periodic output is as shown in Fig. 4.22. Let $R_1 = 0\ \Omega$ and $V_S = -1\ \text{V}$.

4.8 Specify a set of reasonable values for R, C, and V_s in the ideal integrator circuit of Fig. 4.6a such that the output voltage has a slope of 0.01 V/μsec.

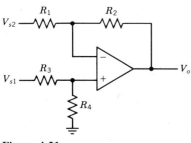

Figure 4.21

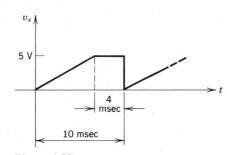

Figure 4.22

72 Mathematical Operators

4.9 The input to the ideal differentiator in Fig. 4.7a is the triangular wave shown in Fig. 4.23. Sketch the waveform of the output voltage if $R = 18$ kΩ and $C = 0.02$ µF.

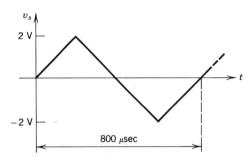

Figure 4.23

4.10 Calculate the currents in the capacitor and resistor of the differentiator described in Problem 4.9.

4.11 Design a practical differentiator (Fig. 4.7b) whose amplitude Bode plot is shown in Fig. 4.24.

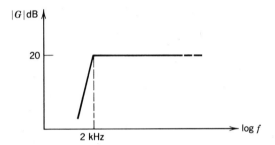

Figure 4.24

4.12 Sketch the output voltage of the practical differentiator of Fig. 4.7b if $R_1 = 2R = 20$ kΩ, $C = 0.01$ µF, and the source is a 10 kHz sinusoidal voltage with a peak amplitude of 2 V.

4.13 Design an ac-to-dc convertor (Fig. 4.9) whose input is described by

$$v_s = \sin 2\pi \cdot 10^3 \, t \text{ V}$$

What is the value of the output voltage? Modify the circuit so that its output voltage is 0.707 V.

4.14 Sketch the output voltage of the half-wave rectifier in Fig. 4.10a if $R_1 = 10$ kΩ and the source is a 10 kHz sinusoidal voltage with a peak amplitude of 2 V. Assume V_F of D_1 is 0.6 V.

4.15 Sketch the transfer characteristic for the half-wave rectifier circuit in Fig. 4.10a if the ideal diode D_1 is reversed.

4.16 How is the transfer characteristic in the half-wave rectifier of Fig. 4.10c changed when the feedback resistor has twice the value of input resistor.

4.17 Design a circuit similar to that of Fig. 4.17d to provide an output voltage proportional to the cube root of its input voltage.

4.18 Design a circuit similar to that of Fig. 4.17d to provide an output voltage proportional to the fourth root of its input voltage.

4.19 Express V_o in terms of V_{s1} and V_{s2} for the noninverting summer of Fig. 4.20a. If $V_{s1} = 2$ V, $V_{s2} = -4$ V, and $R = 10$ kΩ, calculate V_o and the voltage at the amplifier's noninverting input terminal. What is the amplifier's output current?

4.20 Sketch the output voltage of the noninverting summer in Fig. 4.20a if $R = 10$ kΩ and

$$V_{s1} = 4 \sin 2\pi \cdot 10^3 t \text{ V} \quad \text{and} \quad V_{s2} = -2 \text{ V}$$

4.21 Derive the input-output relationship for the averaging amplifier in Fig. 4.20b for $n = 4$.

4.22 Derive the input-output relationship for the adder-and-subtractor circuit in Fig. 4.20c.

4.23 Find V_o in the adder-and-subtractor circuit of Fig. 4.20c if $V_{s1} = 1$ V, $V_{s2} = 2$ V, $V_{s3} = 3$ V, and $V_{s4} = 4$ V.

4.24 The circuit in Fig. 4.20d is a bridge amplifier. All resistances are equal except R_1, which may be an element in a control system that is influenced by such factors as light or temperature. The voltage V_R is a constant. Prove that

$$V_o = \frac{V_R}{2R}(R - R_1)$$

4.25 Derive $V_o(s)/V_s(s)$ for the double integrator in Fig. 4.20e.

4.26 Find v_o in terms of v_{s1}, v_{s2}, v_{s3}, and circuit constants in the summing integrator of Fig. 4.20f.

4.27 Find $V_o(s)$ in terms of $V_{s1}(s)$, $V_{s2}(s)$, and circuit constants in the differencing integrator of Fig. 4.20g.

4.28 The input signal for the absolute value amplifier in Fig. 4.20h is

$$v_s = \sin 2\pi \cdot 10^3 t \text{ V}$$

Sketch the output voltage assuming ideal diodes.

4.29 The circuit in Fig. 4.20i is an analog computer simulation of the differential equation

$$A \frac{d^2x}{dt^2} + B \frac{dx}{dt} + Cx + D = 0$$

Find A, B, C, and D in terms of circuit constants. Let $R_1C_1 = 1$ sec.

4.30 Derive an expression relating V_o and V_s for the log amplifier in Fig. 4.20j. For the transistor, assume

$$I_c = I_S \exp \frac{V_{be}}{V_T}$$

4.31 The circuit in Fig. 4.20k is a capacitance multiplier. Prove that

$$Z(s) = \frac{1}{s(k+1)C} + \frac{kR}{k+1}$$

$$\cong \frac{1}{sC_{eq}} + R$$

where $C_{eq} = (k+1)C$.

4.32 Find the input-output relationship for the analog divider in Fig. 4.20l. Assume the diodes are all matched and

$$I_d = I_S \exp \frac{V_d}{V_T}$$

Chapter 5

Convertors

5.1 PROLOGUE

Convertors, typically, are amplifier circuits that *convert a nonvoltage input variable to a proportionate output voltage*. These circuits translate or change the output signals of transducers to a voltage that can be further processed. Transducers are devices that convert energy forms from one type to another and are often found in the input and output ends of large electronic systems. The output of most input transducers is not a voltage; examples include microphones, photovoltaic cells, and temperature sensors. Carbon microphones are mechanical energy (vibrating columns of air)-to-resistance transducers while photovoltaic cells are light energy-to-electrical current transducers. Operational amplifier circuits are used to convert, and usually amplify, these nonvoltage signals for further processing and conditioning.

A stereo is a common example of an electronic system with input and output transducers. For this case, the input transducer is the phonograph needle or cartridge whose output is best viewed as a current source. A high input impedance, high gain amplifier circuit converts this low level, input current to a voltage that can be further amplified. Ultimately, a power amplifier is used to drive the speaker. The speaker is the output transducer that converts electrical energy to mechanical energy that produces sound.

Most electronic systems today are a combination of analog and digital electronics.

75

76 Convertors

In the system, signals are converted from one form to the other, often times using analog-to-digital and digital-to-analog convertors. These specific convertors are discussed in Chapter 10, but a key component in their design, the comparator, is introduced in this chapter.

5.2 CURRENT-TO-VOLTAGE CONVERTORS

A common application requirement is to monitor and measure a given current signal, and then digitize the result for further processing. For example, in a diode test instrument, it may be necessary to produce a digital readout of diode current for a given bias voltage. In such a case, a current-to-voltage convertor would be used to convert the current to a voltage. Then an analog-to-digital convertor will digitize the analog voltage for display and analysis purposes.

5.2.1 Current-to-Voltage Convertor

The current-to-voltage convertor, Fig. 5.1, is one of the simplest amplifier applications. The circuit contains an amplifier and a single resistor connected from the output back to the inverting input. The input current I is delivered by a current source. The source can be a laboratory generator, a Norton equivalent source of another circuit, a current transducer, or the leakage current of a transistor or other active device.

If we apply the current constraint and Kirchhoff's Current Law (KCL) to node a in Fig. 5.1,

$$I = I_R$$

Since the inverting terminal of the operational amplifier is at virtual ground,

$$V_R = -V_o = IR$$

or

$$V_o = -IR \tag{5.1}$$

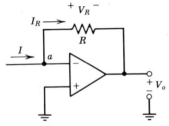

Figure 5.1 Current-to-voltage convertor.

5.2.2 Difference Current-to-Voltage Convertor

The circuit in Fig. 5.2, an extension of the basic current-to-voltage convertor, produces an output voltage proportional to the difference of the two input currents I_{s2} and I_{s1}. The resistor R connected from the amplifier's noninverting terminal to ground converts the current I_{s2} to a voltage that is summed at the output with the voltage drop developed by I_{s1} and the resistor R in the negative feedback loop. Because of the current constraint, the voltage at the noninverting input terminal of the amplifier shown in Fig. 5.2 is

$$V_{ni} = I_{s2}R$$

Because of the voltage constraint, the voltage at the inverting input terminal of the op amp is

$$V_i = V_{ni} = I_{s2}R$$

The amplifier output voltage V_o will be the sum of the voltage drop across R in the feedback loop and the voltage at the inverting input,

$$V_o = -I_{s1}R + V_i = -I_{s1}R + I_{s2}R$$

or

$$V_o = (I_{s2} - I_{s1})R \tag{5.2}$$

Example

In the difference current-to-voltage convertor in Fig. 5.2, $V_o = -1.0$ V, $R = 20$ kΩ, and $V_i = +0.20$ V. Determine the differential input current $I_{s2} - I_{s1}$, I_{s2}, and I_{s1}.

Solution

Using (5.2), we obtain

$$I_{s2} - I_{s1} = \frac{-1 \text{ V}}{20 \text{ k}\Omega} = -50 \text{ }\mu\text{A}$$

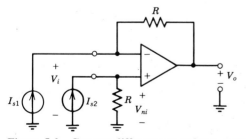

Figure 5.2 Current difference-to-voltage convertor.

If $V_i = +0.2$ V, then

$$V_i = V_{ni} = +0.2 \text{ V}$$

and

$$I_{s2} = \frac{V_{ni}}{R} = \frac{0.2 \text{ V}}{20 \text{ k}\Omega} = 10 \text{ }\mu\text{A}$$

From $I_{s2} - I_{s1} = -50$ μA

$$I_{s1} = 60 \text{ }\mu\text{A} \qquad \blacksquare$$

5.3 CHARGE-TO-VOLTAGE CONVERTOR

The conversion of very small currents to a proportionate voltage requires high value resistors. High value resistors are typically bulky, expensive, and susceptible to noise. An alternate scheme, which overcomes this problem, is to convert the charge that corresponds to the sample of current to a proportionate voltage.

The charge amplifier, or charge-to-voltage convertor, in Fig. 5.3 uses a capacitor in the feedback loop to integrate the input current. If we apply the voltage and current constraints,

$$v_o = -v_c$$

and

$$i = i_c$$

For the capacitor,

$$v_c = v_c(0) + \frac{1}{C}\int_0^t i_c \, dt$$

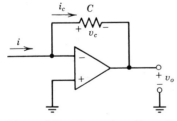

Figure 5.3 Charge-to-voltage convertor.

The output voltage of the circuit is equal to the voltage across the capacitor but with the opposite polarity, that is,

$$v_o = -v_c = -v_c(0) - \frac{1}{C}\int_0^t i_c\, dt$$

If we let q equal the net charge corresponding to the current flow during the time interval from 0 to t, then

$$v_o = -v_c(0) - \frac{q}{C} \tag{5.3}$$

When the input current is a constant (dc), that is, $i = I$, during the time interval under consideration, the output voltage is

$$v_o = -v_c(0) - \frac{1}{C}\int_0^t I\, dt$$

or

$$v_o = -v_c(0) - \frac{It}{C}$$

Thus, the output voltage is linear with a slope of

$$\frac{dv_o}{dt} = -\frac{I}{C}$$

Example

The nonideal op amp's input (bias) current in the charge-to-voltage convertor in Fig. 5.3 is 100 nA. What is the slope of the output voltage for this circuit if the circuit integrates the bias current only and the value of the feedback capacitor is $C = 0.01\ \mu F$?

Solution

$$\frac{dv_o}{dt} = \frac{I}{C} = \frac{100\ \text{nA}}{0.01\ \mu F} = 0.01\ \text{V/msec}$$

The error due to bias current is minimized if the signal current is made much larger than the bias current. If the signal current, in this example, was 50 μA, the error due to the bias current would be 0.2%. ∎

5.4 COMPARATORS

Comparators or level detectors are bistable devices or circuits whose output assumes one of two possible voltage levels depending on the relative values of the signals at the input terminals,

$$V_o = \text{HIGH} \quad \text{for} \quad V_{ni} > V_i$$

and

$$V_o = \text{LOW} \quad \text{for} \quad V_{ni} < V_i$$

Comparators convert the relative values of the two input signals into a digitized output. IC comparators are designed as specialized amplifiers and sold under that name.

The comparator differs from the operational amplifier in a number of respects. The comparator is typically operated open loop; the amplifier seldom is. The comparator is also a much faster device. The slew rate (the maximum rate of change of the output voltage) of a comparator is typically 25 V/μsec while that of the amplifier is 0.5 V/μsec (101A). The comparator also performs a different function. It is a device that has two output states that indicate the relative values of two input voltages. The comparator, essentially, converts the relative values of two analog voltages to a digital decision. The two output states are discrete and usually the levels are compatible with some digital logic family. Occasionally, general-purpose op amps are used as comparators but their slow speed, compared to digital circuits, prohibits their extensive use.

The schematic symbol for a comparator is the same as for the operational amplifier. Both devices have two inputs, an output, and two dc supply pins. However, many comparators also have *ground* and *strobe* pins. The ground pin is used to bias the output stage for digital logic compatibility. The strobe pin is used to enable the comparator at a given time to make a decision based on the voltage at its inputs, or to disable the comparator and have its output go to a predetermined quiescent state. The comparator, like the op amp, may also have two pins to control the input offset voltage.

Two of the more popular comparators are the 710 and the 111. Of the two, the 710 is much faster. However, the 111 is a more versatile device and is frequently used in applications where speed is not of paramount importance. The *output* pin, Fig. 5.4, of this comparator is connected to the open collector of the comparator's output transistor, and the *ground* pin is connected to the emitter. This allows the designer the flexibility of connecting the output, through an external pull-up resistor, to any positive voltage (including zero volts) and the emitter to any negative voltage. Thus, the comparator output can be biased to permit interfacing to various logic families, mechanical and solid-state switches, and output transducers.

The reason for the open collector and open emitter of the output transistor in the 111 is to enhance the ability of the comparator output to interface to various loads. If the comparator must drive a 5 V logic device, a resistor is connected from the collector, or the comparator's output, to +5 V. Its value is in the 1 to 5 kΩ range. The transistor's

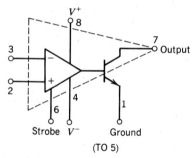

Figure 5.4 111-type comparator.

emitter is connected to ground. When $V_i > V_{ni}$, the output transistor is driven into saturation and sinks a collector current (1–5 mA) from the 5 V supply through the external resistor. When $V_i < V_{ni}$, the output voltage is "pulled-up" to +5 V through the external resistor.

Comparators are used in decision making circuits, and they serve as a building block in digital-to-analog and analog-to-digital circuits. In these circuits, continuously varying analog signals are converted into digital signals of discrete levels.

5.4.1 Detectors

Zero-crossing Detector The zero-crossing detector of Fig. 5.5, ideally, will assume one of two output voltage levels depending on the polarity of the input voltage V_s. The output of the amplifier will be either in negative saturation $V_o(\min)$, or in positive saturation $V_o(\max)$. The input and output are related by

$$V_o = V_o(\max) \quad \text{for} \quad V_s < 0 \tag{5.4a}$$

and

$$V_o = V_o(\min) \quad \text{for} \quad V_s > 0 \tag{5.4b}$$

The amplifier, or comparator, is operated open-loop. This circuit is a nonlinear application of the op amp, and the input constraints do not apply. The potential difference between the amplifier input terminals V_{in} is the same as the source voltage V_s.

In the case of a nonideal amplifier or comparator, there will be a small range of

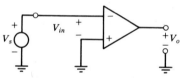

Figure 5.5 Zero-crossing level detector.

values of V_s where V_o and V_s are linearly related because of the finite value of the amplifier's open-loop gain. This range, called a threshold region, is about ± 150 μV. Other amplifier nonidealities, like input offset voltage, extend the region where the amplifier output and input are not related by (5.4).

The block diagram of the phase meter in Fig. 5.6 illustrates the application of two zero-crossing detectors. The detectors, in conjunction with a time interval counter and a time-to-phase convertor, provide an output that is proportional to the phase difference between two sinusoidal input signals. The top detector is used to initiate the time counting when v_{s1} just starts its positive half cycle, and the other detector stops the counting when v_{s2} begins its positive half cycle. The time measured between these two zero crossings is then converted into a phase difference to complete the design of the phase meter.

Nonzero Crossing Detectors A comparator can make a decision at a value other than zero volts. In the case of the circuit in Fig. 5.7a, the comparison or decision is make at the reference voltage connected to the noninverting input. The transfer characteristic for this circuit is shown in Fig. 5.7b,

$$V_o = V_o(\max) \quad \text{for} \quad V_s < V_R \quad (5.5a)$$
$$V_o = V_o(\min) \quad \text{for} \quad V_s > V_R \quad (5.5b)$$

If the amplifier input terminals are reversed, the polarity of the output signal will be reversed. This circuit is appropriate for applications where V_R is a programmable or fixed source, or if a zener or reference diode of the desired value is available.

A second technique of making a nonzero detector is shown in Fig. 5.8. The reference voltage V_R is summed with the signal voltage through a voltage divider composed of R_1 and R_2. The output voltage is a function of the differential input voltage V_{in}, which in turn is a function of V_s, V_R, R_1, and R_2. We can find V_{in} by using superposition,

$$V_{\text{in}} = V_s \frac{R_2}{R_1 + R_2} + V_R \frac{R_1}{R_1 + R_2}$$

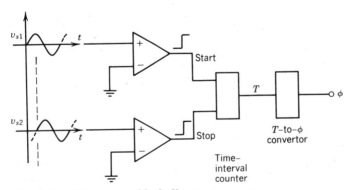

Figure 5.6 Phase meter block diagram.

Comparators   83

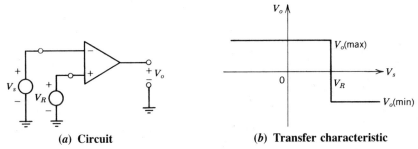

(a) Circuit   (b) Transfer characteristic

Figure 5.7 Nonzero crossing level detector.

The output will assume one of two possible levels depending on whether V_{in} is positive or negative. The polarity of V_{in} will depend on the signal and reference sources, and the resistors:

$$V_o = V_o(\text{max}) \quad \text{for} \quad V_{in} < 0 \quad \text{or} \quad V_s < -\frac{R_1}{R_2} V_R \quad (5.6a)$$

$$V_o = V_o(\text{min}) \quad \text{for} \quad V_{in} > 0 \quad \text{or} \quad V_s > -\frac{R_1}{R_2} V_R \quad (5.6b)$$

Window Detector The use of two comparators and two reference voltages, connected as shown in Fig. 5.9a, produces a window detector. This circuit provides a low output voltage if the signal source V_s is more positive than the upper limit (V_{UL}) reference voltage or less positive than the lower limit (V_{LL}) voltage. The circuit output is high if the signal source is more positive than the lower limit and less positive than the upper limit. The circuit detects a range or window of input voltage and its transfer characteristic is shown in Fig. 5.9b. Mathematically,

$$V_o = \text{HIGH} \quad \text{for} \quad V_{LL} < V_s < V_{UL} \quad (5.7a)$$

and

$$V_o = \text{LOW} \quad \text{for} \quad V_s > V_{UL} \quad \text{or} \quad V_s < V_{LL} \quad (5.7b)$$

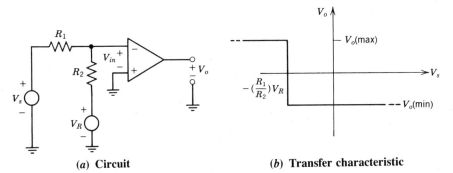

(a) Circuit   (b) Transfer characteristic

Figure 5.8 Another nonzero level detector.

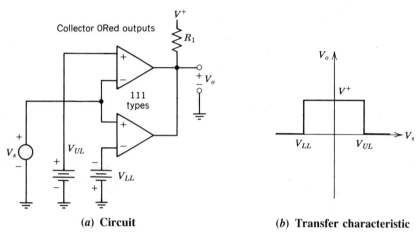

(a) Circuit (b) Transfer characteristic
Figure 5.9 Window detector.

Many comparators (111-type) have open-collector outputs and can be readily interfaced to various loads. In the window detector, the open collectors of each comparator are tied together to form a wire ORed configuration. They are pulled up to V^+ through R_1. When interfacing this circuit, V^+ would be chosen to match the voltage requirements of the load, and R_1 would be chosen to satisfy the current or speed requirements.

Example

What is the maximum magnitude of V_{in} for the nonzero crossing detector in Fig. 5.8 if V_s varies from -10 to $+10$ V, $R_1 = 10$ kΩ, $R_2 = 30$ kΩ, and $V_R = +2$ V. What is the output state for $V_s = -5$ V?

Solution

For $V_s = +10$ V,

$$V_{in} = V_R + V_{R2}$$
$$= 2\text{ V} + \frac{30\text{ k}\Omega}{40\text{ k}\Omega}(10 - 2)\text{ V}$$
$$= +8\text{ V}$$

For $V_s = -10$ V,

$$V_{in} = V_R + V_{R2}$$
$$= 2\text{ V} + \frac{30\text{ k}\Omega}{40\text{ k}\Omega}(-10 - 2)\text{ V}$$
$$= -7\text{ V}$$

The inputs of the amplifier see a voltage that varies from -7 to $+8$ V depending on the specific value of V_s. The maximum magnitude of V_{in} is 8 V. The amplifier or comparator must have an input differential voltage rating of this value or greater.

The crossover (high to low) voltage for the circuit is $(-R_1/R_2)(V_R) = -0.67$ V. For $V_s = -5$ V, the output will be low or at $V_o(\min)$. ∎

5.4.2 Interfacing to Various Loads

The versatility of the open-collector output of the 111-type comparator is illustrated when interfacing the comparator to various loads. Figure 5.10 shows the comparator driving six different loads; a LED (light emitting diode), a mechanical relay, an incandescent lamp, MOS logic, and shunt and series switches.

The LED and the current limiting resistor R_1 are the pullup circuitry for the output transistor of the comparator circuit shown in Fig. 5.10a. The LED, typically rated at 1.5 V and 20 mA, is on or off depending on the on or off status of the output transistor in the comparator, which in turn depends on the comparator's input voltages.

The output of the comparator in Fig. 5.10b drives the coil of a mechanical relay; diode D_1 limits the reverse voltage across the coil during turnoff. The comparator, typically, can be used to interface to loads that require up to 40 V and 40 mA. However, for high loads, a transistor must be used as in the case of the circuit shown in Fig. 5.10c. The transistor in this circuit drives the high voltage, high current incandescent lamp, and the comparator output is used to control the *pnp* driver. R_3, with V^-, establishes the base current.

The high and low levels of the MOS logic block in Fig. 5.10d are 0 and -12 V. The output stage of the comparator converts the relative values of the comparator's input voltages to logic levels compatible with the MOS logic. This technique can be used to interface to almost any logic family.

The circuits in Figs. 5.10e and 5.10f control JFETs (junction field-effect transistors) that function as solid-state switches. Q_1 and R_L simulate the application of a switch and load in shunt and in series with the source V_s. In the shunt switch case, a 10 kΩ load resistor is connected in shunt with V_s when Q_1 is on. The $r_{ds}(\text{on})$ of the JFET, typically 50 Ω, is in series with the load but introduces a negligible error when compared to the 10 kΩ. JFETs are depletion-mode devices, which means the transistor is on when V_{gs} is 0 V and off when V_{gs} is at some "large" negative voltage. Transistor Q_1 is on, or the switch is closed, when the comparator output is at 0 V. When the comparator output is at -15 V, V_{gs} (or V_{gd}) is more negative than the pinchoff (turn off) voltage and the JFET is off, thus simulating an open switch.

In the series switch case, Fig. 5.10f, the voltage source V_s is connected and disconnected from the load by the series switch Q_1. Q_1 is on, or the switch is closed, when the comparator output is at $+15$ V. For this case, diode D_1 is reverse-biased and V_{gs} is 0 V ($I_{R2} = 0$ A) because of the dc path provided by the 1 MΩ from the transistor's source (S) to the high impedance gate (G). When the comparator is at -15 V, D_1 is forward-biased. The gate is at approximately -14.5 V insuring that V_{gs} is more negative than the pinchoff or turnoff voltage, independently of the value of V_s (± 10 V). Integrated circuit versions of solid-state switches are commercially available.

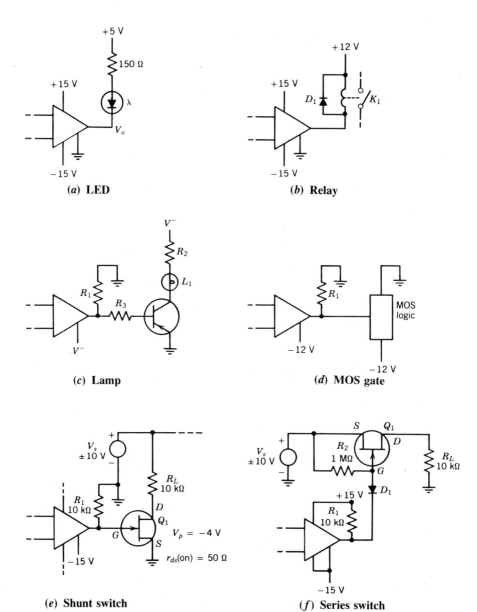

Figure 5.10 Interfacing to various loads.

In certain, select applications, an op amp can be used instead of the comparator when interfacing to various loads. However, the faster speed and the flexibility of the open-collector output usually makes the comparator a better choice.

5.4.3 Schmitt Triggers

A Schmitt trigger is a comparator with hysteresis. Most comparators make a decision by comparing a time-varying signal with a predetermined or programmable voltage reference. The reference for the Schmitt trigger is generated by the circuit itself and is usually a function of the output voltage of the amplifier or comparator and discrete resistors in the circuit. The comparison decision will be made at one of two references or trip voltages and will also depend on the direction of the variation of the incoming voltage.

Inverting Schmitt Trigger   An inverting Schmitt trigger is shown in Fig. 5.11a. This circuit has the first appearance of a noninverting amplifier. However, a closer inspection of the input terminals reveals positive feedback. Because of the positive feedback, the comparator output will be at one saturation limit or the other. This limit is approximately one volt less in magnitude than the corresponding supply voltage and is load sensitive,

$$V_o(\text{max}) \cong (V^+ - 1\text{ V}) \tag{5.8a}$$

$$V_o(\text{min}) \cong (V^- + 1\text{ V}) \tag{5.8b}$$

The voltage at the noninverting input is a function of the maximum or minimum output voltage and the R_1–R_2 divider. The upper limit reference voltage is

$$V_{UL} = \frac{R_1}{R_1 + R_2} V_o(\text{max}) \tag{5.9}$$

and the lower limit reference voltage is

$$V_{LL} = \frac{R_1}{R_1 + R_2} V_o(\text{min}) \tag{5.10}$$

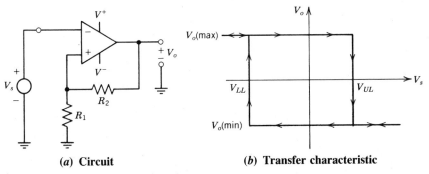

(a) Circuit   (b) Transfer characteristic

Figure 5.11   Inverting Schmitt trigger.

88 Convertors

These limits will be unsymmetric about the ordinate if the V^+ and V^- supplies have different magnitudes.

The transfer characteristic for the inverting Schmitt trigger is shown in Fig. 5.11b. The specific reference voltage (V_{LL} or V_{UL}), at which the output changes state, is a function of the path the input signal is traversing, that is, from low to high or high to low. For large negative values of V_s, the output of the operational amplifier will be at $V_o(\max)$ and the threshold voltage will be at V_{UL}. When V_s reaches V_{UL}, the output switches to $V_o(\min)$ and the threshold changes to V_{LL}. This output state and reference are maintained for all increasing values of V_s. If V_s is then decreased, the output state will change when V_s reaches V_{LL}. The output then switches to $V_o(\max)$ and the reference voltage will be V_{UL}. This circuit behavior produces a hysteresis effect and is graphically illustrated in the circuit transfer characteristic.

Noninverting Schmitt Trigger The noninverting Schmitt trigger of Fig. 5.12a differs from the inverting version in that one comparator input terminal is grounded. When the amplifier's input voltage V_{in} goes positive, the output goes positive to $V_o(\max)$. When V_{in} goes negative, the output goes to $V_o(\min)$. Using superposition, we can express V_{in} in terms of V_o and V_s,

$$V_{in} = V_s \frac{R_2}{R_1 + R_2} + V_o \frac{R_1}{R_1 + R_2}$$

If

$$V_{in} = \frac{V_s R_2 + V_o R_1}{R_1 + R_2} > 0$$

or

$$V_s > -\frac{R_1}{R_2} V_o$$

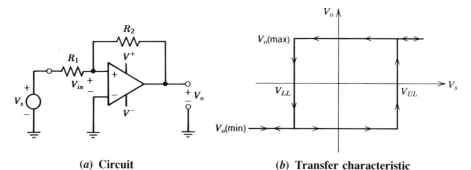

(a) Circuit (b) Transfer characteristic

Figure 5.12 Noninverting Schmitt trigger.

then
$$V_o = V_o(\text{max})$$

If
$$V_{in} = V_s + \frac{R_1}{R_1 + R_2}(V_o - V_s) < 0$$

or
$$V_s < -\frac{R_1}{R_2} V_o$$

then
$$V_o = V_o(\text{min})$$

Thus,
$$V_{LL} = \frac{R_1}{R_2} V_o(\text{min}) \qquad (5.11)$$

and
$$V_{UL} = \frac{R_1}{R_2} V_o(\text{max}) \qquad (5.12)$$

The transfer characteristic for the noninverting Schmitt trigger is shown in Fig. 5.12b. For large negative values of V_s, the output of the circuit is at $V_o(\text{min})$. This state will be maintained until V_s goes positive and reaches the upper limit V_{UL}. For values of V_s more positive than V_{UL}, the output goes to $V_o(\text{max})$. As V_s is decreased, the output remains at $V_o(\text{max})$. When V_s reaches the lower limit V_{LL}, the output switches to $V_o(\text{min})$.

Example

Calculate the upper and lower limits for the inverting and noninverting Schmitt trigger circuits of Figs. 5.11a and 5.12a if, for each circuit, $R_1 = 5$ kΩ, $R_2 = 20$ kΩ, $V_o(\text{min}) = -10$ V, and $V_o(\text{max}) = +10$ V.

Solution

Using (5.9) and (5.10), we find that the upper and lower limits for the inverting Schmitt trigger are
$$V_{UL} = \frac{R_1}{R_1 + R_2} V_o(\text{max})$$
$$= \tfrac{1}{5}(10 \text{ V}) = +2 \text{ V}$$
and $V_{LL} = -2$ V.

90 Convertors

When we use (5.11) and (5.12), the upper and lower limits for the noninverting Schmitt trigger are

$$V_{UL} = \frac{R_1}{R_2} V_o(\text{max})$$
$$= \tfrac{1}{4} (10 \text{ V}) = +2.5 \text{ V}$$
$$\text{and } V_{LL} = -2.5 \text{ V}. \qquad\blacksquare$$

5.5 PHASE DETECTORS

The phase detector model shown in Fig. 5.13a consists of an ac voltage source v_s, an electronic switch S_1, and a load R_L. The source and the switch are operated at the same frequency but are out of phase by θ degrees. When the switch control signal v_1 is high (Fig. 5.13b), the switch is closed and the source v_s is applied across R_L. When v_1 is low, the switch is open and the output is zero. The average value V_o of the instantaneous output voltage v_o is

$$V_o = \frac{1}{2\pi} \int_{\theta}^{\theta+\pi} V_s \sin(\omega t)\, d(\omega t)$$
$$= \frac{V_s}{2\pi} [-\cos(\omega t)] \Big|_{\theta}^{\theta+\pi}$$
$$= \frac{V_s}{2\pi} [-\cos(\pi + \theta) + \cos \theta]$$
$$= \frac{V_s}{\pi} \cos \theta$$

Normally, a low-pass filter is located after the phase detector, so that the high-frequency components of V_o are attenuated and the dc component ($V_s \cos \theta / \pi$) is the dominant portion of the output voltage. Thus, the filtered output of the phase detector, also called a phase comparator or synchronous detector, is directly proportional to the cosine of the phase difference between v_s and v_1. The output of the detector has its maximum value when $\theta = 0°$, its minimum value when $\theta = 180°$, and is zero when $\theta = 90°$.

An alternate scheme of implementing a phase detector is shown in Fig. 5.14. The circuit has two sinusoidal inputs of the same frequency displaced by a phase angle θ, and consists of an analog multiplier, and an op amp low-pass filter. The output of the multiplier is

$$v_1 = k V_{s1} V_{s2} \cos(\omega t) \cos(\omega t + \theta)$$

or

$$v_1 = \frac{k V_{s1} V_{s2}}{2} [\cos \theta + \cos(2\omega t + \theta)]$$

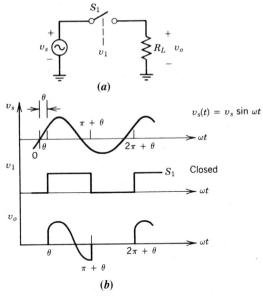

Figure 5.13 Phase detector model.

The low-pass filter (discussed in Section 8.3) that follows will attenuate or eliminate the cos $(2\omega t)$ term and produce a dc output voltage proportional to the phase angle θ,

$$V_o = \frac{kV_{s1}V_{s2}}{2} \cos \theta \tag{5.13}$$

Example

What type of an output is obtained when the filter in Fig. 5.14 is a high-pass filter instead of a low-pass filter?

Solution

A high-pass filter in Fig. 5.14 will reject the dc component at the output that is proportional to phase. Instead, the output will be a sinusoidal signal at twice the frequency of its inputs. The circuit is a frequency doubler. ∎

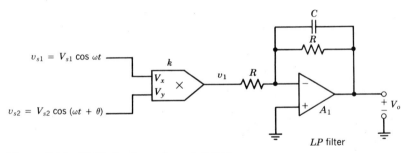

Figure 5.14 Multiplier-type phase detector.

5.6 SUMMARY

In the strictest sense, most operational amplifier circuits are convertors. The output signal is, in general, different from the input. However, the convertors discussed in this chapter convert an input variable to, typically, an output voltage. The current-to-voltage, charge-to-voltage, and phase-to-voltage circuits certainly fall into this category.

A somewhat different convertor is the device called a comparator. The comparator is a specially designed amplifier and has two discrete output states that are related to the relative values of the voltages at the input terminals. The comparator is used in detectors, Schmitt triggers, and analog-to-digital and digital-to-analog convertors.

The detector is the circuit-level equivalent of the comparator and, through the external circuitry, has the added flexibility of shifting and modifying the transfer characteristic. A Schmitt trigger is a comparator with hysteresis and is typically used in noisy environments as a squaring circuit.

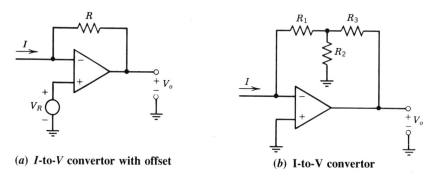

(a) I-to-V convertor with offset

(b) I-to-V convertor

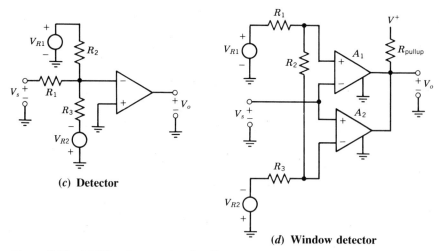

(c) Detector

(d) Window detector

Figure 5.15 Additional convertor circuits.

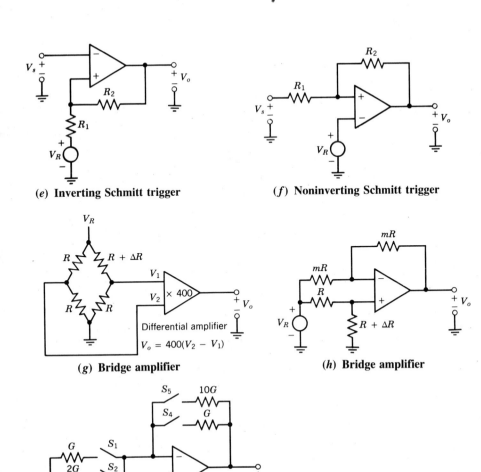

Fig. 5.15 (*Continued*)

PROBLEMS

5.1 The circuit in Fig. 5.15a is a current-to-voltage convertor with an offset. The reference V_R can represent an external source, or it could represent the amplifier's input dc offset voltage called V_{OS}. (a) For the circuit, find V_o in terms of I and circuit constants. (b) If $V_R = V_{OS} = 1$ mV and $I = 1$ μA, what minimum value of R is required so that V_{OS} introduces an error of no more than 1% at the output?

5.2 Calculate V_o for the current difference-to-voltage convertor in Fig. 5.2 if $I_{s1} = 400$ nA, $I_{s2} = 800$ nA, $R = 1$ MΩ, and the amplifier input currents (noninverting and inverting) are sinking 100 and 150 nA, respectively.

5.3 For the I to V convertor in Fig. 5.15b, (a) prove

$$V_o = -I\left(R_1 + R_3 + \frac{R_1 R_3}{R_2}\right)$$

(b) Find the approximate expression for V_o if $R_1 >> R_2$.

5.4 Calculate the circuit currents and voltages for the I to V convertor in Fig. 5.15b if $I = 100$ μA, $R_1 = 10$ kΩ, $R_2 = 20$ kΩ, and $R_3 = 40$ kΩ.

5.5 Sketch the transfer characteristic for the circuit shown in Fig. 5.15c if $R_1 = R_2 = R_3 = 10$ kΩ, $V_{R1} = 8$ V, and $V_{R2} = 6$ V. Label all key points on the graph.

5.6 Derive the expressions for V_s (in terms of R_1, R_2, R_3, V_{R1}, and V_{R2}) in the detector of Fig. 5.15c when V_o is high [$V_o(\max)$] and when V_o is low [$V_o(\min)$].

5.7 Sketch the transfer characteristic for the circuit shown in Fig. 5.15d if $R_1 = R_2 = R_3 = 10$ kΩ, $V_{R1} = 8$ V, and $V_{R2} = 6$ V. Label all key points on the graph.

5.8 Derive the expressions for V_s (in terms of R_1, R_2, R_3, V_{R1}, and V_{R2}) in the window detector of Fig. 5.15d when V_o is high ($\cong V^+$) and when V_o is low ($\cong 0$ V).

5.9 For the Schmitt trigger in Fig. 5.15e, prove that

$$V_{UL} = V_R + \frac{R_1}{R_1 + R_2}[V_o(\max) - V_R]$$

and

$$V_{LL} = V_R + \frac{R_1}{R_1 + R_2}[V_o(\min) - V_R]$$

Sketch the transfer characteristic.

5.10 Sketch the transfer characteristic for the inverting Schmitt trigger in Fig. 5.15e if $R_1 = 10$ kΩ, $R_2 = 20$ kΩ, $V_R = 2$ V, $V_o(\max) = 10$ V, and $V_o(\min) = -10$ V.

Problems 95

5.11 Determine V_{LL} and V_{UL} for the Schmitt trigger in Fig. 5.15f. Sketch the transfer characteristic.

5.12 Sketch the transfer characteristic for the noninverting Schmitt trigger in Fig. 5.15f if $R_1 = 10$ kΩ, $R_2 = 20$ kΩ, $V_R = 2$ V, $V_o(\text{max}) = 10$ V, and $V_o(\text{min}) = -10$ V.

5.13 Find V_o in terms of ΔR, $\Delta R/R$, and V_R for the bridge amplifier in Fig. 5.15g.

5.14 Calculate the circuit currents and voltages for the bridge amplifier in Fig. 5.15g if $V_R = 2$ V, $R = 10$ kΩ, and $\Delta R = 100$ Ω.

5.15 The bridge amplifier in Fig. 5.15h is used for high current transducers. Prove that

$$V_o = \frac{\Delta R}{2R + \Delta R} V_R$$

5.16 Calculate the currents and voltages for the bridge amplifier in Fig. 5.15h if $V_R = 12$ V, $R = 100$ Ω, $\Delta R = 10$ Ω, and $m = 100$.

5.17 What must the 150 Ω resistor be changed to in Fig. 5.10a if a $+15$ V supply is used instead of a $+5$ V supply?

5.18 Redesign the interface circuit of Fig. 5.10a if a -15 V supply is used instead of a $+5$ V supply. Assume that V_F of the LED is 2 V, I_D is 20 mA, and V_o (low) is -15 V.

5.19 Redesign the interface circuit of Fig. 5.10c if an *npn* transistor is used instead of a *pnp* transistor.

5.20 Specify V^+ and V^- in the inverting Schmitt trigger of Fig. 5.11 if $V_{UL} = +2$ V, $V_{LL} = -4$ V, $R_1 = 10$ kΩ, and $R_2 = 30$ kΩ.

5.21 Sketch the transfer characteristic for the noninverting Schmitt trigger in Fig. 5.12a if $V^+ = 15$ V, $V^- = -15$ V, $R_1 = 10$ kΩ, and $R_2 = 30$ kΩ.

5.22 Calculate v_1 in the phase detector of Fig. 5.14 if

$$v_{s1} = 0.1 \cos(2\pi \cdot 10^3 t) \text{ V},$$
$$v_{s2} = 0.2 \cos(2\pi \cdot 10^3 t + 30°) \text{V, and } k = 10.$$

5.23 Figure 5.15i is a circuit model for a digital-to-analog convertor. The circuit provides output voltages from 0 to 0.7 V in 0.1 V steps or from 0 to 7 V in 1 V steps ($V_R = 1$ V) depending on the position of the various switches. Show in tabular form the position (open or closed) of the switches S_1 through S_5 to obtain the following values of V_o: 0 V, 0.1 V, 0.2 V, 0.3 V, 4 V, 5 V, 6 V, and 7 V.

5.24 The circuit in Fig. 5.15j is a V-to-I convertor. The output or source current I_s is proportional to the input or control voltage V_c. Prove that

$$I_s = -\frac{V_c}{R_1}$$

Chapter 6

Oscillators

6.1 PROLOGUE

Oscillation is an effect that repeatedly and regularly fluctuates about some mean value; oscillators are circuits that produce oscillation.

The oscillator, a circuit with an output but no signal input, is basically described by the waveshape, frequency, and amplitude. The most common oscillator waveshapes are sinusoidal, rectangular, and sawtooth.

Secondary characteristics include distortion, stability, programmability, range, and those characteristics unique to the components and circuit implementing the oscillator. For certain applications, the secondary characteristics can be of prime importance; an example is stability in communication circuits.

6.2 OSCILLATOR THEORY

Oscillators are circuits that have a periodically varying output. These circuits or sources, typically, do not have an input signal and are used as signal generators in a wide range of applications that vary from test equipment to communications. There are a number of similarities (and distinct differences) between an amplifier and an oscillator.

A typical amplifier circuit is an example of a negative-feedback control system; it has an input signal whose effect, at the output, is reduced because of the circuit's negative feedback. The typical oscillator is an example of a positive-feedback control system; it does not have an external input signal, but will generate an output signal if it has positive feedback and certain conditions are met. Positive feedback is typically introduced by providing a path from the output of the amplifier back to its noninverting input. Feedback oscillators, and amplifiers, must have at least one active or gain-producing element; thus we concentrate on those oscillators that use the operational amplifier as the active device.

The block diagram of a positive-feedback control system model is shown in Fig. 6.1. The technique of finding the gain for the system is similar to that developed for the negative-feedback control system discussed in Section 1.5.2. A is the amplifier gain and F the feedback factor. The input voltage of the amplifier is called the error voltage V_e:

$$V_e = V_s + V_f$$

The output of the amplifier and the output of the feedback network are related to their inputs by A and F,

$$V_o = AV_e$$

and

$$V_f = FV_o$$

If we substitute for V_f and V_e in the expression $V_e = V_s + V_f$, we end up with the circuit gain,

$$\frac{V_o}{V_s} = \frac{A}{1 - AF} \quad \text{positive feedback}$$

The relationship for V_o/V_s is the same for positive and negative feedback control systems except for the sign of the loop gain AF. An oscillator does not have an input signal; thus, if $V_s = 0$, then

$$\frac{V_o}{V_s} = \infty$$

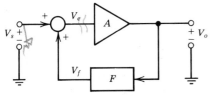

Figure 6.1 Positive feedback control system model.

For the above to be true,

$$1 - AF = 0$$

The quantity AF is, in general, complex and thus contains magnitude and phase information. F may be complex if the feedback network contains reactive components. A may be considered complex if we are operating the amplifier outside its passband where the gain varies with frequency. The condition $1 - AF = 0$ is met when

$$|AF| = 1 \quad \text{(magnitude)} \qquad (6.1a)$$

and

$$\angle AF = 0 \quad \text{(phase)} \qquad (6.1b)$$

These expressions represent the required conditions for oscillations to occur.

If the magnitude of the loop gain is greater than one and the phase is zero, then the amplitude of oscillation will increase exponentially until, in an actual system, something (the supply voltage) restricts the growth. If the magnitude is less than one (and the phase is zero), the amplitude of the oscillation will exponentially decrease to zero.

The frequency of oscillation is the frequency that satisfies the conditions for oscillation. A specific frequency for a given set of values of the components in the feedback network F and gain A will establish this condition and hence the frequency of oscillation. If the magnitude and phase of the loop gain AF are correct, sustained oscillations are ensured.

The feedback network of an oscillator is typically an RC or an LC network. Oscillators with RC feedback networks are normally used in low-frequency applications, and LC oscillators are used in high-frequency cases. The most frequently used active device in low-frequency oscillators is the operational amplifier; in high-frequency circuits, the transistor is used almost exclusively.

Another circuit configuration that produces sinusoidal oscillations is the negative conductance (resistance) oscillator. This type of oscillator uses an LC tank circuit and a device with a negative conductance (resistance) region in its I–V characteristic. Examples of devices with this type of characteristic are the tunnel diode, the four-layer switch or *pnpn* diode, and the unijunction (UJT) transistor.

6.2.1 Oscillation in Amplifiers

One of the most practical problems in amplifier circuits is oscillation (another is grounds). The supposedly negative feedback of the amplifier circuit may end up behaving effectively like positive feedback. A detailed and quantitative description of oscillation in amplifiers is covered in Part III. Our treatment here is qualitative and related to understanding how an amplifier can end up behaving like an oscillator.

Most IC operational amplifiers are designed to have a single-pole response. If the amplifier is surrounded with resistors to form some closed-loop function, the function will also have a single-pole response. If the amplifier is surrounded with resistors and reactive components, additional poles (and zeros) will be introduced into the circuit's gain expression. If the poles introduce the necessary amount of phase shift with a loop gain of one, the circuit will oscillate. When the circuit oscillates, some of the response poles are located on the $j\omega$ axis. In an actual system, the physical connection may be from the output back to the inverting input but, if the net effect produces positive feedback, then oscillation will occur. In many practical circuits, the additional phase shift is due to distributed capacitance and inductance, or reactive loads.

6.3 OSCILLATOR CHARACTERISTICS

The primary characteristics of the oscillator are waveshape, frequency, and amplitude. Other characteristics pertinent to sinusoidal oscillators include distortion, stability, programmability, and those characteristics unique to the components and configuration implementing the oscillator. Linearity, rise and fall time, and output impedance and power are secondary characteristics describing nonsinusoidal oscillators.

Waveshape is the characteristic that describes the form of the output signal. The most common output voltage waveforms are sinusoidal, rectangular, sawtooth, and triangular. Frequency is the number of cycles of the repetitive waveform per unit time and is measured in radians per second or in Hertz for sinusoidal oscillators and in Hertz for the nonsinusoidal cases. Amplitude, for sinusoidal voltages, is measured in RMS or peak-to-peak volts, while for the nonsinusoidal cases, it is measured, typically, in peak-to-peak volts.

Distortion is the characteristic that describes the purity of a desired sinusoidal voltage. Harmonic distortion describes a measure of the amount of a harmonic relative to its fundamental frequency content, and is expressed as a percentage of the latter. Stability describes the ability of a circuit to maintain one or more of its characteristics constant for some given condition(s).

6.4 SINUSOIDAL OSCILLATORS

6.4.1 Wien-Bridge Oscillator

The Wien-Bridge oscillator, Fig. 6.2, is a popular low-frequency (to 1 MHz) oscillator and is characterized by two feedback networks. The negative feedback network, R_3–R_4, establishes the voltage gain of the amplifier, and the R_1–C_1 and R_2–C_2 positive-feedback network establishes the phase shift and attenuation of V_o from the output to the noninverting input terminal.

Conditions for oscillation require that the signal, V_i, must be returned around the loop in phase and with the same magnitude. Assuming an ideal op amp, the voltage gain from the noninverting amplifier terminal to its output is determined by R_3 and R_4,

$$\frac{V_o}{V_i} = 1 + \frac{R_4}{R_3} \qquad (6.2a)$$

100 Oscillators

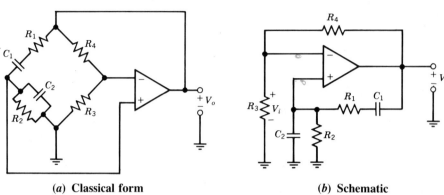

(a) Classical form (b) Schematic

Figure 6.2 Wien-bridge oscillator.

Thus, the output of the op amp will be greater than V_i. The R_1–C_1 and R_2–C_2 feedback network functions as an ac voltage divider and will attenuate the output signal in its transition from the output back to the noninverting input. When the circuit oscillates, the degree of attenuation through the positive-feedback network will equal the gain provided by the negative-feedback portion of the circuit. The phase shift from V_i to V_o through R_3 and R_4 is zero degrees because of the zero phase shift in the noninverting mode. At the frequency of oscillation, the phase through R_1–C_1 and R_2–C_2 will also be zero degrees. The voltages V_i and V_o are related through the positive-feedback path by

$$\frac{V_i}{V_o} = \frac{Z_2}{Z_1 + Z_2} \tag{6.2b}$$

where Z_1 is the series combination of R_1 and C_1 and Z_2 is the parallel combination of R_2 and C_2. Thus, using (6.2b), V_i and V_o are related by

$$\frac{V_i}{V_o} = \frac{1}{1 + \dfrac{R_1}{R_2} + \dfrac{C_2}{C_1} + j\left(\omega C_2 R_1 - \dfrac{1}{\omega R_2 C_1}\right)} \tag{6.2c}$$

For oscillation to occur, the phase through this network must be zero degrees. The frequency of oscillation, in radians per second, is determined from this condition. Thus,

$$\omega_o = \frac{1}{\sqrt{R_1 R_2 C_1 C_2}} \tag{6.3}$$

With the coefficient of j equal to zero, (6.2c) reduces to

$$\frac{V_i}{V_o} = \frac{1}{1 + \dfrac{R_1}{R_2} + \dfrac{C_2}{C_1}} \tag{6.4}$$

For a loop gain of one, the gain through the negative-feedback loop must equal the attenuation through the positive-feedback portion of the circuit,

$$1 + \frac{R_4}{R_3} = 1 + \frac{R_1}{R_2} + \frac{C_2}{C_1}$$

For the special case where $R_1 = R_2 = R$ and $C_1 = C_2 = C$,

$$\omega_o = \frac{1}{RC} \tag{6.5}$$

and the minimum noninverting gain for the circuit is three. Consequently,

$$R_4 \geq 2R_3$$

will ensure that the circuit will oscillate. It is difficult to predict the amplitude of oscillation in this circuit and, to establish a known amplitude, an automatic control scheme must be used.

Example

In the Wien-Bridge oscillator in Fig. 6.2, $R_1 = R_2 = R_3 = 10\ \text{k}\Omega$, $R_4 = 20\ \text{k}\Omega$, and $C = 0.01\ \mu\text{F}$. The output voltage is 6 V RMS. Calculate the frequency of oscillation and the voltage v_i.

Solution

Using (6.5), we obtain

$$f_o = \frac{1}{2\pi RC} = \frac{1}{2\pi(10\ \text{k}\Omega)(0.01\ \mu\text{F})}$$
$$= 1.59\ \text{kHz}$$

The voltage v_i will be one-third of the magnitude of the output voltage [see (6.4)] or

$$v_i = 2\ \text{V RMS} \qquad\blacksquare$$

Amplitude Control The amplitude of the Wien-Bridge oscillator may be established by replacing R_3 with a tungsten lamp, as shown in Fig. 6.3a. When power is first applied to the circuit, the tungsten lamp has a low resistance and the loop gain AF is greater than one. This causes oscillations to build up. As the oscillations increase in magnitude, the tungsten lamp heats up and its resistance also increases. When the loop gain is one, the amplitude of the oscillator output remains constant and equilibrium is reached. For the case where $R_1 = R_2$ and $C_1 = C_2$, the loop gain will be one when the resistance of the lamp is one-half of R_4. Specification sheets for the tungsten lamp provide a graph of lamp resistance versus voltage (Fig. 6.3b). The output voltage of

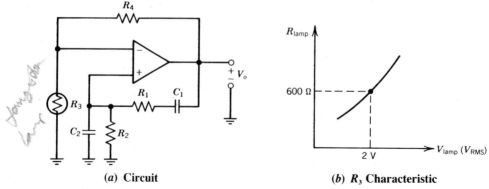

(a) Circuit (b) R_3 Characteristic

Figure 6.3 Tungsten lamp amplitude-stabilized oscillator.

the oscillator will be three times the lamp voltage whose corresponding value of lamp resistance satisfies the loop-gain requirement. The graph in Fig. 6.3b shows, that at a lamp voltage of 2 V, the resistance is 600 Ω. If the circuit is designed with a lamp with this characteristic and $R_4 = 2R_3 = 1.2$ kΩ, then the output voltage of the oscillator will be 6 V.

A second scheme of controlling the amplitude of a Wien-Bridge oscillator is shown in Fig. 6.4. An important advantage of this circuit is that the tungsten lamp is eliminated together with its inherent time constant, linearity, and reliability problems. The amplitude in this circuit is controlled by a negative-feedback path that uses a JFET and capacitor C_4. The circuit gain A_1, from the noninverting amplifier terminal to its output is basically established by R_3 and the drain-to-source resistance (r_{ds}) of the JFET transistor Q_1.

$$A_v = 1 + \frac{R_3}{r_{ds}}$$

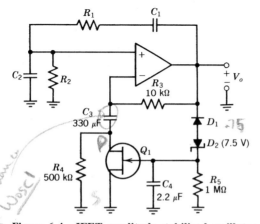

Figure 6.4 JFET amplitude-stabilized oscillator.

When the output of the oscillator exceeds about -8.25 V, D_1 and D_2 conduct and charge C_4. The voltage across C_4 biases Q_1, which establishes the amplifier gain A_v. Capacitor C_3 has a low impedance at the frequency of oscillation and acts as a dc blocking capacitor to prevent the amplifier's offset voltage and offset current errors from being multiplied by the loop gain. Distortion is determined by the amplifier's open-loop gain and the response time of the negative-feedback loop filter, R_5 and C_4. R_4 is chosen to adjust the negative-feedback loop so that the JFET is operated at a small negative gate bias.

The frequency of oscillation is still determined by the positive-feedback path provided by the R_1–C_1 and R_2–C_2 network,

$$\omega_o = \frac{1}{\sqrt{R_1 R_2 C_1 C_2}} \tag{6.6}$$

6.4.2 Quadrature Oscillator

Quadrature oscillators are similar to Wien-Bridge circuits in that they are most attractive for fixed frequency applications in the 10 Hz-to-1 MHz range. Frequencies outside this range are obtainable, but the circuit must be carefully designed.

A quadrature oscillator is shown in Fig. 6.5. As with the Wien-Bridge oscillator, some means of amplitude stabilization is normally required.

The outputs of the quadrature oscillator are two sinusoidal signals. The output of A_1, called the sine output, lags the output of A_2, the cosine output, by 90°.

One approach to determine the frequency of oscillation is to find the loop gain and apply the conditions for oscillation. The oscillator circuit consists of a noninverting amplifier A_1, an inverting integrator A_2, and an ac voltage divider composed of R_1 and C_1.

The loop gain of the circuit, $AF(s)$, is the product of the transfer functions of A_1, A_2, and the voltage divider,

$$AF(s) = -\frac{1 + sR_2C_2}{sR_2C_2} \frac{1}{sR_3C_3} \frac{1}{1 + sR_1C_1}$$

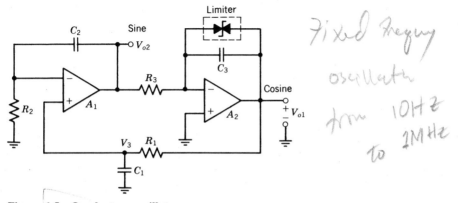

Figure 6.5 Quadrature oscillator.

104 Oscillators

Oscillations will occur when the magnitude of $AF(j\omega)$ equals one and its phase equals zero degrees. If $R_1C_1 = R_2C_2 = RC$, then $AF(s)$ reduces to

$$AF(s) = \frac{-1}{s^2 RR_3CC_3}$$

Thus,

$$AF(j\omega) = \frac{1}{\omega^2 RR_3CC_3} \tag{6.7}$$

and the frequency that will satisfy the unity-gain requirement is

$$\omega_o = \frac{1}{\sqrt{RR_3CC_3}}$$

In a practical circuit, a slight mismatch of the components causes the oscillation to converge or diverge. If the R_1C_1 product is slightly greater than the R_2C_2 product, the oscillator output will diverge but can be limited and stabilized by a limiter circuit.

Zener diodes D_1 and D_2 limit the amplitude of the output signal without significantly affecting the waveform at the cosine output. Because of the filtering the signal receives from the R_1C_1, R_2C_2, and R_3C_3 networks, the two outputs will be nearly distortionless.

Alternate Approach The quadrature oscillator of Fig. 6.5 may be also viewed as the implementation of a differential equation of the form

$$\frac{d^2x}{dt^2} + k^2 x = 0$$

where x is v_{o1} or v_{o2}. Assuming that $R_1C_1 = R_2C_2 = RC$, we find that the transfer function from the output of A_2 to the output of A_1 is given by

$$\frac{V_{o2}}{V_{o1}} = \frac{1}{1 + sRC} \cdot \frac{1 + sRC}{sRC} = \frac{1}{sRC}$$

This is the transfer function of a noninverting integrator. The corresponding time-domain relationship is

$$v_{o2} = \frac{1}{RC} \int v_{o1} \, dt$$

The transfer function from the output of A_1 to the output of A_2 is that of an inverting integrator,

$$\frac{V_{o1}}{V_{o2}} = -\frac{1}{sR_3C_3}$$

which corresponds to

$$v_{o1} = -\frac{1}{R_3C_3} \int v_{o2}\, dt$$

or

$$v_{o2} = -(R_3C_3)\frac{dv_{o1}}{dt}$$

in the time domain. Equating the two time-domain expressions for v_{o2}, we have

$$-R_3C_3 \frac{dv_{o1}}{dt} = \frac{1}{RC}\int v_{o1}\, dt$$

or differentiating,

$$\frac{d^2 v_{o1}}{dt^2} + \frac{1}{RR_3CC_3} v_{o1} = 0$$

The general form of the solution of this differential equation is given by

$$v_{o1} = V_{m1} \cos(\omega_o t + \theta)$$

If we try this solution in the differential equation implemented by the quadrature oscillator in Fig. 6.5, we get

$$-\omega_o^2 V_{m1} \cos(\omega_o t + \theta) + \frac{V_{m1}\cos(\omega_o t + \theta)}{RR_3CC_3} = 0$$

or after simplification,

$$\omega_o = \frac{1}{\sqrt{RR_3CC_3}}$$

This is identical to the expression obtained previously for the frequency of oscillation using loop-gain considerations.

Since

$$v_{o2} = -R_3C_3 \frac{dv_{o1}}{dt}$$

$$= -R_3C_3 \frac{d}{dt} V_{m1} \cos(\omega_o t + \theta)$$

the corresponding equation for v_{o2} is

$$v_{o2} = V_{m1}R_3C_3\omega_o \sin(\omega_o t + \theta)$$

or

$$v_{o2} = V_{m1}R_3C_3\omega_o \cos(\omega_o t + \theta - 90°)$$

Thus v_{o1} leads v_{o2} by 90°, or v_{o1} and v_{o2} are in quadrature. Since a cosine function leads a sine function of the same frequency by 90°, v_{o1} and v_{o2} are called cosine and sine outputs, respectively, in the oscillator of Fig. 6.5.

Example

The RMS value of the sine output voltage V_{o1} of the quadrature oscillator in Fig. 6.5 is 1 V RMS. For this circuit, $R_1 = R_2 = R_3 = R = 51$ kΩ and $C_1 = C_2 = C_3 = C = 820$ pF. Calculate the frequency of oscillation and the magnitude and phase of V_{o1} and V_3 with respect to V_{o2}.

Solution

For the R_1–C_1 network,

$$\frac{V_3}{V_{o1}} = \frac{1}{j\omega_o RC + 1} = \frac{1}{1 + j}$$

$$= 0.707 \angle -45°$$

V_3 will be 0.707 V RMS and will lag V_{o1} by 45°. The A_2 circuit is a noninverting ac amplifier with a gain of

$$\frac{V_{o2}}{V_3} = \frac{j\omega_o RC + 1}{j\omega_o RC} = 1 - j = \sqrt{2} \angle -45°$$

The gain from V_{o1} to V_{o2} is

$$\frac{V_{o2}}{V_{o1}} = \frac{V_{o2}}{V_3}\frac{V_3}{V_{o1}} = 1.00 \angle -90°$$

and confirms that V_{o2} is also 1 V RMS but lags V_{o1} by 90°. ∎

6.5 RECTANGULAR WAVE OSCILLATOR

The low-frequency rectangular wave oscillator in Fig. 6.6 contains two feedback paths. The negative-feedback path consists of the R–C network; the positive-feedback path consists of resistors R_1, R_2, and R_3. The amplifier, which functions as a comparator,

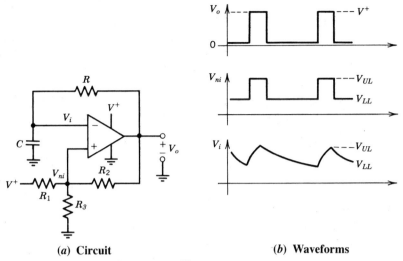

(a) Circuit (b) Waveforms
Figure 6.6 Rectangular wave oscillator.

does not operate in its linear region. The output voltage is at one saturation limit or the other except for the brief time during the transition between the two limits.

The amplifier shown is biased with a single dc supply connected to the amplifier's supply terminals. This voltage establishes the approximate amplitude of the oscillator output.

The voltage at the amplifier's inverting input will be an exponentially varying waveform caused by the charging and discharging of the capacitor C through the series resistor R. The charging and discharging of the capacitor is caused by the output of the oscillator changing from one output state to the other (about V^+ and 0 V in this case).

The voltage at the noninverting input is a rectangular wave whose limits are established by the R_1–R_2–R_3 divider in conjunction with V^+ and the oscillator output.

The amplifier basically functions as a comparator whose reference alternates between two values. The reference voltages at the noninverting input terminal of the op amp are established by the R_1–R_2–R_3 circuit. The time-varying exponential voltage at the inverting input terminal attempts to charge between the two output levels of the op amp causing it to change states when the reference voltages are slightly exceeded. When the output changes states, a new reference voltage is established, and the exponential waveform changes direction toward the new output level. This process is continuously repeated.

If the output is high, that is near V^+, then the voltage V_{ni} at the noninverting input of the operational amplifier is given by

$$V_{ni} \cong V^+ \frac{R_3}{R_3 + R_1 \| R_2} = V_{UL} \tag{6.8}$$

108 Oscillators

If the output is low, that is, near 0 V, then

$$V_{ni} \cong V^+ \frac{R_2 \| R_3}{R_1 + R_2 \| R_3} = V_{LL} \quad (6.9)$$

If $R_1 = R_2 = R_3$, the output of the rectangular wave oscillator will be a square wave (50% duty cycle) and the comparison limits will be

$$V_{UL} = \frac{2V^+}{3} \quad (6.10a)$$

and

$$V_{LL} = \frac{V^+}{3} \quad (6.10b)$$

The voltage at the inverting terminal of the op amp is an exponential waveform. For $V_o = V^+$, the capacitor attempts to charge toward V^+. When it slightly exceeds V_{UL}, the amplifier output will switch to the low state and change the comparison voltage to the lower limit. The capacitor, at this time charged to V_{UL}, starts discharging toward 0 V. This is described by

$$v_i(t) = V_{UL} \exp \frac{-t}{RC} \quad (V_o = 0 \text{ V}) \quad (6.11)$$

The capacitor voltage $v_i(t)$ will then attempt to discharge toward 0 V but, when it is slightly less than V_{LL}, it causes the circuit output to switch to the high state. The time T_1 it takes to discharge from V_{UL} to V_{LL} is found from

$$V_{LL} = V_{UL} \exp \frac{-T_1}{RC}$$

which produces

$$T_1 = t_{\text{disch}} = RC \ln \frac{V_{UL}}{V_{LL}} \quad (6.12)$$

The exponentially growing voltage at the inverting input during the next half cycle is

$$v_i(t) = V^+ - (V^+ - V_{LL}) \exp \frac{-(t - T_1)}{RC} \quad (V_o = V^+) \quad (6.13)$$

When $v_i(t)$ reaches V_{UL}, the output will switch again. For this case,

$$V_{UL} = V^+ - (V^+ - V_{LL}) \exp \frac{-T_2}{RC}$$

Solving for t_{chg} or T_2, we get

$$T_2 = t_{chg} = RC \ln \frac{V^+ - V_{LL}}{V^+ - V_{UL}} \qquad (6.14)$$

The period of oscillation (T) is the sum of the charge and discharge times T_1 and T_2,

$$T = t_{chg} + t_{disch} = T_1 + T_2 \qquad (6.15a)$$

For the special case when $R_1 = R_2 = R_3$,

$$T = 2RC \ln 2 = 1.386 RC \qquad (6.15b)$$

The frequency of oscillation, in Hertz, is the reciprocal of this period.

Basically, R and C establish the frequency of oscillation, and the relative values of R_1, R_2, and R_3 establish the duty cycle. R_1, R_2, and R_3 do affect the frequency of oscillation but, for a given duty cycle, the factor introduced is constant. In the general case, the peak-to-peak amplitude is a function of the voltage(s) connected to the dc supply pins and is not precise. The technique of using back-to-back zener diodes, connected at the output to ground, is frequently used to more accurately establish the output voltage levels.

Example

The supply voltage V^+ for the rectangular wave oscillator in Fig. 6.6 is $+10$ V, $R = 60$ k$\Omega = 2R_1 = 3R_2 = 4R_3$, and $C = 0.01$ µF. Assuming that the output varies between 0 and 10 V, calculate the frequency of oscillation and the duty cycle.

Solution

Using (6.8) and (6.9), we get

$$V_{UL} = 10 \text{ V} \frac{15}{15 + 20||30} = 5.55 \text{ V},$$

and

$$V_{LL} = 10 \text{ V} \frac{20||15}{30 + 20||15} = 2.22 \text{ V}$$

From (6.12) and (6.14),

$$T_1 = RC \ln 2.5 = RC(0.916)$$

and

$$T_2 = RC \ln 1.75 = RC(0.559)$$

The period is the sum of T_1 and T_2 or

$$T = RC \quad (1.475)$$

For the R and C values given, this period produces a frequency of oscillation of

$$f_o = 1.13 \text{ kHz}$$

The duty cycle is the ratio of the T_2 to the period or

$$\text{Duty cycle} = \frac{T_2}{T} = 0.38 = 38\% \qquad \blacksquare$$

6.5.1 Amplitude and Duty Cycle Control

The circuit in Fig. 6.7 is an extension of the rectangular wave oscillator and contains amplitude and duty-cycle control features. The amplitude of the output rectangular wave is established by the back-to-back zener diodes D_3 and D_4; the current through these diodes is limited by resistor R_5. The duty cycle is established by R_1 and R_2 with the diodes D_1 and D_2 providing the directional current for the charging and discharging of capacitor C. The frequency of oscillation is a function of all the circuit components and is determined in the following manner.

When the output signal is high, the *peak* amplitude of the output voltage is V_{op}, a function of the voltage drops across D_3 and D_4. The voltage V_{ni} at the noninverting terminal of the op amp will be

$$V_{ni} = \frac{R_3}{R_3 + R_4} V_{op} = kV_{op} \qquad (V_o = V_{op}) \qquad (6.16)$$

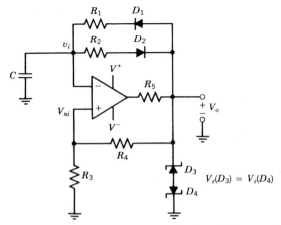

Figure 6.7 Controlling the amplitude and duty cycle of the rectangular wave oscillator.

where

$$k = \frac{R_3}{R_3 + R_4} \qquad (6.16a)$$

Similarly, when the output is low,

$$V_{ni} = \frac{R_3}{R_3 + R_4}(-V_{op}) = -kV_{op} \qquad (V_o = -V_{op}) \qquad (6.17)$$

The voltage at the inverting input $v_i(t)$ will charge and discharge between these limits (kV_{op} and $-kV_{op}$). At the moment the output switches high, the capacitor is at $-kV_{op}$ volts and will start charging, through R_1 and D_1, to $V_{op} - V_F$ volts, where V_F is the forward-voltage drop across D_1. This is expressed by

$$v_i(t) = (V_{op} - V_F) - [(V_{op} - V_F) + kV_{op}]\exp\frac{-t}{R_1 C} \qquad (V_o = V_{op})$$

When $v_i(t)$ reaches kV_{op}, the oscillator output will switch to its low state. The time it takes $v_i(t)$ to go from $-kV_{op}$ to kV_{op} is T_2, which is the same as the time the output is high. Substituting kV_{op} for $v_i(t)$ and solving for T_2 produces

$$T_2 = R_1 C \ln \frac{V_F - (k+1)V_{op}}{V_F + (k-1)V_{op}} \qquad (6.18)$$

Time T_1, during which the output is low, is determined in an analogous manner and given as

$$T_1 = R_2 C \ln \frac{V_F - (k+1)V_{op}}{V_F + (k-1)V_{op}} \qquad (6.19)$$

The sum of the charge and discharge times produces the period of oscillation,

$$T = (R_1 + R_2)C \ln \frac{V_F - (k+1)V_{op}}{V_F + (k-1)V_{op}} \qquad (6.20)$$

The duty cycle of the output is

$$\text{Duty cycle} = \frac{T_2}{T} \qquad (6.21)$$

$$\text{Duty cycle} = \frac{R_1}{R_1 + R_2} \qquad (6.22)$$

112 Oscillators

The voltage $v_i(t)$ may be buffered by a unity-gain amplifier and used as a pseudo-ramp generator. For low values of k, this signal will be nearly linear. The maximum output frequency is, essentially, limited by the slew rate of the amplifier or the maximum rate of change of the amplifier's output voltage. R_1 or R_2 may be used to control the duty cycle, while zeners D_3 and D_4 may be used to control the amplitude of the output voltage.

Example

Design a rectangular wave oscillator, using Fig. 6.7, to meet the following criteria:

1. $V_o = 10 \text{ V}(p\text{-}p)$
2. $f_o = 10 \text{ kHz}$ and
3. Duty cycle $= 25\%$

Solution

In Fig. 6.7, resistors R_1 and R_2 control the duty cycle. Using (6.22), we get

$$\text{Duty cycle} = \frac{R_1}{R_1 + R_2} = \frac{1}{4}$$

or

$$3R_1 = R_2$$

Letting $R_2 = 15 \text{ k}\Omega$, we then obtain,

$$R_1 = 5 \text{ k}\Omega$$

The oscillator will have a 5 V (peak) output if D_3 and D_4 have a zener voltage of 4.4 V assuming the forward-voltage drop of these diodes is 0.6 V.

From (6.16a),

$$k = \frac{R_3}{R_3 + R_4}$$

The constant of proportionality k is set to one half so that the voltage at the noninverting input of the amplifier is half of the output voltage. This condition is met if

$$R_3 = R_4 = 10 \text{ k}\Omega$$

Using (6.20) and letting $V_F = 0.6 \text{ V}$ and $V_{op} = 5 \text{ V}$, we obtain

$$T = 100 \text{ }\mu\text{sec} = (20 \text{ k}\Omega)(C) \ln \frac{0.6 - 1.5(5)}{0.6 - 0.5(5)}$$

Solving the above expression for C produces

$$C = 3.9 \text{ nF}$$ ∎

6.6 SQUARE AND TRIANGLE WAVE GENERATOR

The block diagram of a generalized nonlinear waveform generator is shown in Fig. 6.8. The output of the integrator is a time-varying voltage. This voltage is compared with a reference. When the magnitude of the integrator's output exceeds the reference, the output state of the comparator and the reference change to the opposite polarity, thus changing the direction of the integrator output. When the integrator output reaches the other polarity of the reference, the comparator forces the integrator input to return to its first state, and the procedure is repeated. The output of the integrator is a triangular wave voltage; the output of the comparator is a square wave. The block diagram illustrates how the two types of waveforms are generated, and the necessary functional blocks, but the specific details of a given implementation vary.

The function generator in Fig. 6.9 consists of an integrator A_2 and a comparator A_1. The output state of the comparator is a function of the polarity of the voltage at its noninverting input. The polarity of this input is a function of the output voltages of the integrator and the comparator. The amplitude of the output voltage of the comparator is established by the back-to-back zener diodes D_1 and D_2. Resistor R_4 is used to limit the output current of A_1.

To understand the operation of the circuit, let the output of A_1 be at its positive saturation level. The input to the integrator will be $V_z + V_F$ volts, where V_z is the zener voltage of D_2 (reversed-biased diode) and V_F is the forward-voltage drop of D_1. The integrator's output will be a negative going (slope) voltage ramp. R_2 and R_3 form a voltage summer whose end points are the output of A_1 ($V_z + V_F$) and the steadily decreasing voltage at the output of A_2. The junction of these two resistors is sensed by the noninverting input of A_1. When the noninverting input senses a slightly negative voltage (originally it was positive), the output switches from $+(V_z + V_F)$ to $-(V_z + V_F)$. The voltage ramp at the output of A_2 now reverses its slope from negative to positive. The voltage ramp continues to increase until the junction of R_2 and R_3 goes to a slightly positive value. At this time, the output of A_1 switches back to its positive level. This procedure is repeated continuously.

The triangular wave frequency is determined by R_1, C_1, and $V_z + V_F$. Amplitude

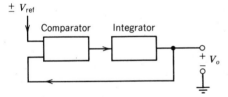

Figure 6.8 Nonlinear waveform generator model.

114 Oscillators

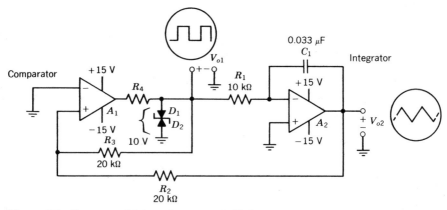

Figure 6.9 Square and triangular wave oscillator.

is determined by the ratio of R_2 to R_3 and $V_z + V_F$. This voltage affects both the amplitude and frequency of the output, but if it is fixed, R_1 and R_3 can be used to control frequency and amplitude, respectively.

Example

Calculate the frequency of oscillation and the magnitude of the outputs of the square and triangular wave oscillator in Fig. 6.9.

Solution

The output voltage of A_1 is established by the back-to-back zener diodes D_1 and D_2. Hence,

$$V_{o1} = \pm 10 \text{ V(peak)}$$

The output voltage of A_1 causes the integrator A_2 to generate, during each half cycle, a linear voltage ramp at its output whose slope is

$$\frac{dv_{o2}}{dt} = \frac{-v_{o1}}{R_1 C_1}$$

$$= \frac{\pm 10 \text{ V}}{(10 \text{ k}\Omega)(0.033 \text{ }\mu\text{F})}$$

$$= \pm 30.3 \text{ V/msec}$$

Since $R_2 = R_3$ and the comparator makes a decision at 0 V, the peak values of v_{o2} are the same as those of v_{o1}. If the output of A_1 is initially at $+10$ V, the comparator will switch states when the output of A_2 is slightly more negative ($\cong 50$ μV) than -10 V. If the output of A_1 is initially at -10 V, the comparator will switch states when the output of A_2 is slightly more positive than $+10$ V. The peak-to-peak value of A_2's output is 20 V and the durations of each half cycle are the same and equal to one-half of the period or

$$\frac{T}{2} = \frac{20\ \text{V}}{30.3\ \text{V/msec}} = 0.660\ \text{msec}$$

Thus, $T = 1.32$ msec and

$$f_o = 758\ \text{Hz}$$ ∎

6.7 VOLTAGE-CONTROLLED OSCILLATORS

6.7.1 Op Amp Sawtooth VCO

The voltage-controlled oscillator, or VCO, in Fig. 6.10 is a sawtooth generator whose frequency is controlled by the input voltage V_c. The circuit is basically an integrator with a voltage-sensitive on/off switch in shunt with the capacitor in the negative-feedback loop. The switch is implemented with a unijunction transistor (UJT) whose on/off status is a function of the gate (g) and anode (a) voltages.

When the gate voltage of the UJT is more positive than the anode or VCO output voltage, the UJT is off and simulates an open circuit. When the anode voltage is more positive than the gate voltage (by a few tenths of a volt), the UJT anode and cathode (c) terminals act like a short circuit. The voltage across the capacitor C will then discharge until the discharge current drops below the UJT's holding current, which is typically a few milliamperes. Then the anode and cathode terminals abruptly act as an open circuit and the circuit integrates once again the constant, but programmable, input voltage.

Since the input or control voltage V_c is a constant, the output will be a positive voltage ramp whose slope is

$$\frac{dv_o}{dt} = \frac{V_c}{RC}$$

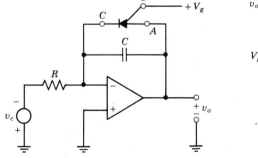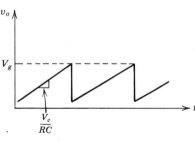

Figure 6.10 Sawtooth VCO.

116 Oscillators

The output will continue to rise linearly until $v_o \cong V_g$ at which time the UJT will turn on and discharge the capacitor. Once the capacitor is discharged, the UJT turns off and the output voltage begins to again increase. This procedure is continually repeated providing a sawtooth at the output of the amplifier. The period of oscillation is a function of the peak output voltage and the integrator's slope,

$$T = \frac{V_g}{\frac{V_c}{RC}} = \frac{V_g}{V_c} RC \qquad (6.23)$$

V_g approximates the peak output voltage and, if V_c is varied, the frequency of oscillation will vary. The circuit can also be viewed as a voltage-to-frequency convertor.

6.7.2 Integrated Triangular VCO

A simplified version of an integrated circuit VCO is shown in Fig. 6.11. The circuit has sawtooth and square wave outputs and is used as an analog building block in the phase locked loop (PLL).

Like the VCO previously discussed, a voltage ramp is generated by charging a capacitor with a constant current. Unlike the previous circuit, the capacitor is not quickly discharged by shunting it with a switch; instead it is discharged linearly by effectively switching in another current source of the same value but of the opposite polarity. The end result is that one current source provides a positive-slope voltage ramp and the other current source provides a negative-slope voltage ramp. The particular current source in Fig. 6.11 that is enabled is a function of the switching transistor Q_3. Transistor Q_3 is driven by a Schmitt trigger (see Section 5.4.3) whose output, a square

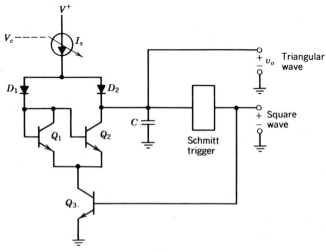

Figure 6.11 *IC triangular VCO.*

wave, is dependent on the relative values of the capacitor voltage and the Schmitt trigger limits.

When the output of the Schmitt trigger is low, Q_3, Q_2, Q_1, and D_1 are off, and the current source I_s charges C through D_2. This produces a voltage ramp across C with a slope of

$$\frac{dv_o}{dt} = \frac{I_s}{C} \qquad (6.24a)$$

When the positive going (slope) capacitor voltage reaches the upper limit of the Schmitt trigger, the output of the Schmitt trigger goes high and turns on Q_3. Transistor Q_3 causes Q_1 to turn on and sink the output current of the source I_s through D_1. Diodes D_1 and D_2, and transistors Q_1 and Q_2, are matched. The base-emitter voltage of Q_1 is impressed across the matched transistor Q_2, which forces Q_2 to sink a current from the capacitor of the same value as I_s. This produces a voltage ramp across the capacitor with a negative slope of

$$\frac{dv_o}{dt} = -\frac{I_s}{C} \qquad (6.24b)$$

When the negative going voltage ramp reaches the lower limit of the Schmitt trigger, the output state of the Schmitt trigger is reversed and the cycle continues.

The current source I_s is externally programmable or voltage controlled, thus providing a variable slope for the voltage ramps and hence a variable frequency of oscillation. If the output is filtered, a voltage-controlled sinusoidal oscillator is produced. The block diagram in Fig. 6.11 is, in principle, the same as for the 566 integrated VCO.

6.8 SUMMARY

Oscillators are regenerative circuits that produce periodic outputs with no signal input. The typical oscillator uses an amplifier and a feedback network. The product of the amplifier's gain A and the transfer function of the feedback network F define the loop gain AF. AF is a complex quantity and, during oscillation, is equal to one. This means that the magnitude and phase of AF must be one and zero degrees, respectively. The frequency that satisfies these conditions is the frequency of oscillation of the circuit.

Typical oscillator waveshapes are sinusoidal, rectangular, and triangular. A commonly used low-frequency sinusoidal oscillator is the Wien bridge. This circuit is characterized by an amplifier with a positive and negative feedback network; the positive-feedback network is composed of an impedance, bridgelike network. Another common sinusoidal oscillator is the quadrature oscillator. This circuit is effectively composed of two integrator circuits. A special characteristic of this circuit is that it has two sinusoidal outputs, 90° out of phase.

The rectangular wave oscillator, with a duty cycle of 50%, produces a square wave output. The amplifier used in this circuit functions as a comparator and operates in its nonlinear region. The positive and negative feedback networks of this circuit use an *RC* and a voltage summing circuit.

Sawtooth and triangular oscillators depend on the constant current-charging-a-capacitor characteristic. The current is, oftentimes, established by a control voltage that makes the circuit a voltage-controlled oscillator (VCO).

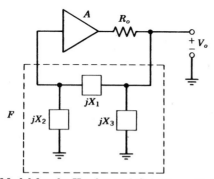

(a) **Model for the Hartley and Colpitts oscillators**

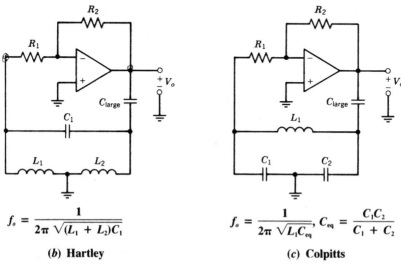

$$f_o = \frac{1}{2\pi \sqrt{(L_1 + L_2)C_1}}$$

(b) **Hartley**

$$f_o = \frac{1}{2\pi \sqrt{L_1 C_{eq}}}, \quad C_{eq} = \frac{C_1 C_2}{C_1 + C_2}$$

(c) **Colpitts**

Figure 6.12 Additional oscillator circuits.

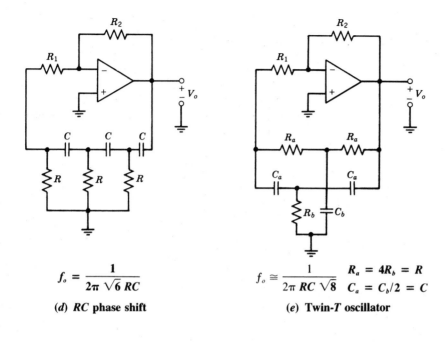

(d) RC phase shift $f_o = \dfrac{1}{2\pi \sqrt{6}\, RC}$

(e) Twin-T oscillator $f_o \cong \dfrac{1}{2\pi\, RC\, \sqrt{8}}$ $R_a = 4R_b = R$ $C_a = C_b/2 = C$

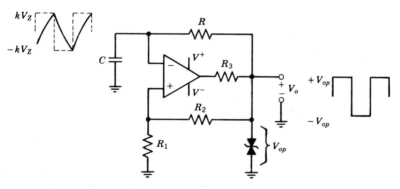

(f) Rectangular wave oscillator

Fig. 6.12 (*Continued*)

PROBLEMS

6.1 The output of the Wien-Bridge oscillator in Fig. 6.2 is described by

$$v_o = 9 \sin(2\pi \cdot 10^3\, t)\ \text{V}$$

Write an expression for the voltage at the amplifier's noninverting and inverting input terminals. Specify a set of values for R, R_3, R_4, and C that will satisfy these relationships if $R_1 = R_2 = R$ and $C_1 = C_2 = C$.

6.2 Determine the relationship between R_3 and R_4 in the Wien-Bridge oscillator of Fig. 6.2 if $R_1 = 2R_2$ and $C_2 = 2C_1$.

6.3 Assuming that the gain of the amplitude-stabilized Wien-Bridge oscillator in Fig. 6.3a is sufficient to produce oscillation, determine a set of values for R_1, R_2, C_1, and C_2 for an output signal frequency of 100 Hz.

6.4 Write the expressions $v_{o1}(t)$, $v_{o2}(t)$, and $v_3(t)$ for the quadrature oscillator in Fig. 6.5 if $R_1 = R_2 = R_3 = R$ and $C_1 = C_2 = C_3 = C$.

6.5 Design a quadrature oscillator (Fig. 6.5) such that

$$v_3(t) = 6 \cos (2\pi \cdot 5 \cdot 10^3 \, t) \text{ V}$$

Also specify the approximate values of the back-to-back zener diodes in the limiter portion of the circuit. Assume that $R_1 = R_2 = R_3 = R$ and $C_1 = C_2 = C$.

6.6 The cosine output of the quadrature oscillator in Fig. 6.5 is defined as $v_{o1}(t) = 2 \sin (2\pi \cdot 10^4 \, t)$ V. Write the expressions for $v_2(t)$ and $v_{o2}(t)$, assuming $R_1 C_1 = R_2 C_2 = R_3 C_3 = RC$.

6.7 The sine output of the quadrature oscillator in Fig. 6.5 is

$$v_{o2}(t) = 2 \sin (2\pi \cdot 10^4 \, t) \text{ V}$$

Write the expressions for $v_3(t)$ and $v_{o1}(t)$.

6.8 Design a quadrature oscillator such that the frequency of oscillation is 10 kHz and V_{o1}(peak) = 5 V.

6.9 Calculate f_o, $V_o(p\text{-}p)$, and the dury cycle for the rectangular wave oscillator in Fig. 6.6 if $R_1 = 2 \text{ k}\Omega$, $R_2 = 4 \text{ k}\Omega$, $R_3 = 8 \text{ k}\Omega$, $R = 10 \text{ k}\Omega$, $C = 0.001 \, \mu\text{F}$, and $V^+ = 10$ V.

6.10 Calculate f_o, $V_o(p\text{-}p)$, and the duty cycle for the rectangular wave oscillator in Fig. 6.6 if $R_1 = 2 \text{ k}\Omega$, $R_2 = 4 \text{ k}\Omega$, $R_3 = 8 \text{ k}\Omega$, $R = 10 \text{ k}\Omega$, $C = 0.001 \, \mu\text{F}$, $V^+ = 10$ V, and $V^- = -10$ V.

6.11 Design a rectangular wave oscillator (Fig. 6.7) to meet the following specifications:

(a) $V_o \cong 12 \text{ V}(p\text{-}p)$
(b) $f_o = 100$ Hz
(c) Duty cycle = 75%

6.12 Sketch $v_o(t)$, $v_i(t)$, and $v_{ni}(t)$ for the rectangular wave oscillator in Fig. 6.6 if $R_1 = R_2 = 10 \text{ k}\Omega$, $R_3 = 20 \text{ k}\Omega$, $R = 2 \text{ k}\Omega$, $C = 0.01 \, \mu\text{F}$, $V^+ = 12$ V.

6.13 Sketch the output waveform of the integrator in the square and triangular wave generator in Fig. 6.9 if R_2 is replaced with a 10 kΩ resistor.

6.14 Derive the expressions for the amplitude and the frequency of the integrator output of the square and triangular wave oscillator in Fig. 6.9.

6.15 What components in the square and triangular wave oscillator of Fig. 6.9 should be made variable to independently control the amplitude and frequency of the triangular wave output?

6.16 What change in the performance of the square and triangular wave oscillator would occur if the inverting input of the comparator was connected to a programmable voltage source (V_R) instead of ground?

6.17 Specify a set of practical values for the components in the sawtooth VCO of Fig. 6.10 so that the period of the output signal is 1 msec, $V_C = 12$ V, and the slope of the sawtooth is 0.01 V/μsec.

6.18 Show that the loop gain for an oscillator of the form shown in Fig. 6.12a is

$$\frac{-AX_2X_3}{-X_3X_1 - X_2X_3 + jR_o(X_1 + X_2 + X_3)}$$

where A is the real and finite open-loop gain and R_o is the finite output resistance of the otherwise ideal amplifier. Prove that the phase and magnitude conditions for oscillation are satisfied if

$$X_1 + X_2 + X_3 = 0$$

and

$$\frac{X_3}{X_2} = -A$$

What determines the value of the ratio R_2/R_1 in the Hartley oscillator in Fig. 6.12b for oscillation to occur? Assume that R_1 is much larger than the reactance of L_1 at the frequency of oscillation.

6.19 What determines the value of the ratio R_2/R_1 in the Colpitts oscillator in Fig. 6.12c for oscillation to occur? Assume that R_1 is much larger than the reactance of C_1 at the frequency of oscillation. *Hint:* Use the results of Problem 6.18.

6.20 The attenuation, at f_o, in the feedback network in the RC phase-shift oscillator in Fig. 6.12d is 1/29. Design an RC phase-shift oscillator for an f_o of 1 kHz. Assume that $R_1 \gg R$ and that the amplifier is ideal.

6.21 The attenuation, at f_o, in the feedback network of the twin-T oscillator in Fig. 6.12e is 1/25. Determine the frequency of oscillation f_o and the approximate values of R_1 and R_2 needed for oscillation if $C_a = 0.01$ μF, $C_b = 0.02$ μF, $R_a = 10$ kΩ, and $R_b = 2.5$ kΩ. Assume that R_1 is large compared to the input impedance of the twin-T network at the frequency of oscillation, and that the op amp is ideal.

6.22 The circuit in Fig. 6.12f is a combination of the rectangular wave oscillator shown in Figs. 6.6 and 6.7. Specify a set of values for the components to satisfy the following specifications:

(a) $V_o(\text{peak}) = \pm 5$ V
(b) $T = 4$ msec
(c) Duty cycle = 50%

Chapter 7

Special-Purpose Circuits

7.1 PROLOGUE

Operational amplifier applications are often categorized into linear and nonlinear sections. While this type of partitioning clearly eases identifying the label for a particular circuit, it doesn't however, vividly bring out the diversity of the op amp's usage. This chapter, a potpourri of op amp circuits, contains those circuits that perform special-purpose functions and cannot be categorized under other major functional headings. Because the circuits are special purpose does not mean they are infrequently used; it just means the circuits' functions are somewhat unique.

7.2 LIMITERS

It is often necessary to limit or bound the maximum and/or minimum excursion of the amplifier output voltage either because its magnitude would be too large for the load or because we would like to avoid operating the amplifier in saturation. Limiting is a circuit technique that uses elements like diodes and zener diodes to control or limit the amplifier's output voltage. A circuit that contains a limiting scheme can be used to perform other functions. There are three types of limiter circuits:

1. Feedback
2. Series
3. Shunt

124 Special-Purpose Circuits

Feedback limiters contain the limiting element(s) in the feedback loop, while series and shunt limiters have the limiting element(s) in series and in shunt with the input network, respectively.

7.2.1 Feedback Limiters

The configuration of Fig. 7.1a limits the minimum value of output voltage V_o to zero volts (ideally) while its positive values are unaffected. For $V_s > 0$, the diode in the feedback loop will be forward-biased and, if we assume a forward-voltage drop of 0 V for the diode, the amplifier output will be zero volts. For $V_s < 0$, the diode will be reverse-biased and

$$V_o = -V_s \frac{R_2}{R_1}$$

The maximum negative value of V_s, before positive saturation occurs, is $-V_o(\max)(R_1/R_2)$ as shown in the transfer characteristic of Fig. 7.1b.

While the amplifier is in its linear region, the gain of the inverting configuration is $-(R_2/R_1)$ and thus the output voltage is proportional to R_2. In the absence of R_2, the circuit becomes an elementary comparator with output levels of zero volts [instead of $V_o(\min)$] and $V_o(\max)$, depending on whether V_s is positive or negative, respectively.

If a constant voltage source V_1 is added in series with the diode in the previous circuit, the result is the modified limiter of Fig. 7.2a and its transfer characteristic of Fig. 7.2b. For this circuit, the diode in the feedback loop will be reverse-biased as long as

$$V_o = -V_s \frac{R_2}{R_1} > -V_1 \tag{7.1}$$

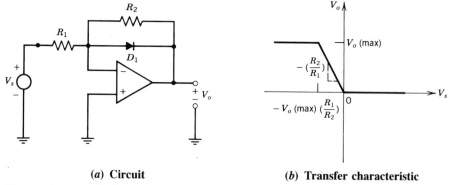

(a) Circuit (b) Transfer characteristic

Figure 7.1 Basic limiter.

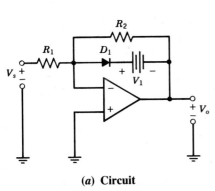
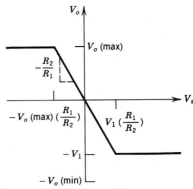

(a) Circuit   (b) Transfer characteristic

Figure 7.2 Modified limiter.

or

$$V_s < V_1 \frac{R_1}{R_2}$$

For $V_s < V_1(R_1/R_2)$, the modified limiter acts like an inverting linear amplifier with a gain of $-(R_2/R_1)$. The maximum possible output voltage is $V_o(\text{max})$ and occurs for $V_s = -V_o(\text{max})(R_1/R_2)$. In the absence of R_2, the resulting circuit provides two output levels $V_o(\text{max})$ and $-V_1$ depending on whether V_s is negative or positive, respectively.

Since a diode, common or zener, in its breakdown region acts effectively like a constant voltage source, the limiter of Fig. 7.3a is a practical implementation of the modified limiter of Fig. 7.2a. When $V_s < 0$, the zener diode will be prevented from conducting in its forward direction by the diode in series with it. For $V_s > 0$ and as long as the output voltage is smaller in magnitude than the zener breakdown voltage V_z, the zener-diode limiter will act like a linear inverting amplifier circuit,

$$V_o = -V_s \frac{R_2}{R_1} \quad \left(V_s < V_z \frac{R_1}{R_2}\right)$$

For $V_s \geq V_z(R_1/R_2)$, the zener diode is conducting in its breakdown region and the output voltage is limited to $-V_z$. The transfer characteristic for the zener diode limiter is shown in Fig. 7.3b.

The feedback loop of the limiter of Fig. 7.4 contains two zener diodes connected, back-to-back, in parallel with resistor R_2. In the transfer characteristic of Fig. 7.4b, zener diode D_1 is conducting in its breakdown region when V_o reaches V_{z1},

$$V_o = +V_{z1} \quad \left(V_s \leq -V_{z1} \frac{R_1}{R_2}\right) \quad (7.2a)$$

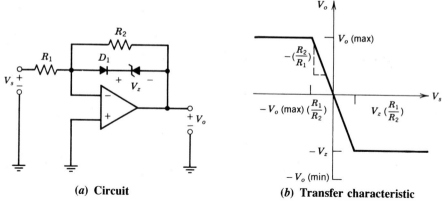

(a) Circuit **(b)** Transfer characteristic

Figure 7.3 Zener-diode limiter.

Zener diode D_2 will be conducting in its breakdown region when V_o reaches $-V_{z2}$,

$$V_o = -V_{z2} \qquad \left(V_s \geqq V_{z2}\frac{R_1}{R_2}\right) \qquad (7.2\text{b})$$

Between these two limits of the output voltage, the limiter of Fig. 7.4a operates like a linear inverting amplifier with a gain of $-(R_2/R_1)$, that is,

$$V_o = -V_s \frac{R_2}{R_1} \qquad \left(-V_{z1}\frac{R_1}{R_2} < V_s < V_{z2}\frac{R_1}{R_2}\right) \qquad (7.3)$$

In the absence of R_2, this circuit will act like a comparator with two output levels, V_{z1} and $-V_{z2}$, depending on whether V_s is negative or positive, respectively.

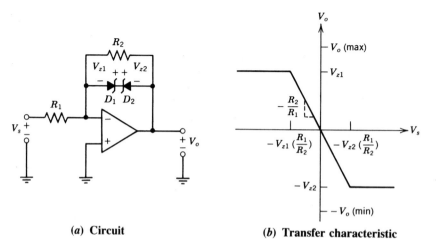

(a) Circuit **(b)** Transfer characteristic

Figure 7.4 Limiter using two zener diodes.

Limiters **127**

Example

Design a shunt limiter circuit (Fig. 7.4) whose input-output relationship is

$$V_o = -3V_{s1}$$

and whose output voltage range is restricted to

$$-10 \text{ V} \leq V_o \leq +10 \text{ V}$$

Solution

The input-output relationship is satisfied for $R_2/R_1 = 3$. Ideally, the output voltage range requirement is satisfied if the zener voltages are

$$V_{z1} = V_{z2} = 10 \text{ V}$$

In practice, a zener diode acts like a signal diode when forward-biased and has a voltage drop of about 0.6 V. One of the zeners is always forward-biased when limiting occurs. Thus,

$$V_{z1} = V_{z2} = 10 \text{ V} - 0.6 \text{ V} = 9.4 \text{ V}$$ ∎

7.2.2 Series Limiter

The diode D_1 is the limiting element in the series limiter shown in Fig. 7.5a. Since its anode is at virtual ground, the potential V_1 between cathode and ground will determine whether the effect of V_s is felt at the amplifier output or not. V_1 is a function of the signal source V_s, fixed reference V_R, and resistors R_1 and R_2. If V_1 is positive, the diode D_1 is reverse-biased and the output voltage of the amplifier is zero because of the feedback connection of R_3 to virtual ground. If V_1 is negative, the circuit will behave like a summing amplifier and the output will be a function of V_s, V_R, and resistors R_1, R_2, and R_3.

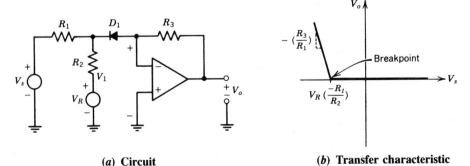

(a) Circuit　　　　　　　　　　　　　(b) Transfer characteristic

Figure 7.5　Series limiter.

The voltage V_1 may be determined using superposition.

$$V_1 = \frac{R_2}{R_1 + R_2} V_s + \frac{R_1}{R_1 + R_2} V_R$$

$$= \frac{R_2}{R_1 + R_2}(V_s + \frac{R_1}{R_2} V_R) \tag{7.4}$$

Voltage V_1 will be positive when

$$V_s + \frac{R_1}{R_2} V_R > 0$$

or

$$V_s > -\frac{R_1}{R_2} V_R$$

When V_1 is positive, the diode is reverse-biased and the output is zero:

$$V_o = 0 \quad \left(V_s > -\frac{R_1}{R_2} V_R\right) \tag{7.5}$$

Similarly, the voltage V_1 will be negative when

$$V_s < -\frac{R_1}{R_2} V_R$$

When V_1 is negative, neglecting the diode drop, the amplifier output will be the inverted and scaled sum of V_s and V_R,

$$V_o = -\frac{R_3}{R_1} V_s - \frac{R_3}{R_2} V_R \quad \left(V_s < -\frac{R_1}{R_2} V_R\right) \tag{7.6}$$

The transfer characteristic in Fig. 7.5b illustrates the relationship between V_o and V_s and the effect of the limiting element. For values of V_s more positive than $-(R_1/R_2)V_R$, the diode is reverse-biased and the output is zero volts. For values of V_s less positive than $-(R_1/R_2)V_R$, the diode is forward-biased and the output and the input are linearly related by R_1 and R_3. The fixed reference V_R has the effect of setting the value of V_s where limiting occurs to $-(R_1/R_2)V_R$.

Example

Design a series limiter (Fig. 7.5a) if (1) $V_R = 2$ V, (2) the slope of the transfer characteristic in the linear region is -10, and (3) limiting occurs at $V_s = -6$ V.

Solution

Let $R_1 = 10 \text{ k}\Omega$. The value of R_2 can be computed from

$$-6 \text{ V} = -\frac{10 \text{ k}\Omega}{R_2}(2 \text{ V})$$

Thus,

$$R_2 = 3.3 \text{ k}\Omega$$

The slope in the linear region is a function of R_1 and R_3,

$$-10 = \frac{-R_3}{10 \text{ k}\Omega}$$

Hence

$$R_3 = 100 \text{ k}\Omega \qquad \blacksquare$$

7.2.3 Shunt Limiter

The signal path in the shunt limiter of Fig. 7.6a is from the source V_s through R_1, R_2, and R_3 to the amplifier output. The limiting element D_1 is in shunt with the signal path and in series with the reference source V_R. The voltage V_1, relative to ground, determines whether the input signal is amplified or bounded by the limiting action of the circuit. If the voltage V_1 tries to go more positive than V_R volts (again neglecting the diode drop), the diode will be forward-biased and V_1 will be clamped to V_R. If the voltage V_1 is less positive than V_R, the diode will be reverse-biased and the V_R source will be disconnected from the circuit, allowing V_o to be linearly related to V_s. The voltage V_1 is a function of the signal source V_s and the resistors R_1 and R_2,

$$V_1 = \frac{R_2}{R_1 + R_2} V_s < V_R \qquad (7.7)$$

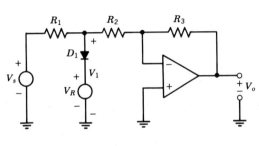
(a) Circuit

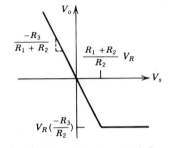
(b) Transfer characteristic

Figure 7.6 Shunt limiter.

For
$$V_s < \frac{R_1 + R_2}{R_2} V_R$$

D_1 is equivalent to an open circuit. Thus,

$$V_o = -\frac{R_3}{R_1 + R_2} V_s \quad \left(V_s < \frac{R_1 + R_2}{R_2} V_R\right) \tag{7.8}$$

For
$$V_s > \frac{R_1 + R_2}{R_2} V_R$$

D_1 is effectively a short, and

$$V_o = -\frac{R_3}{R_2} V_R \quad \left(V_s > \frac{R_1 + R_2}{R_2} V_R\right) \tag{7.9}$$

The transfer characteristic in Fig. 7.6b illustrates the relationship between V_o and V_s and the effect of the limiting element. For values of V_s more positive than $[(R_1 + R_2)/R_2]V_R$, the diode is forward-biased and the output is constant. For values of V_s less positive than $[(R_1 + R_2)/R_2]V_R$, the diode is reverse-biased and the output and input are linearly related.

Example

Define the linear region of operation in terms of V_s for the shunt limiter in Fig. 7.6a if (1) $R_1 = R_2 = R_3 = 10 \text{ k}\Omega$, (2) $V_R = 2$ V, and (3) $|V_o(\text{max})| = |V_o(\text{min})| = 10$ V.

Solution

The maximum negative output voltage is

$$-\frac{R_3}{R_2} V_R = -2 \text{ V}$$

For a linear-region circuit gain of

$$-\frac{R_3}{R_1 + R_2} = -\frac{1}{2}$$

the maximum positive value of V_s will be $+4$ V. The input voltage V_s is at its maximum negative value when $V_o = V_o(\text{max})$. This value is determined by

$$V_s = \frac{10\text{ V}}{-1/2} = -20\text{ V}$$

Hence for

$$-20\text{ V} < V_s < +4\text{ V}$$

the input and output will be linearly related. ∎

7.3 FUNCTION SYNTHESIZER

A nonlinear output may be approximated by a series of linear outputs of varying slope. This synthesizing technique may be accomplished using the circuit in Fig. 7.7a. The circuit is basically an inverting amplifier with a number of series limiter inputs. Each input generates a different slope m for a specified region of the input voltage V_s. The transfer characteristic for the circuit is shown in Fig. 7.7b.

The breakpoints in the characteristic occur at $V_s = \pm V_1$ and $V_s = \pm V_2$. At the cathode side of D_1 in Fig. 7.7a, the voltage v_1 can be expressed (while D_1 is off) in terms of V_{REF} and V_s by using superposition;

$$v_1 = \frac{R_3}{R_3 + R_4} V_s + \frac{R_4}{R_3 + R_4} V_{\text{REF}} \quad (D_1\text{: off})$$

This portion of the circuit is active, and D_1 is conducting only for

$$v_1 < 0$$

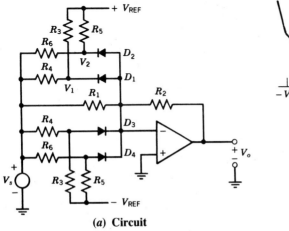

(a) Circuit

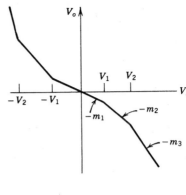

(b) Transfer characteristic

Figure 7.7 Function synthesizer.

or

$$V_s < -\frac{R_4}{R_3} V_{REF}$$

If we repeat this procedure for the anode side of D_3, D_3 is conducting when

$$V_s > +\frac{R_4}{R_3} V_{REF}$$

If we assume that when D_1 or D_3 turns on, D_2 or D_4 turns off, then there will be breakpoints in the transfer characteristic of Fig. 7.7b at

$$V_s = \pm V_1$$

where

$$V_1 = \pm \frac{R_4}{R_3} V_{REF} \tag{7.10}$$

The second set of breakpoints, determined in the same manner, occurs at $\pm V_2$ where

$$V_2 = \frac{R_6}{R_5} V_{REF} > V_1 \tag{7.11}$$

When D_2 or D_4 turns on, D_1 or D_3, respectively, will be already conducting since $V_2 > V_1$.

In the region of operation,

$$-V_1 < V_s < V_1$$

all diodes will be off and

$$V_o = m_1 V_s \tag{7.12}$$
$$= -\frac{R_2}{R_1} V_s$$

where

$$m_1 = -\frac{R_2}{R_1} \tag{7.12a}$$

In the region of operation

$$-V_2 < V_s < -V_1$$

only D_1 will be conducting and

$$V_o = m_2 V_s \tag{7.13}$$
$$= -\left(\frac{R_2}{R_1} + \frac{R_2}{R_4}\right) < V_s$$

where

$$m_2 = -\left(\frac{R_2}{R_1} + \frac{R_2}{R_4}\right) > m_1 \tag{7.13a}$$

In this region, the gain of the circuit is increased because we are effectively shunting the input resistances R_1 and R_4.

For

$$V_s > V_2 \quad (D_3, D_4: \text{on})$$

and

$$V_s < -V_2 \quad (D_1, D_2: \text{on})$$

$$V_o = m_3 V_s$$
$$= -\left(\frac{R_2}{R_1} + \frac{R_2}{R_4} + \frac{R_2}{R_6}\right) V_s \tag{7.14}$$

where

$$m_3 = -\left(\frac{R_2}{R_1} + \frac{R_2}{R_4} + \frac{R_2}{R_6}\right) \tag{7.14a}$$

The circuit shown produces a symmetric transfer characteristic in the second and fourth quadrants because of the equal values in the R_3-R_6 network.

Example

Design a function synthesizer that satisfies the following criteria:

1. $V_o = -2V_s$ for $-2 \text{ V} < V_s < 2 \text{ V}$
2. $V_o = -4V_s$ for $-4 \text{ V} < V_s < -2 \text{ V}$ and $2 \text{ V} < V_s < 4 \text{ V}$
3. $V_o = -8V_s$ for $V_s < -4 \text{ V}$ and $V_s > +4 \text{ V}$.

The reference voltages for the circuit are ± 1 V.

Solution

Initially, we will establish the slope in the three regions of operation. Using (7.12a), we obtain

$$m_1 = -\frac{R_2}{R_1} = -2$$

We will let $R_2 = 20$ kΩ; thus, $R_1 = 10$ kΩ. Using (7.13a) gives us

$$m_2 = -\left(2 + \frac{20 \text{ k}\Omega}{R_4}\right) = -4$$

The above relationship is satisfied for $R_4 = 10$ kΩ. Using (7.14a), we get

$$m_3 = -\left(2 + 2 + \frac{20 \text{ k}\Omega}{R_6}\right) = -8$$

The above relationship is satisfied for $R_6 = 5$ kΩ.

Resistors R_3 and R_5 are used to satisfy the breakpoint requirements. Using (7.10),

$$V_1 = \frac{10 \text{ k}\Omega}{R_3} 1 \text{ V} = 2 \text{ V}$$

produces $R_3 = 5$ kΩ. Using (7.11),

$$V_2 = \frac{5 \text{ k}\Omega}{R_5} 1 \text{ V} = 4 \text{ V}$$

produces $R_5 = 1.25$ kΩ.

After an initial design attempt, the designer always has to go back and check the validity of the component values relative to other criteria. These "other criteria" are identified in Part III but, based on what we know thus far, the resistor values in this design are reasonable. ∎

7.4 SAMPLE-AND-HOLD CIRCUITS

A sample-and-hold circuit tracks an input voltage signal and then holds or stores its value at a given instant when commanded by a logic control signal. The model for a sample-and-hold circuit, Fig. 7.8, consists of a switch and a capacitor. The ideal capacitor, when the switch opens under the command of a control signal, stores the value of the analog input signal v_s.

Two basic sample-and-hold circuits are shown in Fig. 7.9. The inverting version, basically, is an amplifier with a gain of -1. In the sample mode of the cycle, the switch S_1 is closed (S_2 open), and the output is the negative of the input voltage. During the hold mode, switch S_1 is opened and the amplifier maintains at its output the last

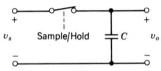

Figure 7.8 Sample-and-hold circuit model.

sample-mode value of v_s. The output voltage is equal to the voltage stored across the capacitor. To repeat another cycle of sample and hold, the capacitor is momentarily discharged by reset switch S_2.

The noninverting version uses a unity-gain, noninverting amplifier as a buffer. When the switch S_1 is closed (S_2 open), v_o vollows v_s. When switch S_1 is opened, the value of v_s stored across the capacitor becomes the input to the buffer amplifier whose output provides a low impedance version of the stored input voltage. Switch S_2 is used to reset the circuit by discharging the capacitor at the end of the sample-and-hold cycle.

Two key parameters of sample-and-hold circuits are acquisition time and decay rate. In time-varying systems, the input signal to a sample-and-hold circuit changes while the circuit is holding a past value. The time required for the circuit to acquire the new value of input signal (to within a given accuracy), when switched from hold to sample, is the acquisition time. The worst case occurs when the output must go from one end of its output range to the other. In the inverting sample-and-hold, acquisition time is primarily a function of the RC product in the feedback loop and the amplifier's slew rate. In the noninverting sample-and-hold, acquisition time is primarily a function of the on resistance of the switch S_1, the amplifier's slew rate, the output impedance of the driving source, and the value of capacitance C. A typical value for the acquisition time of a 20 V signal ($C = 0.01$ μF) is 5 μsec.

Decay rate is measured in volts per second. This parameter measures the ability of the circuit to hold a voltage. In both noninverting and inverting circuits, decay rate is

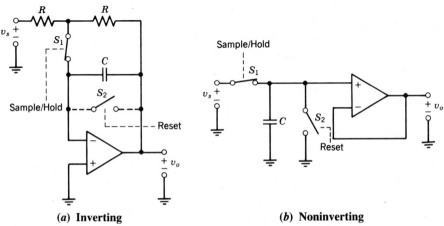

(a) Inverting   (b) Noninverting

Figure 7.9 Basic sample-and-hold circuits.

primarily a function of the impedance of the normally open switch S_2, the input or bias current of the amplifier, and the quality of the capacitor. This parameter is heavily dependent on the physical characteristics of the components. A typical value for decay rate ($C = 0.01$ μF) is 5 mV/sec.

Example

The resistance of switches S_1 and S_2 in Fig. 7.9b while in the open position is 10^{10} Ω; the amplifier's bias current is 10 nA; and the value of the capacitor (ideal) is 0.47 μF. If the capacitor is charged to $+1.000$ V, estimate the time it will take the output voltage to change by 10 mV.

Solution

We have to find the decay rate. The component of the decay rate due to the resistances of the open switches is negligible; initially the current in each will be $1\text{ V}/10^{10}$ Ω $= 0.1$ nA $\ll 10$ nA. Since the input current of the amplifier is constant, the decay rate will also be a constant determined by

$$\frac{dv_o}{dt} = \frac{I}{C} = \frac{10\text{ nA}}{0.47\text{ μF}} = 0.0212\text{ V/sec}$$

For a change in v_o of 10 mV,

$$\Delta t = \frac{10\text{ mV}}{0.0212\text{ V/sec}} = 470\text{ msec}$$

In terms of electronic processing activity, this is indeed a very long time. In commercial sample-and-hold microelectronic circuits, operational amplifiers with ultralow bias currents ($\ll 1$ nA) and smaller capacitors are used. ∎

7.5 PEAK DETECTORS

A peak detector is a special-purpose sample-and-hold circuit. The input signal is tracked until the input reaches a maximum value and then the peak detector automatically holds the peak value until another maximum occurs. Peak detectors are used in transient and repetitive waveform analysis; examples are gas chromatographs and mass spectrometers.

The model for a peak detector is shown in Fig. 7.10. The circuit has three modes of operation: detect, hold, and reset. In the detect mode, S_1 is closed and S_2 is open. When the input signal V_s is more positive (by 0.6 V) than the voltage across the capacitor, the diode D is forward-biased and allows current to flow to charge the holding capacitor C. When the input signal V_s is less positive (by 0.6 V) than the voltage across the capacitor, the diode will be reverse-biased. The discharge path for the capacitor will be a very high impedance, causing the capacitor to maintain its charge.

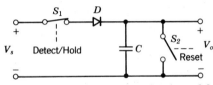

Figure 7.10 Peak detector circuit model.

In the hold mode, both switches are open. For this case, the capacitor will hold the last positive peak value prior to the opening of switch S_1. In the reset mode, switch S_1 is open and S_2 is closed. The closure of S_2 provides a low impedance path to ground and causes the capacitor to rapidly discharge.

The model has its drawbacks and is only used to illustrate basic ideas associated with peak detectors. The major limitations are the forward-voltage drop of the nonideal diode and the effects of a finite load impedance connected across the capacitor.

7.5.1 Noninverting Positive Peak Detector

The circuit in Fig. 7.11 will detect and hold the most positive peak value of V_s. A_2 is a voltage-follower amplifier whose output, after a peak is detected, is the magnitude of this peak. The positive peak of V_s is stored across C, in the hold mode, as a result of the high input impedance of A_2 and the high impedance of the reverse-biased diode D_1.

When $V_s > V_c$, V_{in} of A_1 will be positive; the output of A_1 will be positive, D_1 will be forward-biased; and A_1 will charge the capacitor C to the new peak value. When D_1 is forward-biased, D_2 is reverse-biased since the output of A_1 is more positive than V_c and thus V_o. The voltage at the inverting input terminal of A_1 will be the same as the voltage V_c across the capacitor. While $V_s > V_c$, A_1 functions like a comparator and compares the input voltage with the value stored across the capacitor C.

When $V_s = V_c$, an equilibrium condition is reached. The voltage V_{in} is ideally zero

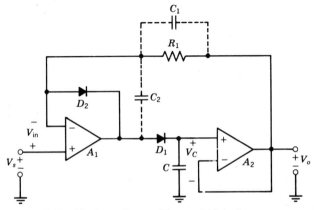

Figure 7.11 Noninverting positive-peak detector.

138 Special-Purpose Circuits

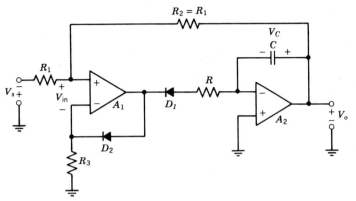

Figure 7.12 Inverting negative-peak detector.

volts because V_s is equal to the voltage stored in C, which is the same as the voltage at the inverting input of A_1.

When $V_s < V_c$, V_{in} of A_1 will be negative; the output of A_1 will be negative; D_1 will be reverse-biased; and D_2 will be forward-biased. The output of A_2 will be $+V_s(\text{peak})$, and the circuit will be in the hold mode until V_s again exceeds V_c. Capacitors C_1 and C_2 may be needed to improve stability and transient response. A switch in parallel with C may be used to discharge the capacitor and initialize the circuit.

7.5.2 Inverting Negative Peak Detector

The circuit in Fig. 7.12 will detect and hold the most negative value of V_s, but will output the value with a positive sign. The negative peak value of V_s is stored across C in the inverting integrator circuit of A_2 in the hold mode because of the high input impedance of A_2 and the high impedance of the reverse-biased diode D_1.

When $V_s > V_c$, V_{in} will be negative; the output of A_1 will be negative; and D_1 will be forward-biased. With D_1 forward-biased, D_2 is reverse-biased. The negative output of A_1 will continually charge C through D_1 and R until $V_s = V_c$. When $V_s = V_c$, an equilibrium condition is reached. The voltage V_{in} is ideally zero volts because $V_s = V_c$ and $R_1 = R_2$.

When $V_s < V_c$, V_{in} will be positive; the output of A_1 will be positive; D_1 will be reverse-biased; and D_2 will be forward-biased. D_2, in both of the detectors discussed, keeps A_1 in its active region while the circuit is in the hold mode. The output will be at $V_s(\text{peak})$, and the circuit will be in a hold mode, until $V_s > V_c$. A switch in parallel with C may be used to discharge the capacitor and initialize the circuit.

7.6 MULTIPLEXER

Systems that combine operational amplifiers with analog switches add a powerful dimension to the data processing capability of the op amp. The switches are often used to control analog operations with digital command signals and the resultant hybrid (analog-digital) circuits, such as multiplexers, and analog-to-digital convertors, are used in a myriad of applications.

An inverting multiplexer is shown in Fig. 7.13a. A multiplexer is a circuit whose output is related to one or more input signals depending on the status of the circuit's switches. The inverting amplifier-type circuit has two signal inputs V_{s1} and V_{s2}, which only affect the output voltage when the analog switches S_1 and S_2 are closed. These switches, controlled by the digital control signals DB_1 and DB_2, provide an output that can be zero, a function of V_{s1} and V_{s2} together, or V_{s1} and V_{s2} individually. If S_1 and S_2 are closed, the circuit functions as an inverting summer. If S_1 is closed and S_2 is open for a brief period of time, and then S_2 is closed and S_1 is opened for a brief period of time, the circuit acts like a multiplexer. If the circuit has n inputs and each one is separately multiplexed to the output in a cyclic fashion, we call the circuit a commutator.

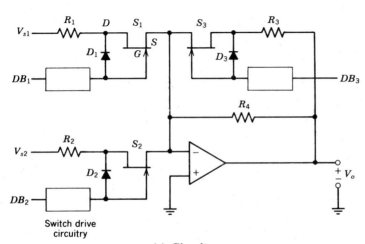

(a) Circuit

S_3	S_2	S_1	V_o
0	0	0	0 V
0	0	1	$-(R_4/R_1)V_{s1}$
0	1	0	$-(R_4/R_2)V_{s2}$
0	1	1	$-(R_4/R_1)V_{s1} - R_4/R_2 V_{s2}$
1	0	0	0 V
1	0	1	$-[(R_4\|R_3)/R_1]V_{s1}$
1	1	0	$-[(R_4\|R_3)/R_2]V_{s2}$
1	1	1	$-[(R_4\|R_3)/R_1]V_{s1} - [(R_4\|R_3)/R_2]V_{s2}$
0-Open 1-Closed			

(b) Input-output relationships

Figure 7.13 Multiplexer.

140 Special-Purpose Circuits

The gain of the circuit is a function of the input and feedback resistors. The feedback resistor is either R_4 (S_3: open) or the parallel combination of R_3 and R_4 (S_3: closed). Using analog switches to change the gain of a circuit is a commonly used circuit technique. The chart in Fig. 7.13b summarizes the relationships between V_o and V_{s1} and V_{s2} for various switch settings.

7.6.1 Analog Switches

At this point it might be appropriate to briefly cover how a transistor is used as an analog switch. Both bipolar and field-effect transistors are used as analog switches; each has its own advantages and drawbacks. Both field-effect transistors, the JFET and the MOSFET, are used more frequently as analog switches than bipolar devices. The device used in the multiplexer circuit is the JFET. JFETs are depletion-mode devices, that is, they require a gate-to-source voltage to turn them off since, for $V_{gs} = 0$ V, they are on. The n channel JFET used in the multiplexer circuit requires a negative V_{gs} to deplete the device's channel of carriers, or have it look like an open switch. If V_{gs} is zero, the switch is closed; if V_{gs} is sufficiently negative, the switch is open.

Two key JFET parameters must be considered when implementing a switch (1) V_{gs}(off) or V_p, and (2) r_{ds}(on). V_{gs}(off) or the pinchoff voltage V_p (numerically equal) is the gate-to-source voltage required to reduce the drain current I_d to near zero (1 nA). In implementing the switch, the output of the switch drive circuitry in Fig. 7.13a must be a negative voltage of magnitude greater than V_{gs}(off) of the JFET. For a closed switch, the drive output circuitry must be zero since the source of the JFET is connected to the summing junction of the amplifier, which is at virtual ground. Once the JFET is on, the resistance from drain-to-source r_{ds}(on) will add to the circuit's input resistance. Normally, the circuit's input resistance is made much larger than r_{ds}(on). Typical values for V_{gs}(off) and r_{ds}(on) are 5 V and 50 Ω. The switch drive circuitry functions as a voltage translator. It converts the high and low levels of the logic input signals to the proper voltage levels to turn the JFET on and off. These translators are available in IC form. Diodes D_1, D_2, and D_3 have a low V_F (0.4 V) and prevent the JFET from turning on gate-to-drain in the switch's off state. Since the JFET is a symmetric device, the diodes limit the gate-to-drain voltage to 0.4 V and insure, in the off condition, that the gate is no more positive than the drain by 0.4 V. Simply, these diodes prevent the gate to drain from behaving like a forward-biased *pn* junction.

7.7 MONOSTABLE MULTIVIBRATOR

The monostable multivibrator in Fig. 7.14 is also called an analog one-shot because it produces a single output pulse each time it is triggered. The circuit has a single stable state and will produce, at its output, a low going pulse of known duration when a positively going trigger pulse is applied to the input. The pulse width is established by R_1, C_1, and V_1.

In the absence of a trigger pulse at v_s and with C and C_1 fully charged, $V_o \cong V_o(\text{max}) = +10 \text{ V}$ because $-V_1$ is applied to the inverting input. The capacitor C_1 is charged to 10 V.

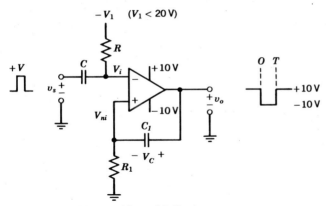

Figure 7.14 Monostable multivibrator.

If the input is triggered at $t = 0$, that is, v_s goes briefly from 0 V to some positive value, the voltage at the inverting input will momentarily go positive and the output will go to $V_o(\min) \cong -10$ V. Since the capacitor C_1 cannot change its voltage instantaneously, the voltage v_{ni} at the noninverting input of the op amp will go to -20 V. With C charged, the inverting input returns to $-V_1$. As long as

$$-V_1 > v_{ni}$$

the output will remain at -10 V.

The voltage at the noninverting input v_{ni} will attempt to go to 0 V as the capacitor C_1 charges toward -10 V. However, this final state is never reached because as soon as v_{ni} is slightly more positive than $-V_1$, the output switches back to the high state or $+10$ V. The voltage v_{ni} is given by

$$v_{ni}(t) = -20 \text{ V} \exp \frac{-t}{R_1 C_1}$$

The time T it takes v_{ni} to reach $-V_1$ described by

$$-V_1 = -20 \text{ V} \exp \frac{-T}{R_1 C_1}$$

is the duration of the output pulse. Solving for T or pulse width produces

$$T = R_1 C_1 \ln \frac{20 \text{ V}}{V_1} \qquad (7.15)$$

In general,

$$T = R_1 C_1 \ln \frac{V_o(\max) - V_o(\min)}{V_1} \qquad (7.16)$$

142 Special-Purpose Circuits

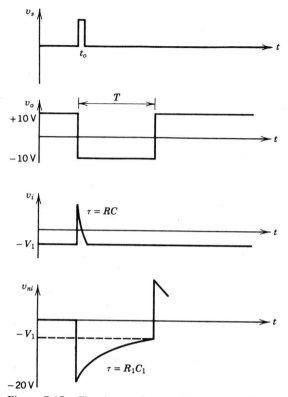

Figure 7.15 Circuit waveforms of the monostable multivibrator.

where $V_o(\max)$ and $V_o(\min)$ are the positive and negative levels of the output of the monostable multivibrator. The various circuit voltages are shown in Fig. 7.15. The RC time constant must be much less than R_1C_1 to allow v_i to return quickly to $-V_1$.

Example

Design a circuit that produces a low going 10 msec output pulse when triggered by a 1 μsec positively going input pulse. The output states must be ± 10 V and a -5 V reference voltage is available in the system.

Solution

The circuit in Fig. 7.14 will fulfull the indicated specifications. Since -5 V is available, then

$$-V_1 = -5 \text{ V}$$

Using (7.15), we obtain

$$10 \text{ msec} = R_1C_1 \ln \frac{20 \text{ V}}{5 \text{ V}}$$

Standard values of R_1 and C_1 that satisfy this equation are

$$C_1 = 0.22 \; \mu F$$

and

$$R_1 = 33 \; k\Omega$$

∎

7.8 SUMMARY

The diversity of the applications of the operational amplifier is clearly illustrated in this chapter. The circuits discussed perform sundry functions that include limiting, waveform synthesizing, sample and holding, peak detecting, multiplexing, and generating a single output pulse.

Limiter circuits utilize diodes to bound or limit the amplifier's output voltage. The function synthesizer, also using diodes, can piecewise simulate or approximate a complex transfer function. The capacitor's ability to store or memorize a voltage is transferred to the circuit level in a circuit called sample and hold. The sample-and-hold feature is also found in peak detectors where the last peak value of the input voltage is stored.

A multiplexer is a circuit whose output voltage is related to two or more input voltages depending on which inputs are multiplexed or switched into the circuit. The switches, implemented with FET transistors, greatly increase the circuit's versatility and functionality. The monostable multivibrator or analog one-shot provides a single output pulse of known duration and amplitude when initiated by an input pulse.

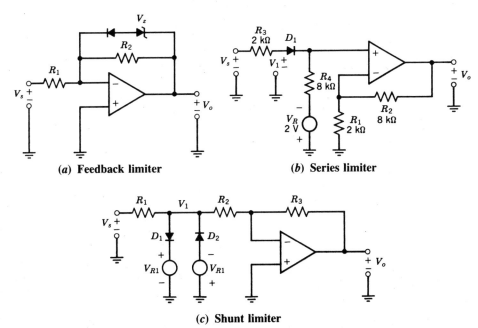

Figure 7.16 Other special-purpose circuits.

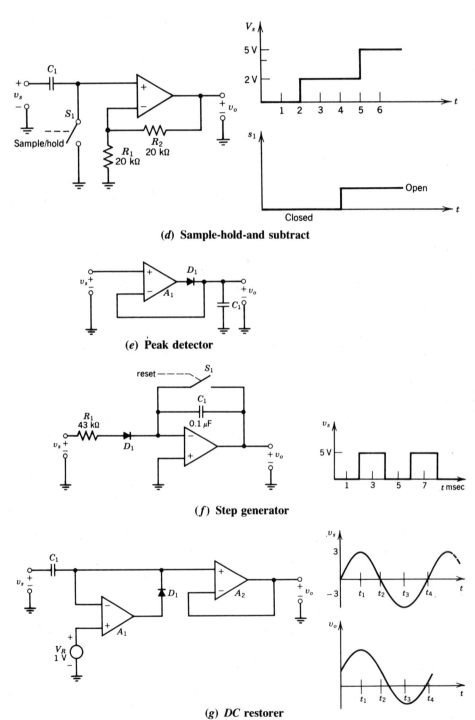

(d) **Sample-hold-and subtract**

(e) **Peak detector**

(f) **Step generator**

(g) **DC restorer**

Figure 7.16 Other special-purpose circuits. (*Continued*)

PROBLEMS

7.1 Sketch the transfer characteristics for the limiters in Figs. 7.1, 7.2, and 7.3 if the diodes in each of these figures are reversed.

7.2 Calculate the breakpoint and the slope of the transfer characteristic of the series limiter in Fig. 7.5 if $R_1 = R_2 = R_3 = 10 \text{ k}\Omega$ and $V_R = 4$ V.

7.3 Calculate the breakpoint and the slope of the transfer characteristic of the shunt limiter in Fig. 7.6 if $R_1 = R_2 = R_3 = 10 \text{ k}\Omega$ and $V_R = 4$ V.

7.4 Sketch the transfer characteristics for the series and shunt limiters in Figs. 7.5 and 7.6 if the polarity of V_R in each of these figures is reversed.

7.5 How would the transfer characteristic of the function synthesizer in Fig. 7.7 be affected if diodes D_1 through D_4 were reversed?

7.6 Express v_o as a function of time for the noninverting sample-and-hold circuit of Fig. 7.9b (in the hold mode) if the amplifier's input circuit can be modeled with a 0.47 μF capacitor in parallel with a 10 MΩ resistor. If the capacitor is initially charged to 10 V, what is the value of v_o 100 msec after the circuit is put in the hold mode?

7.7 Express v_o as a function of time for the noninverting sample-and-hold circuit of Fig. 7.9b (in the hold mode) if the amplifier's input circuit can be modeled with a 0.47 μF capacitor in parallel with a 100 nA current source pointing to ground. If the capacitor is initially charged to 10 V, what is the value of v_o 100 msec after the circuit is put in the hold mode?

7.8 Modify the peak detector model in Fig. 7.10 to detect and hold negative voltage peaks. Illustrate the practical problems associated with this circuit.

7.9 Modify the positive peak detector in Fig. 7.11 to make it a negative peak detector.

7.10 What changes are required in the negative peak detector of Fig. 7.12 to make it a positive peak detector.

7.11 Specify a set of practical values for the resistors and capacitor for the negative peak detector in Fig. 7.12. Show how the values of R and C limit the circuit's response to rapidly changing values of input voltage.

7.12 Specify a set of values for R_1, R_2, R_3, and R_4 of the multiplexer in Fig. 7.13 to satisfy the following relationships:

$$V_o = -2V_{s1} - 4V_{s2} \quad (S_1 = S_2 = 1, S_3 = 0,)$$

and

$$V_o = -V_{s1} - kV_{s2} \quad (S_1 = S_2 = S_3 = 1) \quad (k > 0)$$

What is the value of k?

7.13 Design a monostable multivibrator (Fig. 7.14) whose pulse width is 10 msec and whose output voltage levels are +8 V and −12 V. The input trigger is a positive going 1 μsec pulse.

7.14 Draw the transfer characteristic (V_o versus V_s) for the circuit in Fig. 7.16a. Identify all key points on the graph.

7.15 Design a feedback limiter whose transfer characteristic is shown in Fig. 7.17.

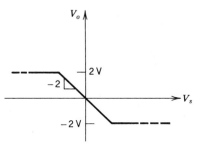

Figure 7.17 Transfer characteristic for Problem 7.15.

7.16 Design a series limiter whose transfer characteristic is shown in Fig. 7.18. Let $V_R = 2$ V.

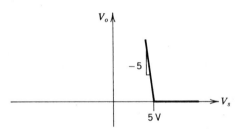

Figure 7.18 Transfer characteristic for Problem 7.16.

7.17 Draw the transfer characteristic for the circuit shown in Fig. 7.16b. Identify the numerical values of all key points on the graph.

7.18 Draw the transfer characteristic for the circuit shown in Fig. 7.16c.

7.19 Design a shunt limiter (Fig. 7.6) whose transfer characteristic is shown in Fig. 7.19. Let $V_R = -2$ V.

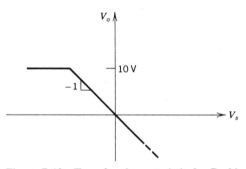

Figure 7.19 Transfer characteristic for Problem 7.19.

7.20 The circuit in Fig. 7.16d is a sample-hold-and-subtract circuit. Sketch the output waveform for the input and switch conditions shown.

7.21 The circuit in Fig. 7.16e is a positive peak detector. Explain its operation. What factors will contribute to the decay rate of the voltage stored by the capacitor? What is a significant disadvantage of this circuit?

7.22 The circuit in Fig. 7.16f is a step generator. Sketch the output waveform for the input shown. Assume that S_1 is opened at $t = 0$ and remains open. What is the function of D_1?

7.23 The circuit in Fig. 7.16g is a dc restorer, a circuit that adds a dc value to a given waveform. Explain the operation of the circuit and identify the values of v_o at t_1, t_2, t_3, and t_4 for the input waveform shown.

Chapter 8

Filters

8.1 PROLOGUE

Filter theory is a topic unto itself. The concepts of this subject are complex because the characteristics and performance of filter circuits are derived abstractly. The introduction of the variable time (or frequency) restricts any intuitive or straightforward approach to analyzing these circuits. In most of the previous applications areas, we followed the input signal and saw how its behavior or waveshape was modified in various portions of the circuit. In filters, we are asked to identify what the circuit will do over the complete frequency spectrum. Typically, filters are analyzed using idealized mathematical and circuit models and the abstract s plane. Since the intent of the next two chapters is to focus on the use of the op amp in filter circuits, we briefly present some filter fundamentals that will serve as background material in support of the applications presented.

Generally speaking, a filter is a circuit designed to pass or amplify input signals of certain frequencies and block or attenuate those of other frequencies. It is a frequency selective circuit used to discriminate input signals of different frequencies. If the magnitude of the input signal is held constant and the frequency is varied, the magnitude and/or phase of the output signal will be constant over some range of frequency and vary outside this range. The all-pass filter, whose magnitude characteristic is constant but whose phase characteristic varies, represents a special case.

If the performance of a filter is frequency dependent, then the circuit must contain capacitors and/or inductors whose impedance varies with frequency. Filters implemented with resistors, capacitors, and inductors are defined as passive. Filters implemented with resistors, capacitors, inductors, and active or gain-producing devices are called active filters. Because of the poor performance of the real inductor, most active filters are implemented with resistors, capacitors, and one or more active devices. Although we focus our attention on active filters that use the operational amplifier as the active device, most of the circuits presented have an equivalent version that uses transistors. The transistor versions are employed when the use of the op amp is not feasible, as in the high-frequency cases.

In our treatment, we concentrate on second-order filters. They are easier to analyze, design, tune, and are frequently used. In addition, second-order filters can be used in cascade to generate higher-order responses when necessary.

8.2 FILTER THEORY

8.2.1 Filter Characteristics

The *primary characteristics* of a filter are:

(A) Nature of the amplitude and/or phase response (low-pass, high-pass, bandpass, band-stop, and all-pass).
(B) Cutoff, corner, or critical frequency (ω_c).
(C) Passband gain (A_o).
(D) Attenuation slope (rolloff) or order.
(E) Characteristic frequency (ω_o).
(F) Figure of merit (Q).

Secondary characteristics include sensitivity to parameter changes, group delay, and drift. Their treatment is beyond the scope of this text.

The *amplitude response* of a filter describes the magnitude of the circuit gain as a function of frequency under sinusoidal excitation conditions. In a typical filter, the amplitude response is nearly constant over a certain range of frequencies, and decreases outside this range. Five typical filter responses are low-pass, high-pass, bandpass, all-pass, and band-stop or band-reject.

The *cutoff,* corner, or critical *frequency* is the frequency where the amplitude response is 3 dB below its passband value. In the ideal amplitude response of first-order, low- and high-pass filters, ω_c represents the frequency where the transition occurs between a constant magnitude and a decreasing magnitude output signal. Because of peaking in the amplitude response of higher order filters, ω_c is, by definition, the -3 dB frequency.

An active filter, which contains a gain-producing or active device, can amplify the input signal in the filter's passband. This *amplification* is described by the passband gain A_o.

The rate of decrease of the gain of the filter circuit outside the passband is described by the attenuation slope or *rolloff* and is typically measured in dB/octave or dB/decade of frequency variation. As the filter order increases, the attenuation slope increases. For first-order (single-pole) filters, the rate of decrease reaches -6 dB/octave or -20 dB/decade. The rate becomes $-6n$ dB/octave or $-20n$ dB/decade for nth-order filters.

The *characteristic frequency* ω_o is the center frequency in bandpass and band-reject filters. It is also the frequency where peaking occurs in second-order, high-Q low and high-pass filters.

The *figure of merit* Q has several interpretations depending on the filter response. In bandpass and band-reject filters, Q measures the ratio of center frequency (ω_o) to bandwidth ($\omega_{cu} - \omega_{cl}$). In second-order low- and high-pass filters, Q measures the degree of peaking at the characteristic frequency ω_o. In a more abstract sense, Q (in conjunction with ω_o) specifies the location of each complex pole-pair of a given filter transfer function.

8.2.2 Transfer Function

The ratio of the Laplace transforms of the output and input signals in a filter is called the *transfer function*. The transfer function $T(s)$ is expressed using the complex variable s. In the general case, $s = \sigma + j\omega$. To obtain the filter's steady-state response under sinusoidal excitation conditions (ac response), s is replaced by $j\omega$ in $T(s)$.

Typically $T(s)$ is the ratio of two polynomials in s,

$$T(s) = \frac{N(s)}{D(s)}$$

Those values of s that make $T(s)$ become infinite are called *poles* of $T(s)$ while those values that make $T(s)$ equal to zero are called zeros. The total number of *poles* and zeros of $T(s)$ must be equal if we include those located at infinity.

An nth-order filter is one whose transfer function has n poles, and to realize these n poles, at least n reactive elements are required. The form of $N(s)$, in conjunction with $D(s)$, identifies the nature of the amplitude response of the filter.

A root locus is a plot of the possible variations of the poles and zeros of $T(s)$ as some parameter is varied. This plot, in the s plane, provides us (indirectly) with a picture of the performance of the filter as a function of various filter component values. As the values of the filter's components change, the location of the poles and zeros in the s plane changes, and so does the performance of the filter.

A biquadratic filter is one whose transfer function is given by the generalized transfer function,

$$T(s) = \frac{a_2 s^2 + a_1 s + a_0}{s^2 + b_1 s + b_0}$$

Filter Theory

The term "biquadratic," in its broadest sense, means any transfer function that has all or part of a quadratic equation in both the numerator and denominator.

The special cases of the biquadratic filter are

1. $a_1 = a_2 = 0$; low-pass

$$T(s) = \frac{a_0}{s^2 + b_1 s + b_0} \quad (8.1a)$$

2. $a_0 = a_2 = 0$; bandpass

$$T(s) = \frac{a_1 s}{s^2 + b_1 s + b_0} \quad (8.1b)$$

3. $a_0 = a_1 = 0$; high-pass

$$T(s) = \frac{a_2 s^2}{s^2 + b_1 s + b_0} \quad (8.1c)$$

4. $a_1 = 0$; band-stop

$$T(s) = \frac{a_2 s^2 + a_0}{s^2 + b_1 s + b_0} \quad (8.1d)$$

5. $a_0/a_2 = b_0$, $a_1/a_2 = -b_1$; all-pass

$$T(s) = a_2 \frac{s^2 - b_1 s + b_0}{s^2 + b_1 s + b_0} \quad (8.1e)$$

For most filters, the coefficients of the numerator and denominator polynomials may be interpreted in terms of the filter's characteristics. Figure 8.1 lists $T(s)$ in terms of the filter characteristics for the four popular first- and second-order filter response types.

First-order, low-pass $\quad T(s) = \dfrac{A_o \omega_c}{s + \omega_c}$

First-order, high-pass $\quad T(s) = \dfrac{A_o s}{s + \omega_c}$

Second-order, low-pass $\quad T(s) = \dfrac{A_o \omega_o^2}{s_2 + (\omega_o/Q)s + \omega_o^2}$

Second-order, high-pass $\quad T(s) = \dfrac{A_o s^2}{s^2 + (\omega_o/Q)s + \omega_o^2}$

Second-order, bandpass $\quad T(s) = \dfrac{A_o s(\omega_o/Q)}{s^2 + (\omega_o/Q)s + \omega_o^2}$

Second-order, band-reject $\quad T(s) = \dfrac{A_o (s^2 + \omega_o^2)}{s^2 + (\omega_o/Q)s + \omega_o^2}$

Figure 8.1 $T(s)$ in terms of A_o, ω_o, ω_c, and Q.

8.2.3 The Frequencies ω_o and ω_c, and Q

The response of the first-order low- and high-pass filters can be graphically described using the straight line approximation of the Bode plot. This plot (magnitude in decibels versus the log of frequency) visually illustrates the behavior of the magnitude of the gain of the filter over the frequency spectrum. Similarly, a linearized version of phase versus frequency approximates the phase behavior. The frequency where the circuit gain makes a transition from a constant magnitude to a decreasing magnitude is called the cutoff, critical, or corner frequency. It is the -3 dB frequency or the frequency where the amplitude of the gain is 0.707 of its value in the passband. This frequency is called ω_c. For first-order low- and high-pass filters, there is no frequency called ω_o.

Bandpass and band-reject filters are at least second order and contain a minimum of two reactive components. The center frequency of the passband or reject band is the characteristic frequency ω_o. The -3 dB frequencies (upper and lower) of the passband or reject band are the frequencies ω_{cu} and ω_{cl}. In these filters, the figure of merit Q measures the ratio of the center frequency ω_o to the 3 dB bandwidth of the pass or reject band,

$$Q = \frac{\omega_o}{\omega_{cu} - \omega_{cl}} \tag{8.2}$$

The center and the cutoff frequencies in bandpass and band-reject filters are related by

$$\omega_o^2 = \omega_{cu} \omega_{cl}$$

This geometric mean tends to become an arithmetic mean for high-Q circuits.

In second-order, low- and high-pass responses, peaking can occur in the vicinity of cutoff. For high-Q ($Q \gg 0.5$) second-order high- and low-pass filters, the frequency ω_o is approximately the frequency where peaking in the response reaches its maximum. Typical responses for these filters are shown in Fig. 8.2. The frequency ω_c is the

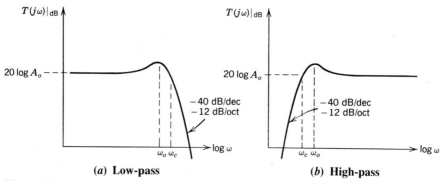

Figure 8.2 Second-order low- and high-pass responses.

frequency where the passband gain is down 3 dB and is lower than ω_o in high-pass filters and higher than ω_o in the low-pass circuits. When Q equals 0.707, the filter has a maximally flat magnitude (MFM) response and ω_c and ω_o are equal. The MFM or Butterworth filter has a minimal ripple in the filter's passband.

The frequency ω_o and the figure of merit Q also have another interpretation. When $T(s)$ has a pair of complex conjugate poles, Fig. 8.3, $D(s)$ can be expressed as

$$D(s) = s^2 + s\frac{\omega_o}{Q} + \omega_o^2 \tag{8.3}$$

The poles in the s plane are expressed as

$$s_{1,2} = -\alpha \pm j\beta \tag{8.4}$$

The coordinates of these poles, in terms of ω_o and Q, are

$$\alpha = \frac{\omega_o}{2Q} \quad \text{and} \quad \beta = \omega_o\sqrt{1 - \frac{1}{4Q^2}}$$

The frequency ω_o corresponds to the distance of the poles from the origin, while Q is a measure of their proximity to the $j\omega$ axis. For the poles of $T(s)$ to be complex, Q must be greater than 0.5. In general, filters with $Q < 0.5$ are of little practical value; hence, we are only interested in filters with complex conjugate poles.

8.2.4 Filter Components

The type and tolerance of the resistors and capacitors used in active filters play an extremely important role in optimizing the filter's performance. In general, ±5% capacitor tolerances and ±1% resistor tolerances are recommended for second-order filters. The best choice of resistors are those made of metal film. Metal film resistors have low noise, very good frequency response characteristics, and fairly low temperature coefficients. Wirewound resistors are also acceptable, but they must be nonin-

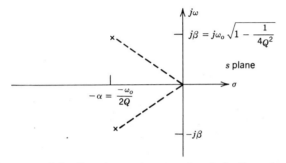

Figure 8.3 Complex conjugate pole-pair in the s plane.

154 Filters

ductively wound. The best capacitors for active filter applications are polycarbonate film, NPO ceramic, and mica. Though capacitors made of these materials are somewhat expensive and large in size, they do have low dissipation factors and low temperature coefficients. For noncritical applications, metalized mylar and polycarbonate capacitors are acceptable.

8.2.5 Butterworth, Chebyshev, and Bessel Responses

Filters are often designed to optimize a certain characteristic: a maximally flat magnitude response, high degree of attenuation in the transition region, or a linear phase response are examples. The names of the special filters that have these characteristics are the Butterworth, Chebyshev, and the Bessel.

The Butterworth or MFM (maximally flat magnitude) filter is one in which the magnitude response is maximally flat in the passband. The Butterworth magnitude monotonically decreases as frequency increases and has the flattest possible response in the vicinity of $\omega = 0$ Hz. This type of filter is used in applications where all frequencies in the passband must have the same gain. The phase response is not linear, that is, the time required for a signal to propagate through the filter is not constant with frequency. An nth-order Butterworth filter will have a $20n$ dB/decade attenuation rate in the stop-band region ($\omega \gg \omega_c$). The magnitude response of a Butterworth filter is expressed as

$$|T(j\omega)| = \frac{|A_o|}{\sqrt{1 + \left(\dfrac{\omega}{\omega_c}\right)^{2n}}}$$

where n represents the number of poles or order of the filter. The poles are uniformly spaced along a semicircle of radius $\omega_c = \omega_o$ in the left half of the s plane.

The Chebyshev filter will have ripple in the passband of the magnitude response, typically 0.5, 1, 2, and 3 dB, but not in the stopband. For higher ripple filters, more attenuation is obtained in the transition region. Chebyshev filters can be of lower order than the Butterworth filters and less complex, for a given attenuation rate, if the amplitude response in the passband need not be constant. The phase response of the Chebyshev filters is less linear than that of the Butterworth filters.

Bessel filters are referred to as linear phase or constant time delay filters. The phase delay of a signal from the input to the output is constant with frequency. Because these filters have almost no overshoot to a step-response input, they are the best choice for filtering rectangular signals without altering the shape of the waveforms.

A Butterworth, Chebyshev, and a Bessel filter might all have the same schematic. The primary difference is Q and the component values. Q sets the shape of the transition region and the overshoot in the magnitude response's passband near the transition region. As an example, a second-order Butterworth low-pass filter will have a Q of 0.707, and a 3 dB ripple, second-order Chebyshev filter will have a Q of 1.305. Figure 8.4 shows a comparison of the second-order, low-pass versions of the three filters.

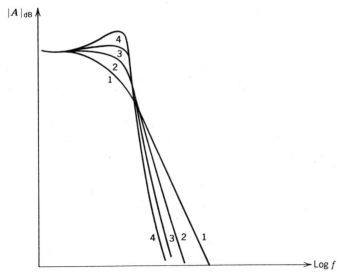

Figure 8.4 Comparing low-pass magnitude responses of second-order Bessel, Butterworth, and Chebyshev filters.

1. Bessell ($Q = 0.577$) 2. Butterworth ($Q = 0.707$) 3. 1 dB ripple Chebyshev ($Q = 0.957$) 4. 3 dB ripple Chebyshev ($Q = 1.305$)

8.2.6 About Active Filters

Active *RC* networks do not use discrete inductors since they are expensive, lossy, heavy, and bulky especially in the low-frequency region. The presence of an ideal inductor in a filter configuration is important, but its behavior can be simulated using a variety of circuit techniques. If we compare the behavior of real inductors and capacitors to their ideal counterparts, the real capacitor is closer to the ideal case. In addition, capacitors have a greater range of values, are smaller in size and weight, and are less expensive to manufacture.

All response types are possible with active filters. The responses can be of various orders from the simple single-pole filters to *n*th-order filters having specialized Butterworth, Chebyshev, and Bessel responses. The moderate to high input impedance and low output impedance of op amp filters allow them to be cascaded without significant deterioration in their performance. The ability to isolate each filter stage allows filters to be individually tuned. Many active *RC* networks can be implemented in microelectronic form where the ultimate in economy can be achieved.

The greatest disadvantage of op amp filters is the frequency limitation of the op amp itself. In general, filters that use an operational amplifier as the active device are used in applications below 100 kHz. A higher bandwidth transistor amplifier would overcome this limitation, but it would do it at the expense of economy or stability from oscillations. Other disadvantages of op amp filters include a limited voltage range, inputs and outputs that are ground referenced, that is, they do not electrically float, and a dependence of the filter's characteristics on the amplifier's parameters.

8.3 FIRST ORDER OP AMP FILTERS

8.3.1 Low-Pass Filter

A low-pass filter allows input signals with frequencies below a given cutoff frequency to be readily passed to its output. In contrast, the filter provides considerable attenuation of those signals above the cutoff frequency. The degree of attenuation of those signals outside the passband is dependent on the order of the denominator of the transfer function under consideration.

The circuit of Fig. 8.5a is a first-order (single-pole), low-pass filter having a cutoff frequency $\omega_c = 1/R_2C$. The voltage gain transfer function of this circuit is

$$T(s) = \frac{(-R_2/R_1)(1/R_2C)}{(s + 1/R_2C)} \tag{8.5}$$

which is of the form

$$T(s) = \frac{A_o \omega_c}{s + \omega_c}$$

The transfer function is obtained from the expression for the inverting amplifier configuration,

$$T(s) = \frac{-Z_2(s)}{Z_1(s)}$$

On an approximate basis, this circuit has a constant amplitude response up to the cutoff frequency ω_c, while outside the passband ($\omega > \omega_c$), the frequency response has a -20 dB/decade (-6 dB/octave) rolloff. The straight line approximation of the magnitude response is shown in Fig. 8.5b. If the rolloff must be increased to improve

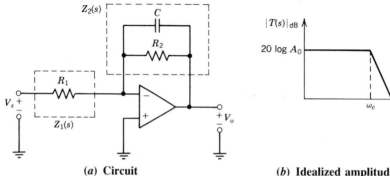

(a) Circuit (b) Idealized amplitude response

Figure 8.5 First-order low-pass filter.

the selectivity (the ability to select the desired frequencies and reject the undesired ones) of the filter, then a higher-order configuration must be used. The higher selectivity circuits have a greater number of poles since each pole produces, eventually, a -20 dB/decade rolloff in the stopband.

The circuit of Fig. 8.5a can be viewed in three different ways. The circuit can be thought of as an inverting amplifier with an upper cutoff frequency controlled by R_2 and C. It can also be visualized as a lossy integrator, or as a low-pass filter. The application and relative values of the circuit components dictate what name is used.

8.3.2 High-Pass Filter

A high-pass filter allows input signals having frequencies above a certain cutoff frequency (ω_c) to be readily passed to its output. Outside the passband, the signals are attenuated. The degree of attenuation is a function of the order of the filter. The greater the number of poles, the higher the order, and the greater the attenuation slope (rolloff) in the stopband.

The circuit of Fig. 8.6a is a first-order (single-pole), high-pass filter having a cutoff frequency $\omega_c = 1/R_1C$. The voltage-gain transfer function of this circuit is

$$T(s) = \frac{\frac{-R_2}{R_1} s}{s + \frac{1}{R_1 C}} \tag{8.6}$$

which is of the form

$$T(s) = \frac{A_o s}{s + \omega_c}$$

The transfer function is obtained from the expression for the inverting amplifier configuration,

$$T(s) = \frac{-Z_2(s)}{Z_1(s)}$$

The idealized magnitude response is shown in Fig. 8.6b. At low frequencies, the capacitor has a high impedance and the gain of the circuit is very low. As ω approaches 0 Hz, the capacitor approaches the behavior of an open circuit and the output goes to zero volts. At frequencies much higher than ω_c, the capacitor may be considered as a short circuit, and the gain becomes constant ($-R_2/R_1$).

The circuit may be viewed three ways. It can be thought of as a modified differentiator, an inverting amplifier with a lower cutoff frequency controlled by R_1 and C, or as a high-pass filter. If the first-order low- and high-pass filters are combined as shown in Fig. 8.7, an elementary bandpass filter results.

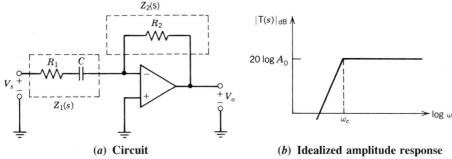

(a) Circuit   (b) Idealized amplitude response

Figure 8.6 First-order high-pass filter.

Example

Find $T(s)$ for the bandpass filter in Fig. 8.7. For the component values shown in the circuit, calculate the lower and upper cutoff frequencies ω_{cl}, ω_{cu}, and the passband gain A_o.

Solution

Since the high-pass and low-pass filters are cascaded and the output impedance of the first filter is essentially zero, the overall transfer function $T(s)$ is equal to the product of the transfer functions of the two filters. Using (8.5) and (8.6), we get

$$T(s) = \frac{A_{o1}s}{s + \omega_{c1}} \frac{A_{o2}\omega_{c2}}{s + \omega_{c2}} \quad (\omega_{c1} \ll \omega_{c2})$$

where

$$\omega_{cl} = \omega_{c1} = \frac{1}{R_1 C_1} = 666.67 \text{ rad/sec } (106 \text{ Hz})$$

$$\omega_{cu} = \omega_{c2} = \frac{1}{R_4 C_2} = 111.1 \text{ krad/sec } (17.7 \text{ kHz})$$

$$A_{o1} = \frac{-R_2}{R_1} = -5$$

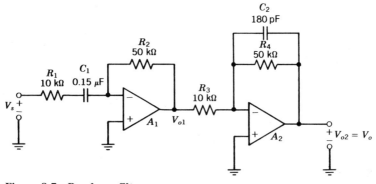

Figure 8.7 Bandpass filter.

and

$$A_{o2} = \frac{-R_4}{R_3} = -5$$

The overall passband gain is

$$A_o = A_{o1}A_{o2} = 25$$

This particular bandpass filter has all the characteristics of an audio amplifier. ∎

8.3.3 All-Pass Filter

An all-pass filter allows input signals of *all* frequencies (within, of course, the bandwidth of the amplifier) to pass to its output. However, the phase relationship between the input and output signals varies as a function of frequency. The first-order all-pass filters in Fig. 8.8, also called phase shifters, use RC lag and lead networks to develop the phase relationship between the input and output.

The transfer function for the circuits in Fig. 8.8 are found by splitting V_s into two equal-valued sources as shown in Fig. 8.9, and then using superposition. For the circuit in Fig. 8.8a,

$$V_o = -V_s + \frac{Z_2}{Z_1 + Z_2} 2V_s$$

$$= -V_s + \frac{\frac{1}{sC}}{R + \frac{1}{sC}} 2V_s$$

After simplification, the resulting transfer function is

$$T_a(s) = \frac{V_o}{V_s} = -\frac{s - \frac{1}{RC}}{s + \frac{1}{RC}} \qquad (8.7a)$$

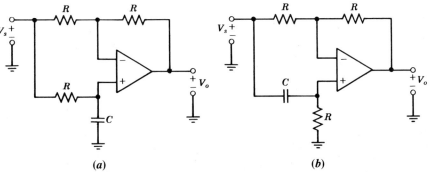

(a)   (b)

Figure 8.8 First-order all-pass filters.

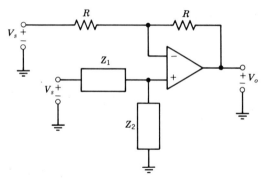

Figure 8.9 Equivalent circuit of the all-pass filters in Fig. 8.8.

The transfer function for the circuit in Fig. 8.8b is found in a similar manner,

$$T_b(s) = \frac{V_o}{V_s} = \frac{s - \frac{1}{RC}}{s + \frac{1}{RC}} \qquad (8.7b)$$

Both responses have symmetrical pole and zero locations with respect to the origin. The magnitude of $T(s)$ for each circuit is one and the phase relationships are

$$\theta_a = -2 \tan^{-1} \omega RC \qquad (8.8a)$$

and

$$\theta_b = -180° - 2 \tan^{-1} \omega RC \qquad (8.8b)$$

The phase of the circuit of Fig. 8.8a varies from 0° to $-180°$ and that of the circuit of Fig. 8.8b varies from $-180°$ to $-360°$.

Example

For the all-pass filters in Fig. 8.8, calculate the phase shift for $\omega = 1/RC$, $\omega = 10/RC$, and $\omega = 1/10RC$.

Solution

Using (8.8a) gives us

$$\theta_a = -2 \tan^{-1} 1 = -90° \qquad \text{at } \omega = \frac{1}{RC}$$

$$\theta_a = -2 \tan^{-1} 10 = -169° \qquad \text{at } \omega = \frac{10}{RC}$$

and

$$\theta_a = -2\tan^{-1}\tfrac{1}{10} = -11° \quad \text{at } \omega = \frac{1}{10RC}$$

Using (8.8b), we get

$$\theta_b = -180° - 2\tan^{-1} 1 = -270° \quad \text{at } \omega = \frac{1}{RC}$$

$$\theta_b = -180° - 2\tan^{-1} 10 = -349° \quad \text{at } \omega = \frac{10}{RC}$$

and

$$\theta_b = -180° - 2\tan^{-1}\tfrac{1}{10} = -191° \quad \text{at } \omega = \frac{1}{10RC}$$

In each case, the range of frequencies in which the phase shift occurs is about a decade above and below $\omega = 1/RC$. ∎

8.4 SECOND-ORDER OP AMP FILTERS

There are a large number of basic filter configurations. We examine four active RC circuit techniques that produce second-order filters with complex conjugate poles:

1. Gyrator filters.
2. Infinite-gain, multiple-feedback filters (IGMF).
3. Infinite-gain, single-feedback filters (IGSF).
4. KRC realizations.

8.5 GYRATOR FILTERS

The gyrator is a two-port circuit or element that may be used to simulate the behavior of an inductor. The gyrator, when its output port is terminated with a capacitor, has the same current and voltage relationship at its input port as the discrete inductor. Thus, gyrator filters are used as active implementations of passive RLC configurations.

8.5.1 Gyrators

The ideal gyrator, Fig. 8.10, is a lossless, passive two-port device for which:

$$V_1 = \alpha I_2 \tag{8.9}$$

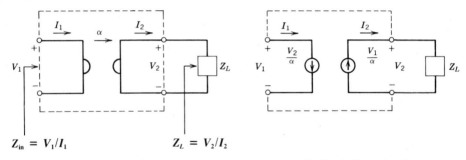

(a) Symbol (b) Equivalent circuit

Figure 8.10 Symbol and equivalent circuit for the gyrator.

and

$$I_1 = \frac{V_2}{\alpha} \tag{8.10}$$

where α is a real constant and is called the gyration ratio.

Using (8.9) and (8.10), we may obtain an expression for the impedance Z_{in} seen across the input port of a gyrator in terms of its load impedance and the gyration ratio:

$$Z_{in} = \frac{V_1}{I_1} = \frac{\alpha I_2}{\frac{V_2}{\alpha}}$$

or

$$Z_{in} = \frac{\alpha^2}{Z_L} \tag{8.11}$$

The input impedance of a gyrator is proportional to the admittance of its load. If the gyrator is terminated with a capacitor $C(Z_L = 1/sC)$, its input impedance is

$$Z_{in} = \frac{\alpha^2}{\frac{1}{sC}} = s(\alpha^2 C)$$

or

$$Z_{in} = sL_{eq} \tag{8.12}$$

where

$$L_{eq} = \alpha^2 C \tag{8.12a}$$

Gyrator Filters

Thus, a gyrator is able to "rotate" (gyrate) a capacitor into an inductor eliminating inductors from network realizations that call for their presence. The gyrator will also rotate inductors into capacitors, resistors into conductors, and open circuits into short circuits.

Figure 8.11 represents a block diagram implementation of a gyrator. This implementation is essentially composed of two noninverting amplifiers with a gain of 2, an inverting amplifier with a gain of one, and a unity-gain summer. All these blocks can be easily implemented using circuits previously described. The voltages at the outputs of each of the amplifiers are

$$V_a = 2V_1$$
$$V_b = -V_a = -2V_1$$
$$V_c = 2V_2$$

and

$$V_d = V_b + V_c = 2(V_2 - V_1)$$

The circuit currents are found using these voltages,

$$I_a = \frac{V_1 - 2V_1}{R_1} = \frac{-V_1}{R_1}$$

$$I_b = \frac{V_1 - V_2}{R_1}$$

$$I_c = \frac{2V_2 - 2V_1 - V_2}{R_1} = \frac{V_2 - 2V_1}{R_1}$$

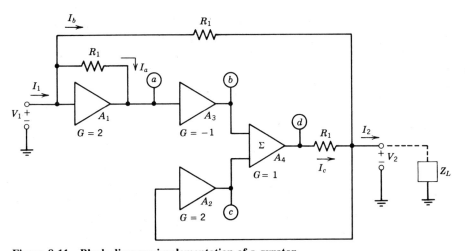

Figure 8.11 Block diagram implementation of a gyrator.

The input and output currents of the gyrator are

$$I_1 = I_a + I_b = \frac{-V_2}{R_1}$$

and

$$I_2 = I_b + I_c = \frac{-V_1}{R_1}$$

Using the expressions for I_1 and I_2, we get

$$V_1 = (-R_1)I_2 \tag{8.13a}$$

and

$$I_1 = \frac{V_2}{-R_1} \tag{8.13b}$$

If we compare the gyrator's defining equations (8.9) and (8.10) with (8.13a) and (8.13b), we see that the circuit of Fig. 8.11 is a gyrator having a gyration ratio $\alpha = -R_1$.

The circuit of Fig. 8.12 is a practical implementation of the gyrator block diagram of Fig. 8.11. In the practical circuit, the A_1 op amp is connected as a noninverting amplifier having a gain of 2, similar to the function of A_1 in Fig. 8.11. Operational amplifier A_2 in conjunction with the two R_3 resistors connected to its inverting input terminal combines the functions of the voltage amplifiers A_2, A_3, and A_4 in the block diagram. A step-by-step analysis of the practical implementation confirms the equivalence of the block diagram and the practical implementation:

$$V_a = 2V_1$$

$$I_a = \frac{V_1 - V_a}{R_1} = \frac{-V_1}{R_1}$$

$$V_d = 2V_2 - V_a = 2V_2 - 2V_1$$

$$I_b = \frac{V_1 - V_2}{R_1}$$

$$I_c = \frac{V_d - V_2}{R_1} = \frac{V_2 - 2V_1}{R_1}$$

$$I_1 = I_a + I_b = \frac{-V_2}{R_1}$$

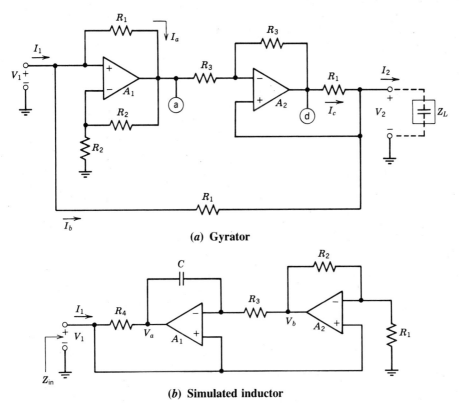

(a) Gyrator

(b) Simulated inductor

Figure 8.12 Active implementations of inductors.

and

$$I_2 = I_b + I_c = \frac{-V_1}{R_1}$$

From the last two steps, we obtain

$$V_1 = (-R_1)I_2$$

and

$$I_1 = \frac{V_2}{-R_1}$$

These expressions are identical to those of (8.13a) and (8.13b), corresponding to the input-output relations of a gyrator having a gyration ratio $\alpha = -R_1$.

Example

The gyrator of Fig. 8.11 is terminated with a capacitor C such that $0.001~\mu F \leq C \leq 0.1~\mu F$. Find the range of values of the equivalent inductance L_{eq} seen across its input, if $5~k\Omega \leq R_1 \leq 100~k\Omega$.

Solution

Using (8.12a) with $\alpha = -R_1$, we obtain

$$L_{eq} = R_1^2 C$$

Therefore, the minimum and maximum values of R_1 and C will produce the minimum and maximum values of L_{eq};

$$L_{eq}(\min) = (5~k\Omega)^2 (0.001~\mu F)$$
$$= 25~mH$$

and

$$L_{eq}(\max) = (100~k\Omega)^2 (0.1~\mu F)$$
$$= 1000~H$$

The range of values for L_{eq}, for nominal values of R_1 and C, is indeed broad. A discrete 1000 H inductor is rather difficult to find.

Example

Find L_{eq} for the simulated inductor circuit of Fig. 8.12b.

Solution

Since the circuit behaves as an inductor, then

$$Z_{in} = \frac{V_1}{I_1} = sL_{eq}$$

To find L_{eq}, we have to find I_1 in terms of V_1 and circuit constants. If we treat A_2 as a noninverting amplifier whose input is V_1, then

$$V_b = \left(1 + \frac{R_2}{R_1}\right) V_1$$

The output of A_1 is found using superposition and is

$$V_a = \left(1 + \frac{1}{sCR_3}\right) V_1 - \frac{1}{sCR_3}\left(1 + \frac{R_2}{R_1}\right) V_1 = \left(1 - \frac{R_2}{sR_1 R_3 C}\right) V_1$$

The current I_1 is

$$I_1 = \frac{V_1 - V_a}{R_4}$$

After eliminating V_a from the last two expressions, we get

$$\frac{V_1}{I_1} = s\frac{R_1R_3R_4C}{R_2} = sL_{eq}$$

where

$$L_{eq} = \frac{R_1R_3R_4C}{R_2}$$ ∎

8.5.2 High-Pass Filter

The circuit of Fig. 8.13a is an elementary second-order passive, high-pass filter with a voltage-gain transfer function given by

$$T(s) = \frac{V_o}{V_s} = \frac{s^2}{s^2 + \dfrac{s}{R_1C_1} + \dfrac{1}{LC_1}} \quad (8.14)$$

The shunt inductor in this passive circuit may be implemented by active means using a gyrator terminated in a capacitor C, as shown in Fig. 8.13b. This is an instance where the inductor is directly replaced by an active RC circuit on a one-to-one basis. Noticing that, in this case, $L = \alpha^2 C$, the corresponding transfer function for the gyrator high-pass filter is

$$T(s) = \frac{s^2}{s^2 + \dfrac{s}{R_1C_1} + \dfrac{1}{\alpha^2 CC_1}} \quad (8.15)$$

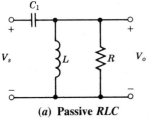
(a) Passive RLC

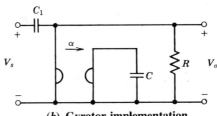
(b) Gyrator implementation

Figure 8.13 High-pass filters.

Example

(a) Design an *RLC* high-pass circuit (see Fig. 8.13a) to satisfy the following criteria:

1. $f_o = 1.59$ kHz
2. $Q = 0.707$

(b) Using the gyrator of Fig. 8.12a in the high-pass active implementation of Fig. 8.13b, specify a set of values for the various resistors and capacitors to duplicate the performance of the *RLC* high-pass configuration.

Solution

(a) Using (8.14), we get

$$T(s) = \frac{s^2}{s^2 + s\dfrac{1}{RC_1} + \dfrac{1}{LC_1}}$$

$$= \frac{s^2}{s^2 + s\dfrac{\omega_o}{Q} + \omega_o^2}$$

Since

$$\omega_o = \frac{1}{\sqrt{LC_1}} = 2\pi f_o$$

and

$$\frac{\omega_o}{Q} = \frac{1}{RC_1}$$

we obtain

$$Q = R\sqrt{\frac{C_1}{L}}$$

Possible values of L and C_1 that satisfy the characteristic frequency f_o requirement are $L = 0.01$ H and $C_1 = 1$ µF. To meet the Q requirement, we must set

$$R = \frac{Q}{\sqrt{\dfrac{C_1}{L}}} = 70.7\ \Omega$$

(b) For the gyrator of Fig. 8.12a, $\alpha = -R_1$. The equivalent inductance implemented by the capacitor-terminated gyrator across R in the active high-pass filter of Fig. 8.13b must then be, using (8.12a),

$$L_{eq} = R_1^2 C = 0.01 \text{ H}$$

This relationship will be satisfied if $C = 0.01$ μF and $R_1 = 1$ kΩ. To complete the rest of the filter circuit design, we arbitrarily set $R_2 = R_3 = 10$ kΩ. ∎

8.5.3 Bandpass

The gyrator configuration of Fig. 8.14 provides a second-order bandpass magnitude response. The gyrator, in this case, is used as a two-port network as opposed to the high-pass active filter of Fig. 8.13b, where the gyrator acted as a one-port replacing an inductor on a one-to-one basis.

To analyze the bandpass gyrator circuit of Fig. 8.14, we use voltage division,

$$V_1 = V_s \frac{Z_{in} \| R_1}{\dfrac{1}{sC_1} + (Z_{in} \| R_1)}$$

or

$$V_1 = V_s \frac{sC_1}{sC_1 + \dfrac{1}{R_1} + \dfrac{1}{Z_{in}}} \quad (8.16)$$

From (8.11),

$$Z_{in} = \alpha^2 \left(sC_2 + \frac{1}{R_2} \right) \quad (8.16a)$$

The output voltage V_o may be found from (8.10),

$$V_o = V_2 = \alpha I_1 = a \frac{V_1}{Z_{in}}$$

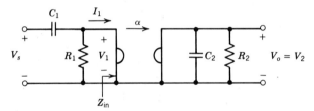

Figure 8.14 Gyrator bandpass filter.

or

$$V_o = V_s \frac{\alpha s C_1}{1 + Z_{in}\left(sC_1 + \dfrac{1}{R_1}\right)} \qquad (8.17)$$

making use of (8.16). Finally, combining (8.16a) and (8.17), we get

$$T(s) = \frac{V_o}{V_s}$$

$$= \frac{\dfrac{s}{\alpha C_2}}{s^2 + s\left(\dfrac{1}{R_2 C_2} + \dfrac{1}{R_1 C_1}\right) + \dfrac{1}{R_1 R_2 C_1 C_2} + \dfrac{1}{\alpha^2 C_1 C_2}} \qquad (8.18)$$

This is the transfer function for a second-order bandpass filter with a center frequency of

$$\omega_o = \sqrt{\frac{1}{R_1 R_2 C_1 C_2} + \frac{1}{\alpha^2 C_1 C_2}}$$

and a 3 dB passband bandwidth BW given by

$$BW = \frac{\omega_o}{Q} = \frac{1}{R_1 C_1} + \frac{1}{R_2 C_2}$$

8.6 SUMMARY

Filters are frequency selective circuits. Their performance and behavior are described by their amplitude and phase response, cutoff and characteristic frequencies, passband gain, attenuation slope, and Q. These characteristics are obtained from the form and coefficients of the transfer function of the filter. The complexity, and to a degree performance, of a filter is measured by its order, which is a function of the number of poles of its transfer function. The simplest filters are of first order with one pole on the negative real axis of the left-half portion of the s plane. Most of the attention is given to second-order filters, which have a pair of complex conjugate poles. One of the simplest implementations of a second-order filter is to replace the discrete inductor in *RLC* circuits with a simulated inductor. A common implementation of a simulated inductor is the gyrator circuit terminated with a capacitor.

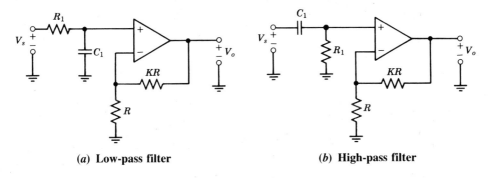

(a) Low-pass filter   (b) High-pass filter

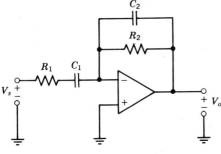

(c) Bandpass filter

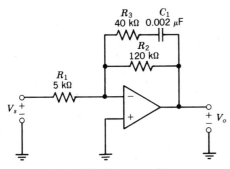

(d) Modified low-pass filter

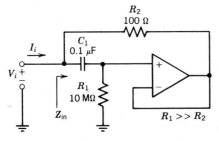

(e) Equivalent LR circuit

Figure 8.15 Additional filter and filterlike circuits.

171

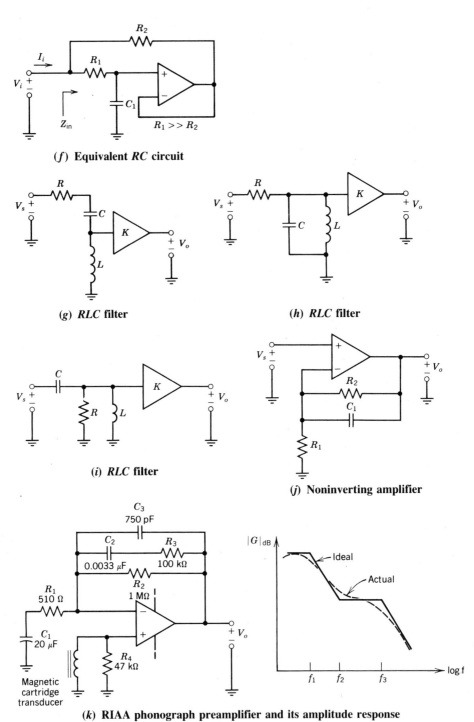

(f) Equivalent RC circuit

(g) RLC filter

(h) RLC filter

(i) RLC filter

(j) Noninverting amplifier

(k) RIAA phonograph preamplifier and its amplitude response

Fig. 8.15 Additional filter and filterlike circuits. (*Continued*)

PROBLEMS

8.1 Using the circuit of Fig. 8.5a, design a first-order, low-pass filter with a 3 dB cutoff frequency of 8 kHz, a gain of -4, and an input resistance of 16 kΩ in the passband.

8.2 Find the voltage-gain transfer function $T(s)$, ω_c, A_o, and the response type for the filter in Fig. 8.5a if the capacitor C is replaced with an inductor L.

8.3 The circuit in Fig. 8.15a is an alternate way to implement an active first-order, low-pass filter. Show that

$$T(s) = \frac{(k+1)\dfrac{1}{R_1C_1}}{s + \dfrac{1}{R_1C_1}}$$

8.4 Using the circuit of Fig. 8.6a, design a first-order, high-pass filter with a cutoff frequency of 200 Hz, a gain of -2, and an input resistance of 20 kΩ in the passband.

8.5 Find the voltage-gain transfer function $T(s)$, ω_c, A_o, and the response type for the filter in Fig. 8.6a if the capacitor C is replaced with an inductor of value L.

8.6 Draw the phase Bode plots for the low- and high-pass filters in Figs. 8.5a and 8.6a.

8.7 Find the voltage-gain transfer function $T(s)$, ω_c, A_o, and the response type for the first-order filter in Fig. 8.15b.

8.8 Find $T(s)$, ω_o, and Q for the bandpass filter in Fig. 8.15c.

8.9 Design a bandpass filter using the circuit of Fig. 8.15c to meet the following specifications:

$$f_o = 1 \text{ kHz}$$
$$A_o = 10$$
$$Q = \text{maximum possible}$$

8.10 Sketch the magnitude Bode plot for the modified low-pass filter in Fig. 8.15d.

8.11 Calculate the frequency where the phase between V_o and V_s is $-45°$ for the all-pass filter in Fig. 8.8a if $R = 10$ kΩ and $C = 0.001$ μF.

8.12 Calculate the frequency where the phase is $-315°$ for the all-pass filter in Fig. 8.8b if $R = 10$ kΩ and $C = 0.001$ μF.

8.13 Specify a set of values for the components in Figs. 8.12a and 8.12b to simulate an inductance of 1 H.

8.14 Find $Z_{in} = V_1/I_1$ for the circuit in Fig. 8.12a if

(A) $Z_L = sL$
(B) $Z_L = R_L$
(C) $Z_L = R_L + 1/sC$

Model the input impedance of this circuit for each of the above cases.

8.15 The circuit in Fig. 8.15e can be modeled as a series RL circuit if the current through the R_1C_1 network is assumed much smaller than the current through R_2. Under these conditions, prove

$$Z_{in} \cong sR_1R_2C_1 + R_2 = sL_{eq} + R_2$$

Find L_{eq} for the values shown.

8.16 The circuit in Fig. 8.15f can be modeled as a series RC circuit if the current through the R_1C_1 network is assumed much smaller than the current through R_2. Under these conditions, prove

$$Z_{in} \cong \frac{1}{sC_{eq}} + R_2$$

where

$$C_{eq} = \frac{R_1}{R_2} C_1$$

8.17 Find $T(s)$ for the RLC filter in Fig. 8.15g. Identify the filter response type, order, characteristic frequency, and passband gain.

8.18 Find $T(s)$ for the RLC filter in Fig. 8.15h. Identify the filter response type, order, characteristic frequency, and passband gain.

8.19 Find $T(s)$ for the RLC filter in Fig. 8.15i. Identify the filter response type, order, characteristic frequency, and passband gain.

8.20 For the circuit in Fig. 8.15j, show that

$$T(s) = \frac{s + \dfrac{R_1 + R_2}{R_1R_2C_1}}{s + \dfrac{1}{R_2C_1}}$$

Sketch the magnitude Bode plot of the filter and show that at low frequencies

$$|T(j\omega)|, \text{dB} = 20 \log\left(1 + \frac{R_2}{R_1}\right)$$

and at high frequencies

$$|T(j\omega)|, \text{dB} = 0 \text{ dB}$$

8.21 Sketch the magnitude and phase Bode plots for the noninverting amplifier in Fig. 8.15j if $R_1 = 1 \text{ k}\Omega$, $R_2 = 9 \text{ k}\Omega$, and $C = 0.01 \text{ }\mu\text{F}$.

8.22 The circuit in Fig. 8.15k is an RIAA (Recording Industry Association of America) phono preamplifier or active filter equalizer. Equalizer circuits, basically, provide the inverse response to that of the signal out of the transducer, so that the frequency response to the next amplifier stage is effectively flat. The ideal and actual amplitude responses, called RIAA curves, are shown in Fig. 8.15k. Estimate the corner frequencies f_1, f_2, and f_3. What causes the circuit gain to decrease below f_1?

Chapter 9

Active Filters

9.1 PROLOGUE

There are a significant number of techniques to implement second-order active filters, some simple and some complex. Fortunately, once the basic principles of any given implementation technique are understood, an engineer can utilize the many aids available to design a filter to a given set of specifications. The second-order infinite-gain, multiple-feedback (IGMF) filter is a fixed, six-component configuration that can provide the more common filter responses. The infinite-gain, single-feedback (IGSF) filter uses an inverting amplifier with two two-ports as the input and feedback networks. The state-variable filter, which is available in integrated circuit form as a "universal" filter, resembles an analog computer circuit and provides several responses simultaneously. The *KRC* filter is composed of an *RC* network and an amplifier with a gain K. The amplifier circuit is a voltage-controlled voltage-source (VCVS).

The objective of this chapter is to introduce the reader to the applications of the operational amplifier in filter circuits. Inasmuch as there are a large number of circuit techniques used to implement second-order active filters, the techniques discussed in this chapter are only a sampling. Additional techniques can be found in books that specialize in this topic. Some of these books are listed in the bibliography.

9.2 INFINITE-GAIN, MULTIPLE-FEEDBACK (IGMF) FILTERS

9.2.1 Theory and Model

The basic multiple-feedback circuit consists of several two-terminal passive elements interconnected to form feedback paths around an operational amplifier; hence the name multiple-feedback filter. The amplifier is used in an inverting configuration. If the circuit is restricted to single resistor and capacitor replacement of the admittances Y_1 to Y_5, and if the transfer function is limited to second order, then a maximum of five elements and two feedback paths, Fig. 9.1, are necessary. At least one low-pass, one high-pass, and four bandpass realizations are possible with this type of circuit. The response of a particular circuit is a function of the location of the resistors and capacitors. However, a minimum of two capacitors is required to achieve a second-order filter.

The two-terminal passive components are described in Fig. 9.1 by their admittances. We will find the voltage-gain transfer function $T(s)$ of the configuration in terms of these admittances.

Applying KCL to node a, we get

$$(V_a - V_s)Y_1 + V_aY_2 + V_aY_3 + (V_a - V_o)Y_4 = 0$$

Another expression is found by noting that the portion of the circuit composed of the amplifier, Y_3, and Y_5 is an inverting amplifier with input voltage V_a,

$$V_o = \frac{-Z_5}{Z_3} V_a = \frac{-Y_3}{Y_5} V_a$$

If we eliminate V_a from the first expression and combine terms, the result is

$$\frac{V_o}{V_s} = T(s) = \frac{-Y_1Y_3}{Y_5(Y_1 + Y_2 + Y_3 + Y_4) + Y_3Y_4} \quad (9.1)$$

The above expression is the mathematical model for the infinite-gain, multiple-feedback

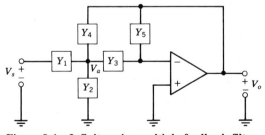

Figure 9.1 Infinite-gain, multiple-feedback filter model.

(IGMF) filter of Fig. 9.1. For a given filter response, we specify the five components and substitute for their admittances in $T(s)$.

9.2.2 Low-Pass Filter

Second-order, low-pass filters have the form

$$T(s) = \frac{A_o \omega_o^2}{s^2 + \frac{\omega_o}{Q} s + \omega_o^2} \tag{9.2}$$

By comparing (9.1) and (9.2), we can deduce what the circuit must look like. Since the numerator is not a function of s, then Y_1 and Y_3 must describe resistors. For a second-order filter, the denominator must contain an s^2, s, and a constant term. If the denominator is to have an s^2 term, then Y_5 and Y_2 or Y_4 must represent capacitors. Y_4 cannot represent a capacitor because the product term of $Y_3 Y_4$ must be the constant term. Hence Y_1, Y_3, and Y_4 are the admittances of resistors and Y_2 and Y_5 are the admittances of capacitors in the low-pass IGMF configuration. The circuit is shown in Fig. 9.2.

Example

Determine the values of f_o, Q, and A_o of the low-pass filter in Fig. 9.2, if $R_1 = R_4 = 82$ kΩ, $R_3 = 39$ kΩ, $C_2 = 560$ pF, and $C_5 = 150$ pF.

Solution

If we substitute for the admittances of the components in (9.1) and express $T(s)$ in the standard form of (9.2), we obtain

$$T(s) = \frac{\frac{-G_1 G_3}{C_2 C_5}}{s^2 + \frac{1}{C_2}(G_1 + G_3 + G_4)s + \frac{G_3 G_4}{C_2 C_5}}$$

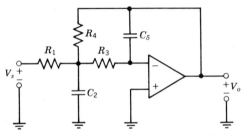

Figure 9.2 Multiple-feedback low-pass filter.

If we compare the corresponding terms of $T(s)$ with (9.2), we get

$$\omega_o^2 = \frac{G_3 G_4}{C_2 C_5}$$

$$\frac{\omega_o}{Q} = \frac{G_1 + G_3 + G_4}{C_2}$$

and

$$A_o \omega_o^2 = -\frac{G_1 G_3}{C_2 C_5}$$

Solving for A_o, f_o, and Q, we get

$$A_o = -\frac{G_1}{G_4}$$

$$f_o = \frac{1}{2\pi} \sqrt{\frac{G_3 G_4}{C_2 C_5}}$$

and

$$Q = \frac{1}{G_1 + G_3 + G_4} \sqrt{\frac{G_3 G_4 C_2}{C_5}}$$

Evaluating the above expressions for f_o, Q, and A_o, we get

$$A_o = -1$$
$$f_o = 9.7 \text{ kHz}$$
$$Q = 0.68$$ ∎

Infinite-gain, multiple-feedback filters make simple low-cost audio filters. An application of the low-pass IGMF filter is as a "scratch" filter. The scratch filter is used to roll off high-frequency noise appearing as hiss, ticks, and pops from worn phonograph records. For this application, the Q of the circuit is designed near 0.707 to provide a maximally flat or Butterworth second-order response. The design procedures for the IGMF low- and high-pass filters are illustrated in a design example in Chapter 14.

9.2.3 High-Pass Filter

A multiple-feedback, high-pass filter is shown in Fig. 9.3. The transfer function for this circuit may be found by substituting for the admittances of the components in the

180 Active Filters

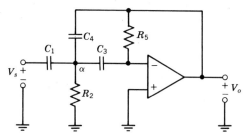

Figure 9.3 Multiple-feedback high-pass filter.

transfer function of the filter's circuit model,

$$T(s) = -\frac{\dfrac{C_1}{C_4}s^2}{s^2 + \left(\dfrac{C_1}{R_5 C_3 C_4} + \dfrac{1}{R_5 C_4} + \dfrac{1}{R_5 C_3}\right)s + \dfrac{1}{R_2 R_5 C_3 C_4}} \qquad (9.3)$$

The expressions for ω_o, Q, and A_o may be interpreted in terms of the circuit's components by comparing (9.3) with the standard form for a second-order high-pass filter

$$T(s) = \frac{A_o s^2}{s^2 + \dfrac{\omega_o}{Q}s + \omega_o^2} \qquad (9.3a)$$

The passband gain A_o of this circuit is the negative of the ratio of C_1 and C_4. This can intuitively be seen by examining the relative values of the impedances of the components as frequency increases. At high frequencies, the impedance of C_3, which is effectively in series with R_5, may be neglected, causing node a to be connected to the inverting input terminal of the op amp. Similarly, the parallel combination of R_5 and C_4 between this terminal and the output of the circuit may be approximated by C_4 at high frequencies, resulting in an inverting configuration with a gain of

$$A_o = -\frac{\dfrac{1}{sC_4}}{\dfrac{1}{sC_1}} = -\frac{C_1}{C_4}$$

R_2, effectively connected across the input of the op amp, does not affect the passband gain of the circuit.

Example

Determine the values of f_o, Q, and A_o of the high-pass filter in Fig. 9.3, if $C_1 = C_3 = C_4 = 0.0033$ μF, $R_2 = 470$ kΩ, and $R_5 = 2$ MΩ.

Solution

By matching the coefficients of (9.3) with those of (9.3a), we get

$$\omega_o^2 = \frac{1}{R_2 R_5 C_3 C_4}$$

$$\frac{\omega_o}{Q} = \frac{C_1}{R_5 C_3 C_4} + \frac{1}{R_5 C_4} + \frac{1}{R_5 C_3}$$

and

$$A_o = -\frac{C_1}{C_4}$$

Solving for f_o and Q, we obtain

$$f_o = \frac{1}{2\pi} \frac{1}{\sqrt{R_2 R_5 C_3 C_4}}$$

and

$$Q = \sqrt{\frac{R_5 C_3 C_4}{R_2}} \frac{1}{C_1 + C_3 + C_4}$$

Evaluating the above expressions for f_o, Q, and A_o, we obtain

$$f_o = 50 \text{ Hz}$$
$$Q = 0.69$$

and

$$A_o = -1$$

The high-pass version of the IGMF filter may be used as an audio "rumble" filter. A rumble filter is used to roll off low-frequency noise associated with worn turntable and tape transport mechanisms and, like the low-pass scratch filter, the Q of the high-pass filter is typically designed for a maximally flat response. If we cascade low- and high-pass filter sections, a broadband bandpass filter is created. This type of filter is often used to limit the audio bandwidth to include only speech frequencies from 300 Hz to 3 kHz. ∎

9.2.4 Bandpass Filter

A two-pole, bandpass filter may be realized by using the multiple-feedback scheme in Fig. 9.4. The transfer function has the form

$$T(s) = \frac{A_o s \dfrac{\omega_o}{Q}}{s^2 + \dfrac{\omega_o}{Q} s + \omega_o^2} \tag{9.4}$$

182 Active Filters

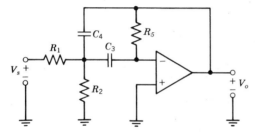

Figure 9.4 Multiple-feedback bandpass filter.

There are three different arrangements of five elements that may be used in implementing the transfer function given by (9.4). These combinations are shown in Fig. 9.5.

9.3 INFINITE-GAIN, SINGLE-FEEDBACK (IGSF) FILTERS

9.3.1 Theory and Model

The circuit model of the infinite-gain, single-feedback (IGSF) filter consists of an inverting amplifier connected to two two-port networks, A and B, as shown in Fig. 9.6. The transfer function for this filter configuration is expressed as the ratio of two short-circuit admittance parameters (y_{ij}). One of the short-circuit admittance parameters is associated with the feedback network and the other is associated with the input network. Before we can develop $T(s)$ for this circuit, we have to briefly discuss network parameters.

In a two-port, or two-terminal pair network, there are four variables of concern; the input and output voltages V_1 and V_2, and the input and output currents I_1 and I_2. Only two of these variables can be independent. These will specify the remaining two variables through an appropriate set of parameters. In the IGSF scheme, we will use the y (short-circuit admittance) parameters for convenience. These parameters correspond to the case where V_1 and V_2 are the independent variables in the two-port under consideration.

The defining equations for the y parameters of a generalized two port, Fig. 9.7, are

$$I_1 = y_{11}V_1 + y_{12}V_2$$
$$I_2 = y_{21}V_1 + y_{22}V_2$$

Y_1	Y_2	Y_3	Y_4	Y_5
G_1	sC_2	sC_3	sC_4	G_5
sC_1	G_2	G_3	G_4	sC_5
sC_1	sC_2	G_3	G_4	sC_5

Figure 9.5 Admittances of additional IGMF bandpass filters (see Fig. 9.1).

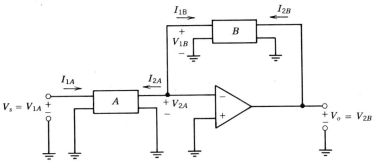

Figure 9.6 Infinite-gain, single-feedback model.

If $V_1 = 0$, then y_{12} can be expressed as the ratio of the current I_1 and the voltage V_2,

$$y_{12} = \frac{I_1}{V_2} \quad \text{for} \quad V_1 = 0$$

On the other hand,

$$y_{21} = \frac{I_2}{V_1} \quad \text{for} \quad V_2 = 0$$

The voltage-gain transfer function $T(s)$ of the circuit of Fig. 9.6 may be found using the above definitions for the y_{12} and y_{21} parameters and the voltage and current constraints of the operational amplifier. From Fig. 9.6,

$$V_{2A} = V_{1B} = 0 \text{ V} \quad \text{(voltage constraint)}$$

and

$$I_{2A} = -I_{1B} \quad \text{(current constraint)}$$

For the feedback network B,

$$\frac{I_{1B}}{V_{2B}} = -\frac{I_{2A}}{V_o} = y_{12B}$$

and for the input network A,

$$\frac{I_{2A}}{V_{1A}} = \frac{I_{2A}}{V_s} = y_{21A}$$

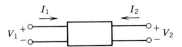

Figure 9.7 Generalized two-port.

184 Active Filters

Expressing $T(s)$ as the ratio of the output and input voltages yields

$$T(s) = \frac{V_o}{V_s} = \frac{\dfrac{I_{2A}}{V_s}}{\dfrac{I_{1B}}{V_o}}$$

or

$$T(s) = \frac{-y_{21A}}{y_{12B}} \qquad (9.5)$$

If the two-ports A and B have RC elements only, the poles of their y parameters will be restricted to the negative real axis of the s plane while the zeros of these parameters may be located off the negative real axis. Making the poles of y_{21A} equal the poles of y_{12B} allows us to produce transfer functions having essentially any desirable pole-zero configuration. For this special case, the zeros of y_{21A} and y_{12B} will establish the zeros and poles of $T(s)$, respectively. Thus, the numerators of the transfer admittances of the passive RC networks determine both the poles and zeros of the filter's transfer function. Since the zeros of these passive networks can be located anywhere in the complex frequency plane, $T(s)$ can have poles and zeros anywhere in the complex plane. Stability, of course, requires the poles to be restricted to the left half plane.

There are a large number of passive RC networks that can be used as input and feedback two-ports. They vary from the simple series RC circuits to the more complex "twin-tee" and "bridged-tee" circuits. Five of the more common networks used in single-feedback filters are given in Fig. 9.8.

If network 4 (a "bridged-tee" circuit) in Fig. 9.8 is used as the feedback network B, networks 1, 2, and 3 can be used as the input network A to produce bandpass, low-pass, and high-pass filter responses, respectively. The zeros of the y parameters of these three networks are located at zero or infinity. The "bridged-tee" network is used in filters that require a high Q and zeros near the $j\omega$ axis.

9.3.2 Bandpass

The input network A of the filter in Fig. 9.9 is a series RC circuit while the feedback network B is a bridged-T circuit. This configuration will produce a bandpass response. For the input network,

$$y_{21A} = \frac{-\dfrac{s}{R}}{s + \dfrac{1}{RC}}$$

Infinite-gain, Single-feedback (IGSF) Filters

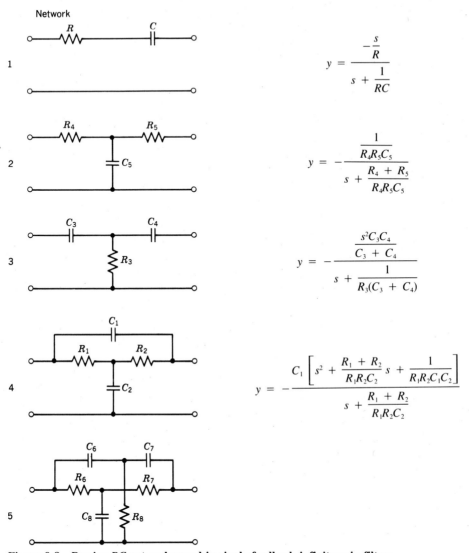

Figure 9.8 Passive *RC* networks used in single-feedback infinite-gain filters.

and for the feedback network,

$$y_{12B} = -\frac{C_1\left(s^2 + \dfrac{s(R_1 + R_2)}{R_1 R_2 C_2} + \dfrac{1}{R_1 R_2 C_1 C_2}\right)}{s + \dfrac{R_1 + R_2}{R_1 R_2 C_2}}$$

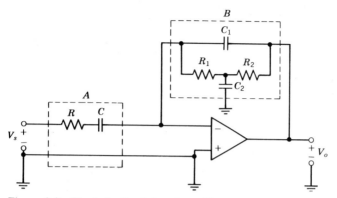

Figure 9.9 Single-feedback bandpass filter.

The y_{21} parameter of network A is determined in a straightforward manner by shorting its output port and finding the expression for I_2/V_1.

Similarly, the y_{12} parameter of network B is obtained by shorting its input port and then finding the expression for I_1/V_2.

The poles for y_{21A} and y_{12B} are at $-1/RC$ and $-1/(R_1 \| R_2)C_2$, respectively. If we force these poles to be equal, the voltage-gain transfer function of the circuit of Fig. 9.9 is

$$T(s) = \frac{V_o}{V_s} = -\frac{y_{21A}}{y_{12B}}$$

or

$$T(s) = -\frac{\dfrac{s}{RC_1}}{s^2 + \left(\dfrac{R_1 + R_2}{R_1 R_2 C_2}\right)s + \dfrac{1}{R_1 R_2 C_1 C_2}}$$

$$= \frac{A_o s \dfrac{\omega_o}{Q}}{s^2 + \dfrac{\omega_o}{Q}s + \omega_o^2}$$

This is a bandpass transfer function for which

$$A_o = -\frac{R_1 R_2 C_2}{R(R_1 + R_2)C_1}$$

$$\omega_o = \frac{1}{\sqrt{R_1 R_2 C_1 C_2}}$$

and
$$Q = \frac{1}{R_1 + R_2}\sqrt{\frac{R_1 R_2 C_2}{C_1}}$$

9.3.3 High-Pass and Low-Pass Filters

In the single-feedback, high-pass filter of Fig. 9.10, the feedback network is a bridged-T circuit. The input two-port network has two capacitors in series with the signal path that contribute zeros at the origin for $T(s)$ of the circuit. In the second-order, high-pass filter, there must be two zeros at dc for maximum rolloff in the stop band and constant gain at high frequencies. The feedback network is used to realize the two poles of $T(s)$.

The y_{12} parameter of network B is

$$y_{12B} = -\frac{C_1\left[s^2 + s\left(\dfrac{R_1 + R_2}{R_1 R_2 C_2}\right) + \dfrac{1}{R_1 R_2 C_1 C_2}\right]}{s + \dfrac{R_1 + R_2}{R_1 R_2 C_2}}$$

and the y_{21} parameter for network A is

$$y_{21A} = -\frac{s^2 \dfrac{C_3 C_4}{C_3 + C_4}}{s + \dfrac{1}{R_3(C_3 + C_4)}}$$

For a second-order high-pass response, we force the poles of the two y parameters to be equal. Hence

$$(R_1 \| R_2)C_2 = R_3(C_3 + C_4)$$

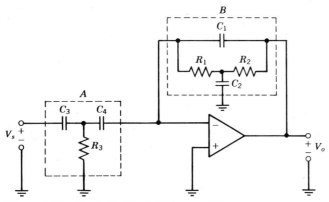

Figure 9.10 Single-feedback high-pass filter.

The resulting transfer function is

$$T(s) = -\frac{y_{21A}}{y_{12B}}$$

$$= -\frac{\dfrac{s^2(C_3 C_4)}{(C_3 + C_4)C_1}}{s^2 + \left(\dfrac{R_1 + R_2}{R_1 R_2 C_2}\right)s + \dfrac{1}{R_1 R_2 C_1 C_2}}$$

$$= \frac{A_o s^2}{s^2 + \left(\dfrac{\omega_o}{Q}\right)s + \omega_o^2}$$

By examining the form of $T(s)$, we can identify the filter's characteristics A_o, ω_o, and Q, noting that

$$A_o = -\frac{C_3 C_4}{(C_3 + C_4)C_1}$$

$$\omega_o = \frac{1}{\sqrt{R_1 R_2 C_1 C_2}}$$

and

$$Q = \frac{1}{R_1 + R_2}\sqrt{\frac{R_1 R_2 C_2}{C_1}}$$

The expressions for ω_o and Q are a function of the resistors and capacitors in the feedback network. The passband gain is a function of the capacitors of both the input and feedback networks. At high frequencies, C_1 dominates in the feedback loop, while C_3 and C_4 dominate in the input network. The gain is established by the series combination of C_3 and C_4, and C_1.

Example

What two RC networks in Fig. 9.8 produce an IGMF second-order low-pass filter? For this configuration, find $T(s)$. What condition is imposed on the RC networks to produce a second-order response?

Solution

A second-order response is produced when circuit 2 of Fig. 9.8 is the input network (A), the bridged-T circuit 4 is the feedback network (B), and the poles of their transfer admittances are set equal to each other. The transfer admittances of the individual networks are

$$y_{21A} = \frac{-1}{R_4 R_5 C_5 \left(s + \dfrac{R_4 + R_5}{R_4 R_5 C_5}\right)}$$

and

$$y_{12B} = \frac{-C_1 \left[s^2 + \left(\dfrac{R_1 + R_2}{R_1 R_2 C_2}\right) s + \dfrac{1}{R_1 R_2 C}\right]}{s + \dfrac{R_1 + R_2}{R_1 R_2 C_2}}$$

When we force the poles of y_{21A} and y_{12B} to be equal,

$$-\frac{R_4 + R_5}{R_4 R_5 C_5} = -\frac{R_1 + R_2}{R_1 R_2 C_2}$$

the circuit will be a second-order, low-pass filter whose transfer function is

$$T(s) = -\frac{\dfrac{1}{R_4 R_5 C_1 C_5}}{s^2 + \left(\dfrac{R_1 + R_2}{R_1 R_2 C_2}\right) s + \dfrac{1}{R_1 R_2 C_1 C_2}}$$

The passband gain A_o is determined by setting $s = 0$,

$$A_o = -\frac{R_1 R_2 C_2}{R_4 R_5 C_5}$$

This is equivalent to

$$A_o = -\frac{R_1 + R_2}{R_4 + R_5}$$

because of the condition imposed when we forced the poles of the admittances to be equal. At low frequencies, the circuit behaves like an inverting amplifier whose gain is established by the resistors in the input and feedback networks. ∎

9.4 STATE-VARIABLE FILTERS

9.4.1 Theory and Model

The expression for a voltage or current time-varying response in an nth-order circuit may be found by solving an nth-order differential equation. It is also possible to obtain this response by solving n first-order differential equations (state equations) in n properly selected unknowns (state variables), provided that the initial conditions of the configuration are specified at some time $t = t_0$.

The second-order state-variable active filter makes use of the state-variable techniques of solving differential equations to implement second-order low-pass, bandpass, and high-pass filter responses. The outputs of this multiple amplifier configuration provide all three responses simultaneously. These state-variable techniques are similar to the analog computer method of solving differential equations. The state-variable filter and the analog computer both use operational amplifiers connected as summers and integrators to implement differential equations by means of time-varying voltages.

Second-order state-variable active filters are commercially available in integrated circuit form and are easy to use because a given response is typically obtained by connecting a few external resistors. The characteristics A_o, ω_o, and Q are independently controlled and their values are relatively insensitive to passive component tolerances and variations. Many of these integrated circuit filters are available with a built-in operational amplifier that can be used as a summer to realize generalized biquadratic transfer functions.

The basic circuit of a second-order state-variable filter is shown in Fig. 9.11. It consists of a summer circuit (A_1) and two inverting integrators (A_2 and A_3). The output V_{o1} of the summer, shown isolated in Fig. 9.12, is found using superposition:

$$V_{o1} = -V_s \frac{R_4}{R_3} - V_{o3} \frac{R_4}{R_5} + V_{o2} \frac{R_7}{R_6 + R_7} \left(1 + \frac{R_4}{R_3 \| R_5}\right) \quad (9.6)$$

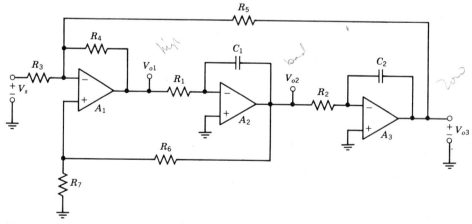

Figure 9.11 Second-order state-variable filter.

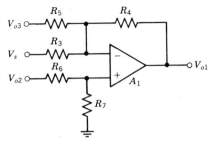

Figure 9.12 Summing amplifier in the state-variable filter of Fig. 9.11.

where the outputs V_{o2} and V_{o3} of the integrators are

$$V_{o2} = -V_{o1}\frac{1}{sC_1R_1} \tag{9.7}$$

and

$$V_{o3} = -V_{o2}\frac{1}{sR_2C_2} = V_{o1}\frac{1}{s^2R_1R_2C_1C_2} \tag{9.8}$$

The transfer function V_{o1}/V_s for the state-variable filter of Fig. 9.11 is found by using (9.6) through (9.8):

$$\frac{V_{o1}}{V_s} = \frac{-\dfrac{R_4}{R_3}s^2}{s^2 + \dfrac{R_7}{R_6 + R_7}\left(1 + \dfrac{R_4}{R_3} + \dfrac{R_4}{R_5}\right)\dfrac{s}{R_1C_1} + \dfrac{R_4}{R_1R_2R_5C_1C_2}} \tag{9.9}$$

$$= \frac{A_{hp}s^2}{s^2 + \left(\dfrac{\omega_o}{Q}\right)s + \omega_o^2}$$

The transfer function given by (9.9) has a high-pass response for which

$$A_{hp} = -\frac{R_4}{R_3} \tag{9.10a}$$

$$\omega_o = \sqrt{\frac{R_4}{R_1R_2R_5C_1C_2}} \tag{9.10b}$$

and

$$Q = \sqrt{\frac{R_1R_4C_1}{R_2R_5C_2}}\frac{1 + \dfrac{R_6}{R_7}}{1 + \dfrac{R_4}{R_3} + \dfrac{R_4}{R_5}} \tag{9.10c}$$

192 Active Filters

The remaining transfer functions, V_{o2}/V_s and V_{o3}/V_s, are found using (9.7) through (9.10). They are

$$\frac{V_{o2}}{V_s} = \frac{\left(\dfrac{\omega_o}{Q}\right)\left(\dfrac{R_4}{R_3}\right) s \left(1 + \dfrac{1 + \dfrac{R_6}{R_7}}{\dfrac{R_4}{R_3} + \dfrac{R_4}{R_5}}\right)}{s^2 + \dfrac{\omega_o}{Q} s + \omega_o^2} \quad (9.11)$$

and

$$\frac{V_{o3}}{V_s} = \frac{-\dfrac{R_5}{R_3}\omega_o^2}{s^2 + \dfrac{\omega_o}{Q} s + \omega_o^2} \quad (9.12)$$

The outputs V_{o2} and V_{o3} provide bandpass and low-pass responses in addition to the high-pass response at V_{o1}. Although the passband values of the gains of the responses are different, the poles of all three responses are identical.

Components C_1, C_2, R_4, R_5, and R_6 are not accessible to the user in some commercial versions of the state-variable filter, since they are part of the integrated circuit. To set the ω_o of the poles at its desired value, resistors R_1 and R_2 are selected using (9.10b). Once these resistors are determined, the value of resistor R_7 may be obtained from (9.10c) to set the Q of the poles. R_1, R_2, and R_7 may also be used to adjust or tune the final values of ω_o and Q. Adjusting ω_o first with the use of R_1 and R_2 insures that once ω_o is set, Q may be independently adjusted by means of R_7. For the low- and high-pass responses, the passband gain may be also adjusted by means of R_3 as the last step in the tuning procedure.

Example 1

Find the expressions for ω_o, Q, and passband gain in terms of R_1, R_2, R_3, and R_7, for the second-order state-variable filter of Fig. 9.11. Let $C_1 = C_2 = 1000$ pF, $R_4 = 10$ kΩ, and $R_5 = R_6 = 100$ kΩ.

Solution

Using (9.10b) and (9.10c), we get

$$\omega_o = \sqrt{\frac{10 \text{ k}\Omega}{(100 \text{ k}\Omega)(10^{-9} \text{ F})^2 R_1 R_2}} \quad (9.13)$$

$$= 10^8 \sqrt{\frac{10}{R_1 R_2}}$$

and

$$Q = \frac{1 + \dfrac{10^5}{R_7}}{1.1 + \dfrac{10^4}{R_3}} \sqrt{\dfrac{R_1}{10R_2}} \qquad (9.14)$$

The passband gain A_{hp} for the high-pass response is found directly from (9.10a),

$$A_{hp} = -\frac{R_4}{R_3} = -\frac{10^4}{R_3} \qquad (9.15)$$

The passband gain for the low-pass response, given by (9.12), is

$$A_{lp} = -\frac{10^5}{R_3} \qquad (9.16)$$

The passband gain A_{bp} of the bandpass response is found by letting $s = j\omega_o$ in (9.11) and is given as

$$A_{bp} = \frac{10^4}{R_3} \frac{1 + \dfrac{10^5}{R_7}}{1.1 + \dfrac{10^4}{R_3}} \qquad (9.17)$$

∎

Example 2

Assuming $R_1 = R_2$ in the universal active filter of Example 1, obtain the values of R_1, R_2, R_3, and R_7 to meet the following specifications:
1. $A_{lp} = -10$
2. $f_o = 1$ kHz
3. $Q = \sqrt{10}$

Solution

Solving (9.13) for R_1 and R_2, we obtain

$$R_1 = R_2 = 10^8 \frac{\sqrt{10}}{\omega_o}$$
$$\cong 50.33 \text{ k}\Omega$$

Similarly, solving (9.16) for R_3, we get

$$R_3 = -\frac{10^5}{A_{lp}} = 10 \text{ k}\Omega$$

194 Active Filters

Since $R_1 = R_2$ and $R_3 = 10$ kΩ, the expression for Q given by (9.14) will provide us with the value of R_7 needed to meet the Q specification of the filter response. Solving for R_7 produces

$$R_7 = \frac{10^5}{(2.1)(10) - 1} = 5 \text{ k}\Omega$$

The state-variable filter of Fig. 9.11 provides simultaneously high-pass and low-pass outputs having inverted forms (with respect to its input) and a noninverting bandpass output. If V_s and R_3 are removed from the inverting input terminal of A_1 and connected instead to its noninverting input terminal, it can be shown that the modified state-variable circuit will still provide high-pass, bandpass, and low-pass outputs at V_{o1}, V_{o2}, and V_{o3}, respectively. This time, though, the high-pass and low-pass outputs will be noninverting while the bandpass output will be inverted. This feature adds to the versatility of the state-variable filter. ∎

9.4.2 Bandpass and Band-Reject Biquad Filters

The circuits in Fig. 9.13 are called biquad filters. They implement second-order filter responses and use inverting amplifiers, summers, and integratorlike circuits in a manner similar to the state-variable filters.

Example

Find (a) the voltage-gain transfer function $T(s)$, A_o, ω_o, and Q for the bandpass circuit in Fig. 9.13a and (b) the values of the circuit components that satisfy the following specifications:

1. $f_o = 5$ kHz
2. $A_o = 5$
3. $Q = 10$

Solution

A summary of the steps to produce $T(s)$ for the bandpass circuit in Fig. 9.13a follows:

1. $V_3 = -\dfrac{V_o}{sR_3C_1}$

2. $V_1 = -\dfrac{\dfrac{R_2}{R_1}V_s}{sR_2C_1 + 1} + \dfrac{\dfrac{R_2}{R_3}V_o}{(sR_2C_1 + 1)sR_3C_1}$

3. $V_o = -V_1 = \dfrac{\dfrac{R_2}{R_1}V_s}{sR_2C_1 + 1} - \dfrac{\dfrac{R_2}{R_3}V_o}{(sR_2C_1 + 1)sR_3C_1}$

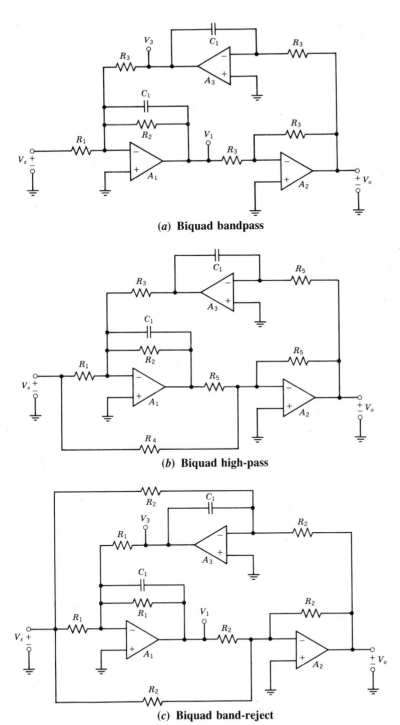

(a) Biquad bandpass

(b) Biquad high-pass

(c) Biquad band-reject

Figure 9.13 Biquad active filter circuits.

Solving for V_o/V_s produces

$$T(s) = \frac{s\dfrac{1}{R_1C_1}}{s^2 + \dfrac{1}{R_2C_1}s + \dfrac{1}{R_3^2C_1^2}}$$

$$= \frac{sA_o\dfrac{\omega_o}{Q}}{s^2 + \dfrac{\omega_o}{Q}s + \omega_o^2}$$

where

$$\omega_o = \frac{1}{R_3C_1}$$

$$A_o = \frac{R_2}{R_1}$$

and

$$Q = \frac{R_2}{R_3}$$

In designing the circuit, C_1 must be specified and then R_3 is used to satisfy the f_o requirement, R_2 is used to satisfy the Q requirement, and R_1 is used to satisfy the A_o requirement. For the specifications given,

$$C_1 = 0.01 \ \mu F$$
$$R_3 = 3.183 \ k\Omega$$
$$R_2 = 31.83 \ k\Omega$$
$$R_1 = 6.366 \ k\Omega$$

∎

Example

Find (a) the voltage-gain transfer function $T(s)$, ω_o, and Q for the unity gain band-reject filter in Fig. 9.13c and (b) the values of the circuit components that satisfy the following specifications:

1. $f_o = 120$ Hz
2. $Q = 10$

Solution

A summary of the steps to produce $T(s)$ for the band-reject circuit in Fig. 9.13c follows:

1. $V_3 = -\dfrac{1}{sR_2C_1}(V_o + V_s)$

2. $V_1 = \dfrac{V_o}{sR_2C_1(1 + sR_1C_1)} + \dfrac{V_s}{sR_2C_1(1 + sR_1C_1)} - \dfrac{V_s}{1 + sR_1C_1}$

3. $V_o = -V_1 - V_s$

$ = -\dfrac{V_o}{sR_2C_1(1 + sR_1C_1)} - \dfrac{V_s}{sR_2C_1(1 + sR_1C_1)} + \dfrac{V_s}{1 + sR_1C_1} - V_s$

Solving for V_o/V_s produces

$$T(s) = -\dfrac{s^2 + \dfrac{1}{R_1R_2C_1^2}}{s^2 + \dfrac{1}{R_1C_1}s + \dfrac{1}{R_1R_2C_1^2}}$$

The second-order, band-reject transfer function with a gain of -1 has the form of

$$T(s) = -\dfrac{s^2 + \omega_o^2}{s^2 + \dfrac{\omega_o}{Q}s + \omega_o^2}$$

where, for the circuit shown in Fig. 9.13c,

$$\omega_o = \sqrt{\dfrac{1}{R_1R_2C_1^2}} \quad \text{and} \quad Q = \sqrt{\dfrac{R_1}{R_2}}$$

For $Q = 10$, $R_1 = 100R_2$. Component values that satisfy the required conditions are

$$R_1 = 1 \text{ k}\Omega$$
$$R_2 = 100 \text{ k}\Omega$$
$$C_1 = 0.133 \text{ }\mu\text{F}$$

■

9.5 KRC REALIZATIONS

9.5.1 Theory and Model

In the preceding sections, some general properties of active *RC* networks were introduced, and several types of active *RC* circuit configurations were analyzed. In the gyrator filters, a simulated inductor was used to replace the discrete inductor in *RLC* circuits. In the state-variable filter, functional blocks like the integrator and summer were used to implement filter transfer functions. In the "infinite-gain" realizations, an operational amplifier was used to provide the required high gain to generate the various filter responses. The *KRC* filter is a unique circuit configuration. It requires an active element, with a relatively low value of gain (K), which we will refer to as a controlled source. This controlled source is realized through the use of one or more operational amplifiers. The filter derives its name from the fact that a voltage-controlled voltage source (VCVS) with a gain of K and an *RC* network are used to implement it.

In general, a controlled or dependent source is an active network element whose output voltage or current is a function of some voltage or current elsewhere in the circuit. There are four types of controlled sources: the voltage-controlled voltage source (VCVS), the voltage-controlled current source (VCCS), the current-controlled voltage source (CCVS), and the current-controlled current source (CCIS). Certain physical devices have characteristics that make them act in a manner similar to that of controlled sources. For example, a bipolar transistor in its active region acts somewhat like a current-controlled current source. Similarly, a field-effect transistor in its pinchoff region acts essentially like a voltage-controlled current source.

KRC filters are conceptually a little more difficult to visualize than the filter topics previously discussed. They also require an extensive amount of manipulation to derive their overall transfer functions.

Circuit Model The *KRC* filter model of Fig. 9.14 consists of an amplifier circuit with a fixed gain K connected to a three-terminal *RC* network. The fixed-gain amplifier circuit is a VCVS. It is treated as an ideal voltage amplifier with infinite input impedance, zero output impedance, and an output voltage that is equal to the input voltage multiplied by some positive or negative constant.

The voltage-controlled voltage source has two implementations. For positive values

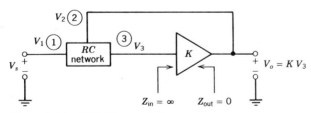

Figure 9.14 *KRC* filter model.

of K, the noninverting circuit of Fig. 9.15a is used. The gain is given by

$$K = 1 + \frac{(K-1)R}{R}$$

The circuit meets the requirements of high input and low output impedance. For negative values of K, the inverting circuit of Fig. 9.15b may be used. The gain is

$$-K = -K\frac{R}{R}$$

This VCVS has a low output impedance, but its input impedance is R. The RC network in Fig. 9.14, when looking into the amplifier from node 3, will see this resistance to virtual ground. Hence, this resistance must be accounted for in the RC network.

The transfer function for the KRC filter will be expressed in terms of the transfer functions of the RC network and the gain K of the amplifier. The voltage at node 3 of the RC network, V_3, is related to the output voltage by K,

$$V_3 = \frac{V_o}{K}$$

Voltage V_3 is also related to the voltages at nodes 2 (V_o) and 1 (V_s) through the transfer functions T_{32} and T_{31}, where

$$T_{32} = \frac{V_3}{V_2}\bigg|_{V_1=V_s=0} = \frac{V_3}{V_o}\bigg|_{V_1=V_s=0}$$

and

$$T_{31} = \frac{V_3}{V_1}\bigg|_{V_2=0} = \frac{V_3}{V_s}\bigg|_{V_2=0}$$

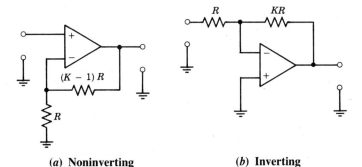

(a) Noninverting   (b) Inverting

Figure 9.15 Voltage-controlled voltage sources (VCVS).

Thus, superposition gives us

$$V_3 = \frac{V_o}{K} = T_{31}V_s + T_{32}V_o$$

Solving for V_o/V_s, we get

$$T(s) = \frac{V_o}{V_s} = \frac{KT_{31}}{1 - KT_{32}} \quad (9.18)$$

Typically, T_{31} and T_{32} will be ratios of two polynomials in s,

$$T_{31} = \frac{N_{31}}{D_{31}} \quad (9.19)$$

and

$$T_{32} = \frac{N_{32}}{D_{32}} \quad (9.20)$$

In general, the denominator expressions of the two transfer functions will be the same,

$$D_{31} = D_{32} = D \quad (9.21)$$

In view of (9.21), if we substitute (9.19) and (9.20) in (9.18), we arrive at the transfer function for the *KRC* filter of Fig. 9.14,

$$T(s) = \frac{V_o}{V_s} = \frac{KN_{31}}{D - KN_{32}} \quad (9.22)$$

The roots of D will be on the negative real axis while those of N_{31} and N_{32} can be located essentially anywhere in the s plane except for the positive real axis. We have to take a closer look at the expression of the denominator of $T(s)$ to see the effects of K on the filter's performance. If the *RC* network is second order, then the denominator of $T(s)$ will have the following form:

$$D - KN_{32} = (s^2 + c_1 s + c_0) - K(d_2 s^2 + d_1 s + d_0)$$

where the c terms are positive while the d terms are nonnegative that is, positive or zero. This expression may also be written as

$$D - KN_{32} = (1 - Kd_2)s^2 + (c_1 - Kd_1)s + (c_0 - Kd_0) \quad (9.23)$$

The gain K is part of each term in the denominator of $T(s)$ and, hence, it can affect its roots. $T(s)$ will have complex poles ($Q > 0.5$), if

$$(c_1 - Kd_1)^2 - 4(1 - Kd_2)(c_0 - Kd_0) < 0 \quad (9.24)$$

KRC Realizations 201

For *K positive,* this inequality may be satisfied by using K to reduce the $(c_1 - Kd_1)$ term. For *K negative,* this inequality may be satisfied by using K to increase the magnitudes of the $(1 - Kd_2)$ and $(c_0 - Kd_0)$ terms, either together or separately. This allows us to generate complex poles for the voltage-gain transfer function of the *KRC* filter of Fig. 9.14. The *RC* networks in *KRC* filters are such that they affect only one of the coefficients in (9.23).

9.5.2 Low-Pass Filter

The *KRC* filter of Fig. 9.16 has a second-order, low-pass response. Transfer function T_{31} is determined from the *RC* network by grounding port 2 as shown in Fig. 9.17a,

$$T_{31} = \left. \frac{V_3}{V_1} \right|_{V_2=0}$$

$$= \frac{\dfrac{1}{R_1 R_2 C_1 C_2}}{s^2 + \left(\dfrac{1}{R_2 C_1} + \dfrac{1}{R_2 C_2} + \dfrac{1}{R_1 C_1}\right) s + \dfrac{1}{R_1 R_2 C_1 C_2}}$$

$$= \frac{N_{31}}{D}$$

Transfer function T_{32} is similarly found by grounding port 1 of the *RC* network in Fig. 9.17b,

$$T_{32} = \left. \frac{V_3}{V_2} \right|_{V_1=0}$$

$$= \frac{\dfrac{s}{R_2 C_2}}{s^2 + \left(\dfrac{1}{R_1 C_1} + \dfrac{1}{R_2 C_1} + \dfrac{1}{R_2 C_2}\right) s + \dfrac{1}{R_1 R_2 C_1 C_2}}$$

$$= \frac{N_{32}}{D}$$

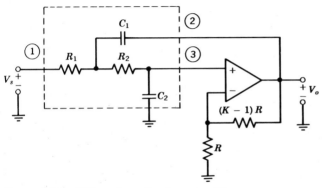

Figure 9.16 *KRC* noninverting low-pass filter.

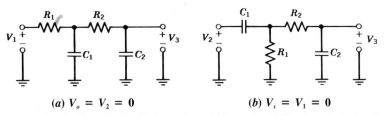

(a) $V_o = V_2 = 0$ (b) $V_s = V_1 = 0$

Figure 9.17 Circuits for finding T_{31} and T_{32} for the RC network of Fig. 9.16.

The coefficients of the s^2 and constant terms in the numerator of T_{32} are zero. This permits us to reduce the $(c_1 - Kd_1)$ term in (9.24) by increasing K without affecting the other two terms. The transfer function of the filter of Fig. 9.16 is found by using (9.22) and the expressions for N_{31}, N_{32}, and D,

$$T(s) = \frac{KN_{31}}{D - KN_{32}}$$

or

$$T(s) = \frac{\dfrac{K}{R_1 R_2 C_1 C_2}}{s^2 + \left(\dfrac{1}{R_1 C_1} + \dfrac{1}{R_2 C_1} + \dfrac{1 - K}{R_2 C_2}\right) s + \dfrac{1}{R_1 R_2 C_1 C_2}}$$

The details in deriving T_{31}, T_{32}, and $T(s)$ are left as an exercise for the reader.

The form of the numerator of $T(s)$ confirms the low-pass response of this circuit. The filter contains two zeros at infinity. The denominator is a second-order polynomial in s, the poles of which can be made complex by means of the gain K of the VCVS.

The numerator and the coefficients in the denominator of $T(s)$ can be interpreted in terms of the filter's characteristics A_o, ω_o, and Q. For the numerator,

$$A_o \omega_o^2 = \frac{K}{R_1 R_2 C_1 C_2}$$

For the denominator,

$$\omega_o^2 = \frac{1}{R_1 R_2 C_1 C_2}$$

and

$$\frac{\omega_o}{Q} = \frac{1}{R_1 C_1} + \frac{1}{R_2 C_1} + \frac{1 - K}{R_2 C_2}$$

These relationships may be used to express the filter characteristics in terms of the various R and C elements in the circuit.

In filters, generally speaking, each component will typically affect several filter characteristics. The design of these circuits by hand, can be an arduous task. However, a number of aides are available to simplify the design procedure. For common filter configurations, nomographs are available that provide design values for the various filter components based on the expected performance.

Computer programs (*c*omputer-*a*ided *d*esign, or CAD) are also available to design engineers. These sophisticated programs are easy to use, and provide the designer with the ability, not only to design a circuit for a given set of primary characteristics, but they also provide information about sensitivity, drift, accuracy, and so on. References on filter nomographs and CAD are provided in the bibliography.

9.6 INTRODUCTION TO SENSITIVITY FUNCTIONS

The sensitivity of a filter characteristic y with respect to a change in value x of one of the elements in a circuit is defined as

$$S(y, x) = \frac{\frac{\partial y}{y_o}}{\frac{\partial x}{x_o}} = \frac{x_o}{y_o} \frac{\partial y}{\partial x}$$

where x_o and y_o are the nominal values of x and y, respectively. As defined above, sensitivity is a measure of the percentage change of y (from its nominal value y_o) caused by a percentage change in x (from its nominal value x_o). All other parameters are assumed to remain constant.

The lower the sensitivity of a characteristic, the less its performance depends on the changes of the elements implementing the filter. Thus, sensitivity may be used as a criterion when comparing different filter configurations. If a small percentage change of a filter element x results in a small percentage change in a characteristic, the sensitivity of y with respect to x may be approximated by

$$S(y, x) \cong \frac{\frac{\Delta y}{y_o}}{\frac{\Delta x}{x_o}} \quad (\Delta x, \Delta y \text{ small})$$

This last expression may be used for quantitative purposes in obtaining the actual percentage change in y due to a given percentage change in x. For example, if $S(x, y)$ is -2, a 1% increase in parameter x will cause the characteristic y to decrease by 2%.

Sensitivity functions commonly found in the discussions of active filter configurations are $S(\omega_o, x)$ and $S(Q, x)$ where x may represent a resistor or capacitor in the

204 Active Filters

Property	IGMF	IGSF	KRC	State Variable
		Implementation		
Minimum number of components	+	−	0	−
Stability of characteristics	+	+	−	−
Ease of adjustment of characteristics	0	−	0	+
Spread of component values	−	+	$-(K < 0)$	+
	−	+	$+(K > 0)$	+
Possibility of high Qs	−	+	−	+
Possibility of high passband gain	−	+	+	+

+ superior
0 average
− inferior

Figure 9.18 Summary of the advantages and disadvantages of the various filter implementations.

filter circuit, or an active device characteristic such as A or GB. Sensitivity and its interpretation as a criterion in comparing active filter designs represents an involved topic and is beyond the scope of this book. The bibliography lists many excellent books on this topic.

9.7 COMPARING THE VARIOUS IMPLEMENTATIONS

There are a significant number of filter configurations that can be used to implement the various frequency responses. How does the designer select the "best" one? The answer to this question, of course, depends on how the word "best" is defined. Numerous properties describe the various techniques, and each technique has its advantages and disadvantages. Properties like number of components, stability of characteristics, ease of adjustment of the characteristics, and spread of component values must be compared with the design criteria. Thus, the so-called best configuration is the one whose positive properties best fit the requirements of the design. To provide a guide in making this choice, some of the advantages and disadvantages of the four techniques discussed in this chapter are listed in Fig. 9.18.

9.8 SUMMARY

The particular implementation configurations examined in this chapter were the infinite-gain single and multiple feedback circuits, the state-variable filters, and the *KRC* realization. All configurations are capable of implementing second-order filters of various responses with a pair of complex conjugate poles. The IGMF technique uses six components, described by their admittances, and provides two feedback loops around the amplifier. This particular circuit is frequently used to implement audio filters. The IGSF technique is quite versatile because a number of two-ports can be used in the input and feedback networks of the filter. The state-variable filter uses summing amplifiers and integrators and provides the low-pass, high-pass, and bandpass responses simultaneously. The *KRC* realization uses the basic noninverting and inverting amplifier circuits in conjunction with an *RC* network to implement second-order filters. The gain K of this technique can be used as an additional degree of freedom in controlling the location of the filter's poles.

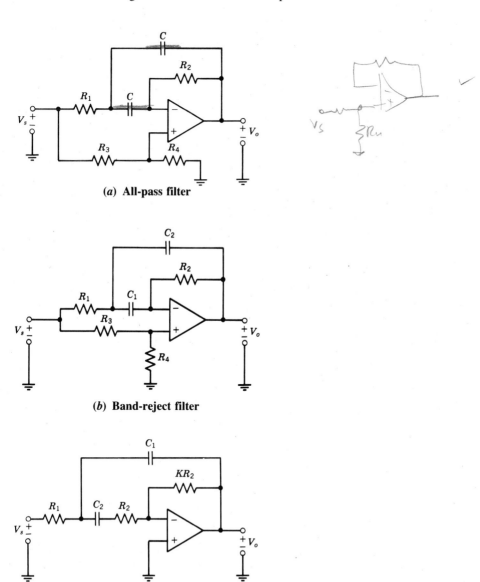

(a) All-pass filter

(b) Band-reject filter

(c) *KRC* filter

Figure 9.19 Additional active filter circuits.

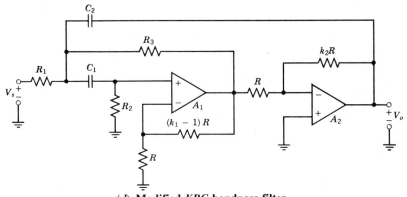

(d) Modified *KRC* bandpass filter

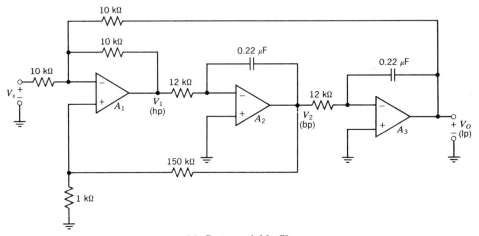

(e) State-variable filter

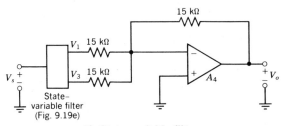

(f) State-variable filter

Fig. 9.19 Additional active filter circuits. (*Continued*)

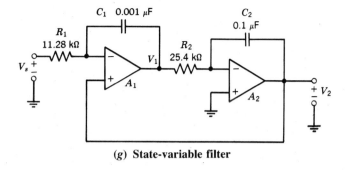

(g) State-variable filter

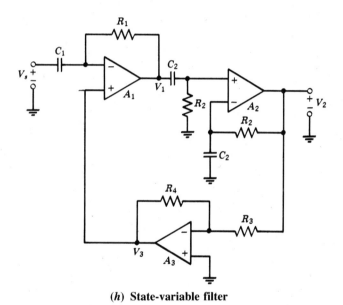

(h) State-variable filter

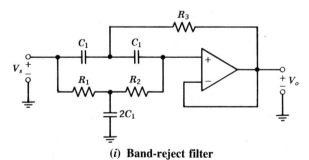

(i) Band-reject filter

Fig. 9.19 Additional active filter circuits. (*Continued*)

PROBLEMS

9.1 What component(s) in the IGMF model in Fig. 9.1 may be left out (open-circuited) and still have the configuration provide second-order bandpass responses with only four components, that is, two resistors and two capacitors? *Hint:* Examine $T(s)$.

9.2 Design a low-pass IGMF filter whose $f_o = 8$ kHz, $Q = 0.707$, and $A_o = -10$.

9.3 Design a high-pass IGMF filter whose $f_o = 100$ Hz, $Q = 0.707$, and $A_o = -10$.

9.4 Derive the expressions for y_{12} and y_{21} for network 1 in Fig. 9.8.

9.5 Derive the expressions for y_{12} and y_{21} for network 2 in Fig. 9.8.

9.6 Derive the expressions for y_{12} and y_{21} for network 3 in Fig. 9.8.

9.7 Derive the expressions for y_{12} and y_{21} for network 4 in Fig. 9.8.

9.8 Derive the expressions for y_{12} and y_{21} for network 5 in Fig. 9.8.

9.9 What type of filter response is produced when the passive networks 2 and 4 in Fig. 9.8 are used as the input and feedback networks, respectively, in the IGSF model of Fig. 9.6? What are the expressions for A_o, ω_o, and Q?

9.10 Design an IGSF bandpass filter whose $f_o = 1$ kHz, $Q = 10$, and $A_o = -10$.

9.11 Design an IGSF high-pass filter whose $f_o = 50$ Hz, $Q = 0.707$, and $A_o = -1$.

9.12 Find $T(s)$, ω_o, A_o, and Q for the high-pass biquad filter in Fig. 9.13b. Let $R_2 R_4 = R_1 R_5$.

9.13 Design a band-reject biquad filter (Fig. 9.13c) such that $\omega_o = 1$ kHz and $Q = 20$.

9.14 Find $T(s)$ for the all-pass filter in Fig. 9.19a. Let $R_2 R_3 = 4 R_1 R_4$.

9.15 Find $T(s)$ for the band-reject filter in Fig. 9.19b. Let $R_1 R_4 (C_1 + C_2) = R_2 R_3 C_1$.

9.16 Derive $T(s)$ for the KRC filter in Fig. 9.19c. Identify the filter's response, order, ω_o, A_o, and Q. Find $T(s)$ for the case where $R_1 = R_2 = R$ and $C_1 = C_2 = C$. Note that if $R_2 = 0$, this circuit becomes an IGMF configuration.

9.17 Prove that $T(s)$ for the modified KRC bandpass filter in Fig. 9.19d is

$$T(s) = \frac{s K_1 K_2 R_2 R_3 C_1}{As^2 + Bs + C}$$

where

$$A = R_1 R_2 R_3 C_1 C_2 (1 + K_1 K_2)$$
$$B = R_3 C_1 (R_1 + R_2) + R_1 R_3 C_2 + R_1 R_2 C_1 (1 - K_1)$$

and

$$C = R_1 + R_3$$

9.18 Calculate the value of f_o, A_o, and Q for the bandpass response of the state-variable filter in Fig. 9.19e.

9.19 What filter response is produced if the outputs V_1 and V_3 of Fig. 9.19e are summed as shown in Fig. 9.19f?

9.20 Find $T(s) = V_2/V_s$ and calculate the characteristics A_o, ω_o, and Q for the filter in Fig. 9.19g.

9.21 Find V_1/V_s, V_2/V_s, and V_3/V_s for the state-variable filter in Fig. 9.19h.

Chapter 10

Analog ICs and Subsystems

10.1 PROLOGUE

The circuits discussed in the previous chapters have centered around the use of *one* operational amplifier. However, the amplifier also plays a key role in implementing higher-level analog functions where it is only *one* of *several* different analog building blocks used to implement these functions. This becomes quite apparent when we examine the block diagrams of integrated circuits like the phase locked loop and the analog-to-digital convertors. The integration of higher-level analog functions is proceeding on a path similar to that in the digital electronics area.

The categorization of circuits discussed in this chapter is not so much by the functions these circuits perform, but by their complexity. The complexity is either due to the amount of circuitry or to the sophistication of the function. While the op amp is considered, in general, as a low-frequency amplifier, simplified integrated-circuit versions of the op amp are found in some high-frequency circuits; examples are specialized TV circuits, modulators, demodulators, and phase locked loops.

10.2 VOLTAGE-CONTROLLED VOLTAGE SOURCES

Voltage sources are low output impedance generators that maintain an output voltage independently of the output or load current. A voltage-controlled voltage source or

Voltage-Controlled Voltage Sources

VCVS is a voltage source whose output voltage is a function of its input voltage and circuit constants.

The two basic amplifier circuits, the noninverting and the inverting, can also be viewed as voltage-controlled voltage sources. They can be considered voltage sources because each of these circuits has a very low output resistance and an output voltage that is independent of the output current. This of course is true as long as the output current is within the rated value for the amplifier. Since the output voltage is proportional to the input voltage, both circuits can be called voltage-controlled voltage sources. As we have seen in the active filter applications area, the two basic amplifier circuits are used to perform VCVS functions.

The basic amplifier circuits have a pronounced limitation if they are going to be used as practical sources; their output current is limited to about 20 mA. This limitation can be overcome by using a discrete or integrated current amplifier in series with the output of the operational amplifier. Low cost, IC current amplifiers, which were designed expressly for this purpose, increase the output current capability of the VCVS from about 20 mA to nearly an ampere. The current and voltage amplifiers are cascaded as shown in Fig. 10.1 and, since the current amplifier is inside the feedback loop, the output voltage is still established by the input voltage and resistors R_1 and R_2. For the noninverting and inverting sources in Figs. 10.1a and 10.1b,

$$V_o = -\frac{R_2}{R_1} V_{REF} \qquad (10.1a)$$

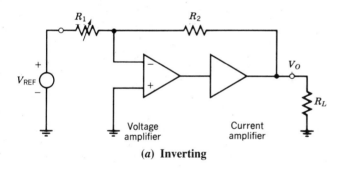

(a) **Inverting**

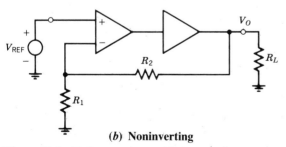

(b) **Noninverting**

Figure 10.1 Voltage sources with increased output current capability.

and

$$V_o = \left(1 + \frac{R_2}{R_1}\right) V_{REF} \qquad (10.1b)$$

10.3 VOLTAGE-CONTROLLED CURRENT SOURCES

Current sources are high-output impedance generators that sink or source a given output current independently of the load voltage. A voltage-controlled current source or VCCS is a current source whose output current is a function of the input voltage and circuit constants.

10.3.1 Three-Amplifier Current Source

The design objective in most constant current sources is to maintain a constant voltage across a "sensing" resistor. If the current through the resistor is the same as the output current and is independent of the output voltage, we call the source a current source. If the value of the current is determined by an input or control voltage, we call the source a voltage-controlled current source or VCCS.

The circuit in Fig. 10.2 consists of three basic amplifier configurations: the summer A_1, the voltage-follower A_2, and the unity-gain inverter A_3. The output of the summer A_1, which establishes the potential at one end of the sensing resistor R_I, is the inverted sum of the control voltage V_c and the output of A_3. The output of A_3 is the negative of the load voltage V_L, which is the voltage at the other end of R_I. We will show that the voltage across R_I, established by V_c, is constant and independent of the load voltage V_L.

The load voltage V_L is the input voltage of the follower amplifier A_2. The output of A_2 is

$$V_2 = V_L$$

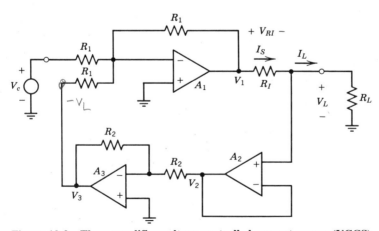

Figure 10.2 Three-amplifier voltage-controlled current source (VCCS).

A_3 is an inverting amplifier with a gain of -1 and, thus, its output voltage is

$$V_3 = -V_2 = -V_L$$

This voltage and the control voltage V_c serve as inputs to the inverting summer circuit of A_1 whose output is

$$V_1 = -(V_c - V_L) = -V_c + V_L$$

The difference in potential across R_I is

$$V_{RI} = (-V_c + V_L) - V_L = -V_c$$

Thus, the current I_s is a function of V_c and R_I or

$$I_s = -\frac{V_c}{R_I} \tag{10.2}$$

The input current of amplifier A_2 is typically very small compared to I_s; hence,

$$I_L \cong I_s = -\frac{V_c}{R_I} \tag{10.3}$$

The feedback loop composed of A_1, A_2, and A_3 is called a "bootstrap" configuration. As R_L changes, V_L changes. The feedback loop causes the voltage at the other side of R_I to change by the same amount. Thus the "bootstrapping" or lifting effect results in a voltage across R_I that is independent of the load voltage.

The maximum output voltage of a current source, beyond which the current regulation deteriorates, is called the compliance voltage. For this implementation, the compliance voltage is a function of the voltage across $R_I(-V_c)$ and the maximum output voltage of A_1. The maximum output voltage of A_1 is a function of the bias voltages connected to the V^+ and V^- amplifier terminals and the voltage drop across the components in the op amp's output stage.

For the polarity of V_c shown in Fig. 10.2, the current generator will "sink" current from the load. If the polarity of V_c is reversed, the generator will "source" current. The circuit can be converted to a digitally programmable current source by letting V_c be the output of a digital-to-analog convertor. If V_c is an ac voltage source, the circuit becomes an ac current source.

10.3.2 Single Amplifier Current Source

In applications that do not require precise values of current, the single-amplifier current source in Fig. 10.3 will suffice. This circuit requires fewer components but is only accurate in the 1% range. The principle of generating a constant current by forcing a known voltage across a resistor is the same in this circuit as in the three-amplifier version.

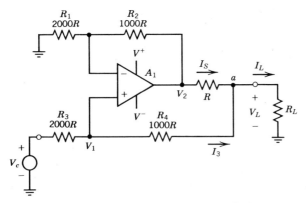

Figure 10.3 Single-amplifier voltage-controlled current source.

Voltage V_1 is related to the control and load voltages V_c and V_L through the resistor divider formed by R_3 and R_4. This voltage, a function of V_c and V_L, is amplified by A_1 and lifts or bootstraps the voltage at the amplifier output end of R. As V_L changes, the amplifier output voltage also changes by the same amount and maintains a constant current in the resistor R. The major error in this circuit is the feedback current in R_3 and R_4. As long as this current is much smaller than I_s, then $I_L \cong I_s$ and the load current will be independent of R_L.

If we neglect the amplifier's input current, the current through R_3 and R_4 is

$$I_3 = \frac{V_c - V_L}{R_3 + R_4} = \frac{V_c - V_L}{3000R} \tag{10.4}$$

The voltage V_1 is the sum of V_L and the voltage drop across R_4,

$$V_1 = V_L + I_3 R_4 = V_L + \frac{V_c}{3} - \frac{V_L}{3}$$

or

$$V_1 = \frac{2V_L}{3} + \frac{V_c}{3}$$

With respect to V_1, the configuration composed of A_1, R_1, and R_2 is a noninverting amplifier with a gain of $[1 + (R_2/R_1)]$ or $3/2$. Thus, the output of A_1 is

$$V_2 = \frac{3}{2}\left(\frac{2V_L}{3} + \frac{V_c}{3}\right) = V_L + \frac{V_c}{2} \tag{10.5}$$

The potential difference across R is

$$V_2 - V_L = \left(V_L + \frac{V_c}{2}\right) - V_L = \frac{V_c}{2}$$

Voltage-Controlled Current Sources

which produces a current through R independent of R_L

$$I_s = \frac{V_c}{2R} \qquad (10.6)$$

If we apply *KCL* to node a, we get a relationship for the load current,

$$I_L = I_s + I_3 = \frac{V_c}{2R} + \frac{V_c - V_L}{3000R}$$

If

$$|I_s| \gg |I_3|$$

as in this case, then

$$I_L \cong I_s = \frac{V_c}{2R} \qquad (10.7)$$

Example

The amplifier in the current source of Fig. 10.3 is a 108A with an input (bias) current of 2 nA. Assuming $R = 1 \text{ k}\Omega$, $V^+ = 15 \text{ V}$, $V^- = -15V$, and $V_c = +2 \text{ V}$, calculate the load current I_L and estimate the largest load resistance that the current source can drive before current regulation is lost. Compare I_L and the amplifier's input current for the case when $R_L = 0 \, \Omega$.

Solution

From (10.7),

$$I_L \cong \frac{2 \text{ V}}{2 \text{ k}\Omega} = 1 \text{ mA}$$

The largest load resistor that the circuit can drive depends on the maximum load voltage before the amplifier reaches saturation. The maximum load voltage depends on the maximum value of the output voltage of the amplifier and the voltage drop across R. As an estimate, we will allow 3 V for the potential difference required for the op amp's output stage. This means that

$$V_2(\text{max}) = V^+ - 3 \text{ V} = 12 \text{ V}$$

Using (10.5), we get

$$V_L(\text{max}) = V_2(\text{max}) - \frac{V_c}{2}$$

$$= 12 \text{ V} - 1 \text{ V} = 11 \text{ V}$$

Since $I_L \cong 1$ mA, then

$$R_L(\text{max}) \cong \frac{V_L(\text{max})}{I_L} = 11 \text{ k}\Omega$$

Under short-circuit conditions,

$$V_L = 0 \text{ V}$$

and

$$V_c = 2 \text{ V} = I_3(R_3 + R_4)$$

The error current I_3 is

$$I_3 = \frac{2 \text{ V}}{3 \text{ M}\Omega} = 667 \text{ nA} \ll I_L$$

The amplifier input current (2 nA) is much smaller than I_3 and has little effect on the accuracy of the circuit. The low input current rating of the 108A is advantageous in such applications. ∎

10.4 VOLTAGE REGULATORS

10.4.1 Theory and Model

The voltage regulator is the second largest selling analog IC in the world. The first, of course, is the op amp. Although it is a relatively simple device to use, the extensive use of the regulator in industrial products more than justifies our treatment here. The voltage regulator may be considered to be an op amp application in the sense that a key element in its design is the integrated voltage amplifier. The regulator is a variable-voltage input, constant-voltage output device. It converts an unregulated voltage to a constant voltage that is also independent of the load.

The simplified diagram for a positive-voltage regulator is shown in Fig. 10.4. This series regulator consists of a pass transistor Q_1, an error amplifier A_1, and a zener diode. The reference voltage developed across the zener diode, together with resistors R_1 and R_2, establish the value of the regulator's output voltage V_O. The output voltage will be proportional to the reference voltage, with R_1 and R_2 establishing the constant of proportionality,

$$V_O = V_{\text{REF}} \frac{R_1 + R_2}{R_1} \tag{10.8}$$

For fixed-voltage three-terminal regulators, resistors R_1 and R_2 are part of the device.

Voltage Regulators   217

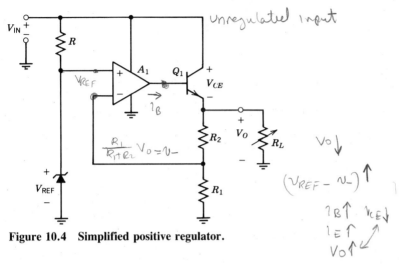

Figure 10.4   Simplified positive regulator.

The "error" amplifier A_1 is a simplified voltage amplifier implemented with a differential amplifier circuit. It compares a portion of the output voltage, divided down by R_1 and R_2, with the reference voltage V_{REF}. When an *error* or difference is sensed, the amplifier drives the base of Q_1 to compensate for this difference. The emitter of Q_1, which is the output of the regulator, will follow the base. If the output voltage drops below the established value given by (10.8), the output of A_1 will go more positive, raising the output voltage back to its proper level. If the output voltage rises above the established value, A_1 will go less positive, lowering the output voltage.

Transistor Q_1 is the regulator "pass" transistor; it *passes* the required output current to the load. Its collector voltage is the unregulated dc input, and its emitter voltage is the regulated dc output. The collector-to-emitter voltage V_{CE} makes up the difference between the two voltages, and it can be relatively large. The load current is supplied by the unregulated dc input voltage V_{IN} through Q_1 and along with V_{CE} determines the power dissipated by the pass transistor. Q_1 is designed as a power transistor and occupies over half of the IC chip in the higher current regulators.

The performance of the voltage reference is crucial to the overall performance of the regulator. It is the basis for comparison by the error amplifier in establishing and regulating the output voltage. The noise, stability, and thermal effects of the reference are directly reflected to the output. The simplest voltage reference is a zener diode. Monolithic voltage regulators use either a zener diode or a complex circuit that can be modeled as a zener diode.

Earlier IC regulators required a reasonably large number of external components to complete the circuit. These circuits were versatile, and they represented a step toward the fully integrated regulators we find today. Two popular IC regulators of that era were the 104 and 723. Most regulators sold today are three-terminal, fixed-voltage integrated circuits. Some of the more popular three-terminal regulators are the 105, 340-xx, 78xx, 320-xx, and 79xx. The 105 is a fixed +5 V regulator while the 340-xx and 78xx are positive regulators that come in a number of different (xx) output voltages values. Common output voltage values are 5, 6, 8, 12, 15, 18, and 24 V.

The negative voltage (with respect to ground) regulator counterparts of the 340 and 78xx are the 320-xx and 79xx.

10.4.2 Self-Protection Features

A more detailed block diagram for the positive-voltage regulator is shown in Fig. 10.5. The circuit reflects the self-protection features built into most regulators; thermal shutdown, current limiting, and protection from excessive power dissipation. Pass transistors Q_1 and Q_2 form a Darlington pair that increases the effective value of β, thus increasing the output current capability of the regulator.

Transistor Q_4 is the thermal shutdown transistor. This transistor turns on and reduces the maximum output current when its V_{BE} drops to 0.3 V because of a high die temperature. The base-emitter junction of Q_4 is used as a temperature sensor where 0.550 V is normal and 0.300 V represents a high temperature. The thermal shutdown circuitry prevents the IC chip or die from overheating in case of excessive load, momentary short, or a large increase in ambient temperature. The normal die temperature is about 80°C, but when the chip reaches a temperature of 165 to 190°C, the maximum output current is reduced to prevent further internal heating. This protection feature uses the temperature coefficient of the base-emitter junction of Q_4 (-2.2 mV/°C). The base of Q_4 is biased to about 0.3 V, too low to turn it on under normal conditions. As the die temperature rises and becomes excessive, the turn-on voltage of Q_4 drops to 0.3 V. Q_4 then begins to conduct and drains current away from the base of the pass transistors, reducing the maximum output current of the regulator.

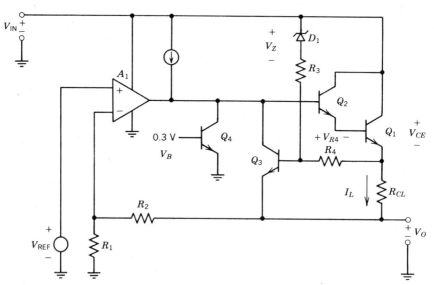

Figure 10.5 Positive regulator with self-protection features.

Example

To illustrate the basic ideas associated with the thermal shutdown provision in the circuit of Fig. 10.5, let us assume that the base-to-emitter voltage of Q_4 is 0.550 V when the die temperature is 75°C. For a temperature coefficient of -2.2 mV/°C, this voltage will drop to 0.300 V when a die temperature increase of

$$\Delta T = \frac{-250 \text{ mV}}{-2.2 \text{ mV/°C}} = 114°C$$

has taken place. Hence, when the die temperature reaches 75°C + 114°C or 189°, Q_4 will turn on and sink base current away from the pass transistors, thus limiting the output current. ■

Transistor Q_3 and resistor R_{CL} limit the maximum regulator output current. When the output current through R_{CL} causes a voltage drop of about 0.5 V, the normally off Q_3 starts to conduct and limits the base current available to the pass transistors. With the base current limited, the regulator output current is also limited. This current limiting scheme is also used in the output stage of the general-purpose op amp.

The purpose of D_1, R_3, and R_4 is to protect the pass transistor Q_1 against excessive power dissipation. This series circuit is activated when V_{CE} of the pass transistor Q_1, which approximately equals $V_{IN} - V_O$, exceeds the zener voltage of D_1. For high values of V_{IN}, the voltage $V_{IN} - V_O$ is distributed between D_1, R_3, and R_4. The voltage drop across R_4 reduces the maximum value of the output current at which the current-limiting circuit will limit, thus maintaining control of the power dissipated by the pass transistor.

Example

The values for the components in the current limiting and excessive power dissipation circuits for the 109 are $R_{CL} = 0.3 \ \Omega$, $R_3 = 10 \text{ k}\Omega$, $R_4 = 230 \ \Omega$, and $V_z(D_1) = 6.3$ V. Calculate the maximum output current when $(V_{IN} - V_O) = 6$ V and $(V_{IN} - V_O) = 9$ V.

Solution

The output of the 109 is rated at 1.5 A maximum. If the current limiting resistor is 0.3 Ω, then the voltage that will cause Q_3 to begin to conduct is 0.45 V. If $(V_{IN} - V_O)$ is 6 V, then D_1 will not conduct and the maximum output current is 1.5 A.

If $(V_{IN} - V_O) = 9$ V, D_1 will break down and the difference between $(V_{IN} - V_O)$ and V_z will be distributed between R_3 and R_4. The drop across R_4 will be

$$V_{R4} = \frac{R_4}{R_3 + R_4} [(V_{IN} - V_O) - V_z]$$

$$= 0.067 \text{ V}$$

220 Analog ICs and Subsystems

If Q_3 turns on at 0.450 V and $V_{R4} = 0.067$ V, then the voltage drop across R_{CL} that will cause current limiting is now reduced to

$$V_{RCL} = 0.450 \text{ V} - 0.067 \text{ V} = 0.383 \text{ V}$$

This corresponds to a maximum output current of

$$I_L = \frac{0.383 \text{ V}}{0.3 \text{ }\Omega} = 1.28 \text{ A} \qquad \blacksquare$$

The negative voltage regulator, in concept, is similar to the positive version. In the negative regulator, the polarity of the reference voltage is reversed and the pass transistor is a *pnp*. Most regulators are three-pin devices with a fixed dc output voltage. They can be made adjustable using external resistors.

In a complex electronic system, it is good design practice to use a voltage regulator for each digital and analog assembly or printed circuit board. The circuitry on a printed circuit board that is locally regulated sees a very low impedance back to the dc supply and is less sensitive to noise and oscillations.

10.4.3 Applications

The monolithic, three-terminal regulator is complete and ready to use. Apply an unregulated input voltage, and the device produces a fixed, regulated output voltage. Figure 10.6a illustrates the simplicity of its application. An input and output capacitor are added to aid frequency stability and transient response of the circuit.

The two most common positive regulator values are +5 and +15 V. Five volts is used to bias digital circuits and +15 V is used for analog circuits. Typical values of the input and output capacitors are 0.22 and 0.1 µF, respectively. The manufacturer's data sheets provide information on the values and types of capacitors recommended for a particular device.

Although positive (and negative) regulators have various output voltage values, odd voltages or variable voltages are often required. The addition of two resistors, R_1 and

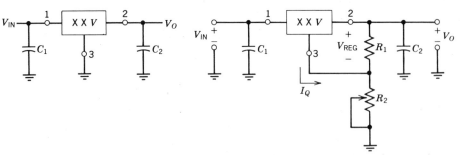

(a) **Fixed output voltage**   (b) **Variable output voltage**

Figure 10.6 Regulator applications.

R_2 in Fig. 10.6b, converts the fixed output regulator to a variable output. The output voltage of the regulator circuit shown is

$$V_O = V_{REG} + \left(\frac{V_{REG}}{R_1} + I_Q\right) R_2$$

where V_{REG} is the fixed regulator voltage and I_Q is the device's quiescent current. Quiescent current is a parameter of monolithic voltage regulators and is defined as that part of the input current that is not delivered to the load. Its value can be found in the data sheet and is moderately constant. If R_2 is a potentiometer, then V_O can be adjusted over a range of values. Capacitors C_1 and C_2 are added for stability.

Example

The typical value of the quiescent current for a 340-8 V regulator is 8 mA. If R_1 in Fig. 10.6b is a fixed 1 kΩ resistor and R_2 is a 250 Ω potentiometer, then

$$V_O = 8 \text{ V} + \left(\frac{8 \text{ V}}{1 \text{ k}\Omega} + 8 \text{ mA}\right) R_2$$

and the output voltage can be adjusted from +8 V ($R_2 = 0$) to +12 V ($R_2 = 250\ \Omega$). ∎

10.5 DIGITAL-TO-ANALOG CONVERTORS

Communicating between the analog and digital worlds requires devices that can translate the language of the two domains. A digital-to-analog (D/A or DAC) convertor accepts a digital word as its input and translates, or converts, it to an analog voltage. An analog-to-digital (A/D or ADC) convertor accepts an analog voltage and converts it to a digital word. A word is a group of digital two-state signals that collectively represent data.

Both convertors are used extensively in microprocessor based systems. They serve as the interface between the analog and digital portions of the system.

10.5.1 Unipolar DACs

The basic elements of a D/A convertor are the voltage reference, resistor network, switches, and an operational amplifier. The basic unipolar (one output voltage polarity) convertor design shown in Fig. 10.7 is an n-bit, parallel-entry, binary-coded DAC. The circuit is similar to a summing amplifier whose inputs are switched on or off as a function of the input digital word. The amplifier also functions as a current-to-voltage convertor because it sums (and converts to a voltage) the binary-weighted currents established by the resistors, voltage reference, and the switches. The switches are

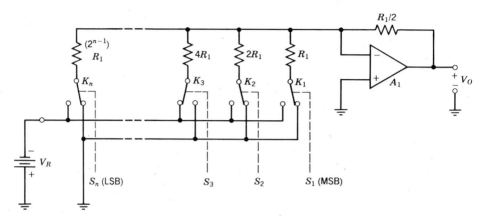

Figure 10.7 Basic D/A convertor.

usually single-pole, double-throw switches and are implemented using field-effect or bipolar transistors.

The input-output relationship of the circuit is

$$V_O = V_R(K_1 2^{-1} + K_2 2^{-2} + \cdots + K_n 2^{-n}) \quad (10.9)$$

where $K_x = 0$ ($x = 1, 2, \ldots, n$) when the switch K_x contact is connected to ground and S_x is a logic zero. $K_x = 1$ when the switch K_x contact is connected to V_R and S_x is a logic one. The input digital word takes the form

$$\text{Digital word} = (S_1 S_2 S_3 \ldots S_n) \quad (10.10)$$

where S_n is the least significant bit (LSB) and S_1 is the most significant bit (MSB). The 0 and 1 state of S_x ($x = 1, 2, \ldots, n$) corresponds to the low and high state of the digital signal that controls the appropriate switch.

The resolution, or smallest programmable voltage, for the DAC is determined by the least significant bit (LSB),

$$V_O(\text{smallest}) = V_R 2^{-n} \quad (10.11)$$

The full-scale value, or largest programmable voltage, occurs when all bits of the input word are one:

$$\begin{aligned} V_O(\text{full scale}) &= V_R(2^{-1} + 2^{-2} + \cdots + 2^{-n}) \\ &= V_R(1 - 2^{-n}) \\ &= V_R - V_O(\text{smallest}) \end{aligned} \quad (10.12)$$

The output voltage of the DAC can be programmed from zero volts to its full-scale value in increments of the value established by the LSB.

Several key parameters of DACs are number of bits, polarity of output voltage,

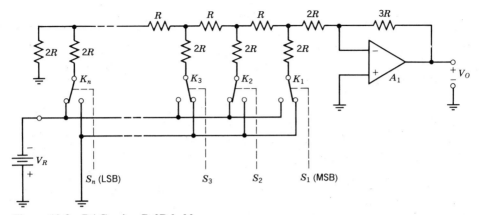

Figure 10.8 DAC using R-2R ladder.

input digital code, and accuracy or linearity. The typical commercial DAC is eight bits, bipolar, two's complement binary, with a 0.1% full-scale linearity accuracy. Other parameters include hysteresis, settling time, slew rate, drift, supply rejection, conversion rate, temperature coefficient, clock rate, and monotonicity. The discussion of these parameters can be found in books that deal with analog-digital convertors.

The disadvantage of the circuit shown in Fig. 10.7 is the wide range of resistance values required. The commonly used R-2R ladder network of Fig. 10.8 overcomes this disadvantage at the expense of an additional resistor per bit. It can be shown that the input-output relationship for this configuration is the same as the one for the previous convertor. Many other resistor network schemes exist, but they all have the same purpose: to produce binary weighted currents (or voltages) that can be switched on or off in accordance with a digital input word.

Example

Design a unipolar DAC, with a R-2R ladder as shown in Fig. 10.8, to meet the following specifications;

1. $V_O(\text{full scale}) \cong +10 \text{ V}$
2. $V_O(\text{smallest}) \leq 10 \text{ mV}$

Solution

From (10.11) and (10.12),

$$\frac{V_O(\text{full scale})}{V_O(\text{smallest})} = \frac{V_R - V_O(\text{smallest})}{V_O(\text{smallest})}$$

or

$$\frac{V_O(\text{full scale})}{V_O(\text{smallest})} = 2^n - 1 \geq 10^3$$

Solving this last expression for n, we obtain

$$n \geq \frac{\log 1{,}001}{\log 2} = 9.97$$

Thus, to resolve 10 mV of 10 V or 1 part of 1000, the DAC must have 10 bits. For a nominal full-scale value of 10 V, let

$$V_R = -10 \text{ V}$$

From (10.11), the LSB will have a weighted value of

$$V_O(\text{smallest}) = \frac{10 \text{ V}}{2^{10}} = 9.766 \text{ mV}$$

and the full-scale value will be exactly

$$V_O(\text{full scale}) = 10 \text{ V} - \text{LSB} = 9.990234 \text{ V}$$

The nominal full-scale value of all DACs is asymptotically approached and is a function of the number of bits of the DAC.

The selection of the value of the resistors in the R-$2R$ ladder depends largely on the input current of the op amp. This current must be much smaller than the minimum current through the feedback resistor ($3R$). Minimum feedback current will flow when the output voltage is $V_O(\text{smallest})$. Thus,

$$I_B \ll \frac{V_O(\text{smallest})}{3R}$$

or, in this case,

$$I_B \ll \frac{9.766 \text{ mV}}{3R}$$

If the maximum value of the amplifier bias current is assumed to be 5 nA, we neglect its effect by arbitrarily making the feedback current caused by the LSB 1000 times larger or at least 5 µA. The value of the feedback resistor for A_1 is given by

$$3R \leq \frac{9.766 \text{ mV}}{5 \text{ µA}} = 1.95 \text{ k}\Omega$$

We will let $R = 0.5$ kΩ, which establishes

$$2R = 1 \text{ k}\Omega$$

and

$$3R = 1.5 \text{ k}\Omega$$

The maximum feedback current in A_1 occurs when all bits are one and will be approximately

$$\frac{10 \text{ V}}{1.5 \text{ k}\Omega} = 6.67 \text{ mA}$$

which is within the rated value of the operational amplifier used in Fig. 10.8. The offset voltage V_{OS} for the amplifier should be very low or nulled out. ∎

10.5.2 Bipolar DACs

Most applications require the output of a DAC to provide both positive and negative values of voltage. The additional hardware requirements to convert a unipolar DAC to a bipolar DAC are usually modest; one more resistor and an additional reference source. There are a number of codes that can be used to represent the two polarities of numbers; however the physical circuit is usually the same. A block diagram of a bipolar DAC is shown in Fig. 10.9a.

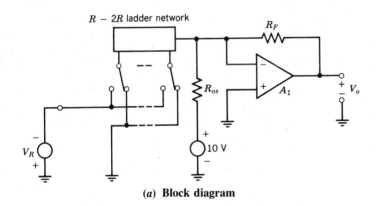

(a) Block diagram

	Unipolar		DAC Bipolar	
V_0		Binary	Two's Complement	V_0
0		000	100	$-V_{FS}$
		001	101	
		010	110	
		011	111	
		100	000	0
		101	001	
		110	010	
$+V_{FS}$		111	011	$+V_{FS}$

(b) Codes

Figure 10.9 Bipolar DAC.

The second reference, 10 V, and the resistor R_{OS} supply a constant current to the inverting input of the amplifier. This current is adjusted to equal in magnitude the current contributed to the summing junction by the most significant bit (MSB). Whether or not the MSB is turned on determines the polarity of the output; therefore it is called the sign bit. Negative numbers in digital systems are frequently represented using the two's complement code. In this code, negative numbers are represented by taking the two's complement of their positive number counterparts. The two's complement of a number is the one's complement plus one where the one's complement is found by changing binary ones to zeros and zeros to one. The three-bit straight binary and two's complement binary codes for the unipolar and bipolar DACs are shown in Fig. 10.9*b*.

10.6 ANALOG-TO-DIGITAL CONVERTORS

The analog-to-digital convertor, or ADC, is more complex than the digital-to-analog convertor. There are a number of different design approaches, but many of them use a parallel entry DAC in a feedback loop. A simplified block diagram of such an ADC will be used to illustrate analog-to-digital conversion principles and the corresponding application of the operational amplifier. The ADC in Fig. 10.10 is intended to operate in conjunction with a computer. It consists of a DAC and a comparator. The inputs to the ADC are the bits to program the DAC and the unknown voltage V_x.

Initially, the output voltage of the DAC is programmed to its most negative value, causing the output of the comparator to be low. As long as the output of the comparator is low, the program stored in the computer will continually increment the value of the digital word, thus incrementing the output voltage of the DAC. When the comparator output just goes high, the computer ceases to increment the DAC any further. The value (digital word) of the DAC that caused the null or the output of the comparator to change state is the same as the unknown voltage V_x. This, of course, is true only within the accuracy and resolution of the system. When the comparator changes state,

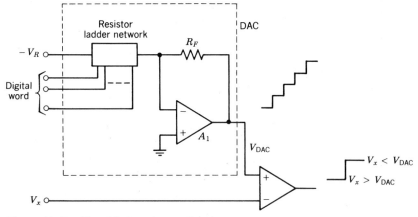

Figure 10.10 Simplified analog-to-digital convertor.

the program that incremented the DAC is stopped, and its last programmed value can be interpreted as the value of the voltage V_x.

The accuracy of the ADC is primarily a function of the accuracy and resolution (number of bits) of the DAC. The speed or conversion rate varies depending on the speed of the computer, the response time of the DAC, the response time of the comparator, and the specific value of the unknown voltage. Modular or self-contained analog-to-digital convertors also have START-TO-CONVERT and CONVERSION-FINISHED signals. For these convertors, a signal must be initiated that instructs the ADC to start the conversion. When the conversion is completed, and the digital bits representing the unknown voltage are available, the ADC initiates a completion signal. These convertors will also have an internal clock to control the conversion rate.

Other ADC design approaches vary greatly. The analog hardware used includes summing amplifiers, integrators, resistor networks, limiters, comparators, and switches. The digital hardware includes registers, counters, clocks, and control logic.

10.7 PHASE LOCKED LOOP

The intent of the coverage of the phase locked loop (PLL) in this section is not to present an extensive and detailed treatment of the device. Rather, the intent of this presentation is to show how the operational amplifier is used in a more complex analog system.

The basic concepts of the PLL have been developed since the early 1930s. However, its current availability in a low-cost, self-contained, monolithic IC package has changed it from a specialized design technique to a general-purpose building block. The device represents an increase in the level of complexity in analog ICs because it contains four rather basic analog building blocks.

A phase locked loop is an electronic feedback system consisting of a phase detector, a low-pass filter, an amplifier, and a voltage-controlled oscillator (VCO) as shown in Fig. 10.11. The phase detector, low-pass filter, and amplifier are in the forward signal path, while the VCO is in the feedback path. When the circuit is in operation or lock, the PLL can be approximated as a linear feedback system where the input signal variable is frequency or phase instead of voltage or current. The popular 565 PLL is a general-purpose circuit designed for highly linear FM demodulation. Other applications of the PLL include frequency shift keying, frequency multiplication or translation, motor-speed control, and background-music decoder. The principles of operation of the PLL are explained from two points of view.

When no input signal $v_{in}(t)$ is present, the voltage $v_d(t)$ driving the VCO is equal to zero. With $v_d(t) = 0$, the VCO operates at a set frequency f_o, which is known as the free-running frequency. If an input signal is applied to the PLL, the phase detector compares the phase and frequency of the input with the phase and frequency of the output of the VCO and generates an error voltage $v_e(t)$ that is related to the phase and frequency difference between the two signals. This error voltage is filtered by the low-pass filter, amplified, and applied to the control terminal of the VCO. In this manner, the control voltage $v_d(t)$ forces the VCO frequency to vary in a direction that reduces

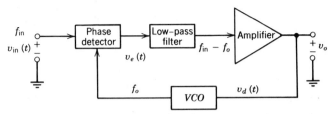

Figure 10.11 Phase-locked loop block diagram.

the difference between f_o and f_{in}. When the input frequency f_{in} is sufficiently close to f_o, the feedback nature of the PLL will cause the VCO to synchronize or lock with the incoming signal. Once the two signals are in lock, the VCO and the input signal will be at the same frequency but with a finite phase difference between them. The net phase difference is necessary to provide the voltage $v_d(t)$ that shifts the VCO from its free-running frequency to the input signal frequency f_{in} and, thus, keep the PLL in lock.

From another viewpoint, the phase detector is a multiplier or mixer that mixes the input and VCO signals producing the sum and difference frequencies $f_{in} \pm f_o$. The loop will servo and lock when the VCO duplicates the input frequency. The output of the multiplier contains ac components that are filtered, and a dc component that is a function of the phase angle between the VCO and the input signal. The dc component is amplified and fed back to the VCO to provide the necessary voltage to keep it oscillating at the proper frequency.

10.8 MODULATORS AND DEMODULATORS

10.8.1 Modulation and Demodulation

In a communications system, information, either audio or data, cannot be directly transmitted efficiently. To increase the efficiency of transmission, the low-frequency information is used to change some *characteristic* of a higher-frequency carrier. This process is called modulation. If the amplitude of the carrier is changed in proportion to the amplitude of the information to be transmitted, the process is called amplitude modulation (AM). Changing the frequency or the phase angle of the carrier results in frequency modulation (FM) or phase modulation (PM), respectively. At the receiver end of the communications system, the audio or data information is extracted or recovered by a process called demodulation or detection.

10.8.2 Balanced Modulator

The amplitude modulator shown in Fig. 10.12 uses an analog multiplier (see Section 4.8). A high-frequency carrier described by

$$v_c = V_{cp} \sin(\omega_c t) \qquad (10.13)$$

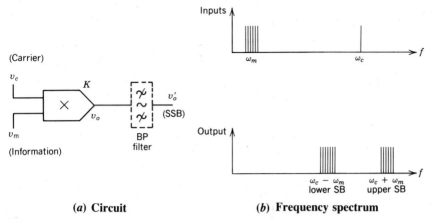

Figure 10.12 Balanced modulator.

is applied to one input of the multiplier, and a low-frequency modulating signal described by

$$v_m = V_{mp} \sin (\omega_m t) \tag{10.14}$$

is applied to the other input. The output of the multiplier is

$$v_o = K V_{cp} V_{mp} \sin (\omega_c t) \sin (\omega_m t) \tag{10.15}$$

where K is the multiplying constant, usually $1/10$.

Equation (10.15) can be expressed as

$$v_o = \frac{K V_{cp} V_{mp}}{2} [\cos (\omega_c - \omega_m)t - \cos (\omega_c + \omega_m)t] \tag{10.16}$$

using the trigonometric identity

$$\sin A \sin B = \tfrac{1}{2}[\cos (A - B) - \cos (A + B)] \tag{10.17}$$

The sum $(\omega_c + \omega_m)$ and difference $(\omega_c - \omega_m)$ frequencies are called side frequencies because they are above and below the carrier frequency ω_c. If the modulating signal varies over a band or range of frequencies (like the audio range), the frequency spectrum of the modulator output will be double sideband (DSB), having upper and lower sidebands instead of a single side frequency. Since v_o does not contain the carrier frequency ω_c, this type of modulation is called balanced or suppressed carrier modulation. The frequency spectrum for a balanced modulator is shown in Fig. 10.12b. The information in the form of the modulated signal is present in both the upper and lower sidebands. Thus only one sideband is necessary during the detection or demodulation process. The addition of a bandpass filter (passing either of the sidebands) to the multiplier output, as shown in Fig. 10.12a, results in a single sideband (SSB) output.

Example

What are the frequencies of the lower and upper sidebands of the balanced modulator circuit in Fig. 10.12, if the carrier is

$$v_c = 1.5 \sin (2\pi \cdot 50 \cdot 10^3 \, t) \text{ V}$$

and the modulating signals are described by

$$v_m = 1.2 \sin (2\pi \cdot f_m \cdot t) \text{ V}$$

where

$$0.5 \text{ kHz} \leq f_m \leq 4 \text{ kHz}$$

The multiplier constant is equal to one.

Solution

Using (10.16), we find that the output signal is

$$v_o = 0.9[\cos 2\pi(50 \cdot 10^3 - f_m)t - \cos 2\pi(50 \cdot 10^3 + f_m)t] \text{ V}$$

which produces a lower sideband frequency range from 46 to 49.5 kHz and an upper sideband frequency range from 50.5 to 54 kHz. ∎

Single-Sideband (SSB) Demodulator Demodulation is the process of recovering a modulating signal ω_m from the modulated output voltage v_o. To demodulate the output voltage, all we have to do is connect a SSB signal, say of frequency $(\omega_c + \omega_m)$ to one multiplier input and a second signal of frequency ω_c to the other. The output of the demodulator, shown in Fig. 10.13, will have a sum $(\omega_c + \omega_m) + \omega_c$ and a difference frequency $(\omega_c + \omega_m) - \omega_c = \omega_m$. If the modulator output signal is of the form

$$v'_o = V_1 \cos (\omega_c + \omega_m)t \qquad (10.18)$$

and the carrier signal is of the form

$$v_c = V_2 \cos (\omega_c t) \qquad (10.19)$$

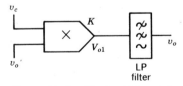

Figure 10.13 Single-sideband demodulator.

then the output of the multiplier will be

$$v_{o1} = \frac{KV_1V_2}{2}[\cos(\omega_m t) + \cos(2\omega_c + \omega_m)t] \quad (10.20)$$

A low-pass filter, connected to the output of the multiplier, removes its high-frequency component and leaves the modulating signal or

$$v_o = \frac{KV_1V_2}{2}\cos(\omega_m t) \quad (10.21)$$

In certain applications, first-order, low-pass op amp filters (see Section 8.3) may be used.

Example

What is the output signal of the SSB demodulator in Fig. 10.13 if the modulator output contains a 50 kHz carrier and a 2 kHz tone and is described by

$$v_o = 0.9 \cos(2\pi 52 \text{ kHz})t \text{ V}$$

The carrier voltage is described by

$$v_c = 1.5 \cos(2\pi 50 \text{ kHz})t \text{ V}$$

For the multiplier, K is one.

Solution

The output of the multiplier is found using (10.20),

$$v_{o1} = 0.68[\cos(2\pi 2 \text{ kHz})t + \cos(2\pi 102 \text{ kHz})t] \text{ V}$$

The output of the low-pass filter is

$$v_o = 0.68 \cos(2\pi 2 \text{ kHz})t \text{ V} \quad \blacksquare$$

Standard AM Modulator The standard AM amplitude modulator is shown in Fig. 10.14. The summer A_1 sums the modulating signal and the peak value of the carrier to produce

$$v_{o1} = -[V_{cp} + V_{mp}\sin(\omega_m t)] \quad (10.22)$$

This signal is then multiplied by the carrier

$$v_c = V_{cp}\sin(\omega_c t) \quad (10.23)$$

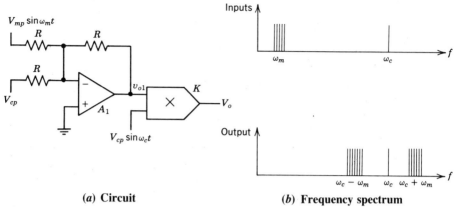

(a) Circuit (b) Frequency spectrum
Figure 10.14 Standard AM modulator.

to produce

$$v_o = -K[V_{cp}^2 \sin(\omega_c t) + 0.5 V_{cp} V_{cm} \cos(\omega_c - \omega_m)t \\ - 0.5 V_{cp} V_{cm} \cos(\omega_c + \omega_m)t] \quad (10.24)$$

The output of the standard modulator contains the carrier frequency and its two sidebands. Its frequency spectrum is shown in Fig. 10.14b.

10.9 TIMER

The IC timer (555) is an analog integrated circuit that can be used as a rectangular wave oscillator (astable operation), or as a single-pulse generator (monostable operation) upon application of an external trigger signal. External components, two resistors and a capacitor, establish the frequency of oscillation in the astable mode as shown in Fig. 10.15. Other applications include time-delay generation, pulse-width modulation, pulse-position modulation, and ramp generation.

The functional block diagram of the 555 timer contains the equivalent of two comparators A_1 and A_2, a flip-flop, an output stage, a voltage divider, and a discharge transistor Q_1. The operation and function of the internal sections of the IC will be explained for the astable condition. If the timer is connected as shown in Fig. 10.15, it will free-run. The operation of the timer in the astable mode is very close to that of the rectangular wave oscillator discussed in Section 6.5.

The voltage divider consists of three equal-valued resistors and establishes the reference voltages for the comparators at

$$V_{UL} = \frac{2V^+}{3} \quad \text{(upper limit)} \quad (10.25a)$$

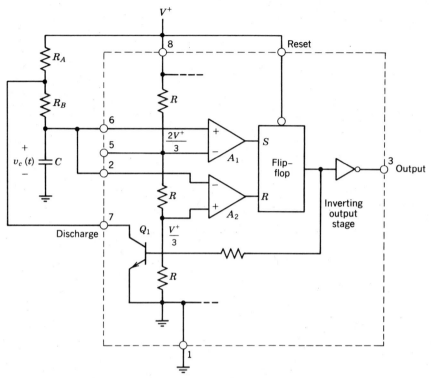

Figure 10.15 555 timer (astable operation).

and

$$V_{LL} = \frac{V^+}{3} \quad \text{(lower limit)} \quad \text{(10.25b)}$$

The voltage at the threshold pin (6) is an exponentially growing and decaying waveform as the capacitor C attempts to charge to V^+ through resistors R_A and R_B and discharge to zero through resistor R_B. The junction of R_A and R_B is connected to the discharge pin (7). Internally, this pin is connected to the discharge transistor Q_1, which is conducting for the discharge portion of the cycle and off for the charge portion.

The flip-flop is set low (timer output is high) when the capacitor voltage $v_c(t)$ drops below $V^+/3$, causing the output of comparator A_2 to go high. The flip-flop output is set high (timer output is low) when the capacitor voltage rises above $2V^+/3$, forcing the output of comparator A_1 high. The two comparators function as a window detector, which was covered in Section 5.4.

The discharge cycle of the capacitor voltage $v_c(t)$ is initiated, say at time $t = 0$, when its magnitude just exceeds V_{UL}. This causes the output of the flip-flop to go high (timer output is low), turning on the discharge transistor Q_1 and thus grounding pin 7. The capacitor voltage during its discharge cycle decreases exponentially toward zero (from the initial value V_{UL}) with a time constant $R_B C$ and is given by

$$v_c(t) = V_{UL} \exp \frac{-t}{R_B C}$$

or

$$v_c(t) = \frac{2V^+}{3} \exp \frac{-t}{R_B C} \tag{10.26}$$

The discharge cycle is terminated at $t = T_1$, when the capacitor voltage just drops below V_{LL}, causing the output of the flip-flop to go low (timer output is high). The duration T_1 of the discharge cycle is found from

$$V_{LL} = \frac{V^+}{3} = \frac{2V^+}{3} \exp \frac{-T_1}{R_B C}$$

Solving for T_1, we get

$$T_1 = R_B C \ln 2 \cong 0.69 \, R_B C \tag{10.27}$$

Immediately after the end of the discharge cycle at $t = T_1$, the low state of the flip-flop turns Q_1 off, initiating the charge cycle of $v_c(t)$. During this time, $v_c(t)$ charges exponentially toward V^+ (from its initial value V_{LL}) with a time constant $(R_A + R_B)C$ and is given by

$$v_c(t) = V^+ - (V^+ - V_{LL}) \exp \frac{-(t - T_1)}{(R_A + R_B)C}$$

or

$$v_c(t) = V^+ - \frac{2V^+}{3} \exp \frac{-(t - T_1)}{(R_A + R_B)C}$$

The charge cycle of $v_c(t)$ is terminated at $t = T_1 + T_2$. Evaluating the above expression when the capacitor voltage becomes $2V^+/3$, we find the duration of the charge cycle T_2 from

$$T_2 = (R_A + R_B)C \ln 2 \cong 0.69(R_A + R_B)C \tag{10.28}$$

The period of oscillation of the astable multivibrator of Fig. 10.15 is the sum of the durations of the discharge and charge cycles and equals

$$T = T_1 + T_2 \cong 0.69(R_A + 2R_B)C \tag{10.29a}$$

The corresponding expressions for frequency and duty cycle are

$$f = \frac{1}{T} \cong \frac{1.44}{(R_A + 2R_B)C} \tag{10.29b}$$

and

$$\text{Duty cycle} = \frac{T_1}{T_1 + T_2} = \frac{R_B}{R_A + 2R_B} \tag{10.29c}$$

The duty cycle for the 555 is, by definition, the ratio of the time the output is low to the total period.

Example

Design a square wave oscillator using a 555 timer with the following specifications;

1. $f = 2$ kHz
2. Duty cycle $= 25\%$

Solution

The period of the oscillating waveform is

$$T = \frac{1}{f} = 500 \ \mu\text{sec}$$

From the duty cycle requirements,

$$T_2 = (0.25)T = 125 \ \mu\text{sec}$$

and

$$T_1 = 375 \ \mu\text{sec}$$

Choosing $R_A + R_B = 10$ kΩ and using (10.28), we get

$$C = \frac{T_1}{0.69(R_A + R_B)} = 0.054 \ \mu\text{F}$$

From (10.27),

$$R_B = \frac{T_2}{0.69C} = 3.34 \text{ k}\Omega$$

Thus, $R_A = (R_A + R_B) - R_B = 6.66 \text{ k}\Omega$. ∎

10.10 SUMMARY

The operational amplifier is not only important in circuits in which it is *the* active device, but it also plays a key role in complex analog systems. Evidence of this can be found in voltage regulators, analog-to-digital and digital-to-analog convertors, phase locked loops, and timers.

The VCVS and VCCS are voltage and current sources whose output values are established and controlled by an input voltage. These sources are often used in active filter circuits and, when current amplifiers are added, they may be used as voltage supplies. A key component in the widely used voltage regulator is the error amplifier implemented in IC form by a simplified voltage amplifier. The regulator is a variable-voltage in, constant-voltage out device and is an important part of a power supply. The analog and digital convertors use both the operational amplifier and comparator to implement circuits that convert discrete voltages to an analog voltage and vice versa.

Modulators, demodulators, and phase locked loops are circuits used in communications systems whose key element is the analog multiplier. The timer uses comparators and digital electronics to provide a functional block that, among other things, can be used as an oscillator or single-pulse generator.

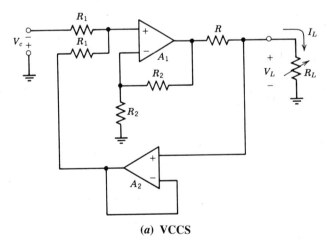

(a) VCCS

Figure 10.16 Additional analog ICs and subsystem circuits.

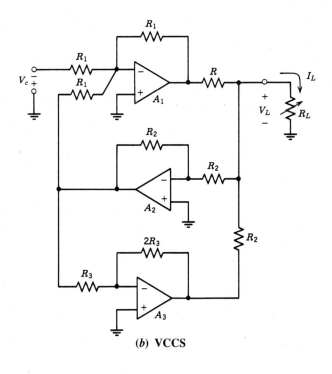

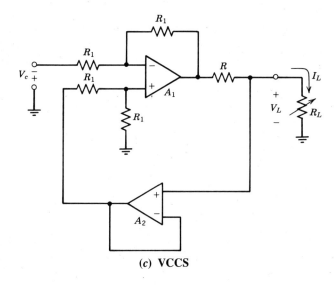

Fig. 10.16 Additional analog ICs and subsystem circuits. (*Continued*)

237

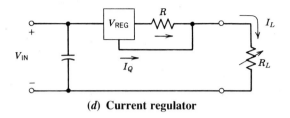

(*d*) **Current regulator**

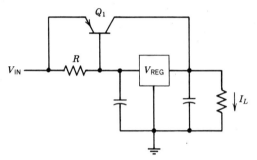

(*e*) **Voltage regulator with current booster.**

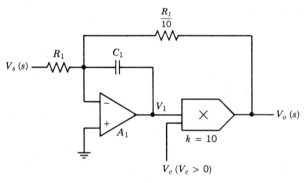

(*f*) **Tunable low-pass filter**

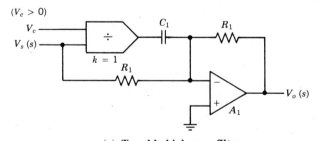

(*g*) **Tunable high-pass filter**

Fig. 10.16 Additional analog ICs and subsystem circuits. (*Continued*)

PROBLEMS

10.1 What is the required input-output voltage phase relationship for the current amplifiers in Fig. 10.1? What transistor amplifier circuit is traditionally used to implement the current amplifier?

10.2 Mathematically describe the load current $i_L(t)$, the load voltage $v_L(t)$, and the output voltages $v_1(t)$, $v_2(t)$, and $v_3(t)$ for all the amplifiers in the circuit of Fig. 10.2, if $v_c(t) = 2 \sin(2\pi \cdot 10^3)t$ V, $R_1 = R_2 = 10$ kΩ, and $R_L = 1$ kΩ.

10.3 The circuit in Fig. 10.16a is a VCCS. Find the input-output relationship I_L/V_c.

10.4 For the VCCS in Fig. 10.16b, prove that

$$I_L = \frac{V_c}{R}$$

10.5 Find the input-output relationship I_L/V_c for the VCCS circuit in Fig. 10.16c.

10.6 Specify the values of R_1, R_2, R_3, R_4, R_{CL}, and V_B for the regulator of Fig. 10.5 to meet the following specifications:
(a) $V_O = 12$ V
(b) $I_L(\text{max}) = 200$ mA
(c) Maximum die temperature = 180°
(d) $P_D(\text{max})$ of $Q_1 = 1.26$ W
The temperature coefficient of all transistors is -2.2 mV/°C, V_{BE}(turn-on) is 0.500 V at 75°C, V_{REF} is 1.8 V, and $V_z = 5.3$ V.

10.7 A fixed 8 V regulator is used in Fig. 10.6b. The circuit must be made adjustable from 8 to 12 V, and the regulator's I_Q is 5 mA. If $R_1 = 1.5$ kΩ, find $R_2(\text{max})$.

10.8 Modify the circuit in Fig. 10.6b to make it an adjustable negative regulator circuit. What is the output voltage range if $V_{REG} = -5.2$ V, $R_1 = 510$ Ω, and R_2 is a 500 Ω potentiometer? Assume that the quiescent current is very small.

10.9 The regulator circuit in Fig. 10.16d is configured as a current source. For this circuit, prove

$$I_L = \frac{V_{REG}}{R} + I_Q$$

10.10 The output current (200 mA) of the regulator in Fig. 10.16e can be increased by using an external booster transistor Q_1. What is the required value of R if V_{BE} of Q_1 is 0.600 V. For large values of load current, a current-limiting circuitry for Q_1 must be added.

10.11 For the D/A convertor in Fig. 10.7, $n = 10$ and $V_R = 10$ V. What is the output voltage V_O for an input digital word of (10 0110 1101)?

240 Analog ICs and Subsystems

10.12 For the D/A convertor in Fig. 10.7, $n = 8$, $V_R = 10$ V, and $R_1 = 2$ kΩ. Calculate the current in the feedback resistor if the input digital word is (0110 1101).

10.13 Show that (10.11) and (10.12) apply to the D/A convertor of Fig. 10.8.

10.14 Extend the unipolar-bipolar table in Fig. 10.9 for a four-bit bipolar D/A convertor.

10.15 The D/A convertor in the A/D circuit shown in Fig. 10.10 contains 10 bits with a nominal full-scale value of 10 V. What will be the value of the digital word after conversion occurs if $V_x = 7.510$ V.

10.16 Define the range of frequencies for the lower and upper sidebands in the balanced modulator of Fig. 10.12, if the 50 kHz carrier is modulated with signals ranging from 100 Hz to 10 kHz.

10.17 The circuit in Fig. 10.16f is a voltage-tunable, first-order, low-pass filter. Prove that

$$\frac{V_o(s)}{V_s(s)} = \frac{-1}{\dfrac{sR_1C_1}{10\,V_C} + 10}$$

and

$$f_c = \frac{100 V_C}{2\pi R_1 C_1}$$

The frequency f_c is the cutoff frequency of the filter.

10.18 Find V_o/V_s for the circuit in Fig. 10.16g. Sketch its amplitude Bode plot.

10.19 Design a 555-type oscillator (see Fig. 10.15) with a period of 100 msec and a duty cycle of 10%.

PART III

PRACTICAL LIMITATIONS

Chapter 11

Voltage Gain and Bandwidth

11.1 PROLOGUE

The ideal operational amplifier, used in the previous chapters to implement a number of circuit functions, enabled us to concentrate on the design philosophy behind these circuits. In real-life circuits, the practical amplifier represents only an approximation of the ideal case. Its characteristics and parameters that previously were assumed to be infinite or zero now have to be taken into account while analyzing a given op amp circuit design.

In this and the following two chapters, the more important limitations of the nonideal operational amplifier are gradually introduced. An exposure to and an understanding of these limitations will enable the designer of op amp circuits to minimize, whenever possible, their adverse effects. When the designer takes into account the limitations of the op amp, the performance of the practical circuit ultimately becomes a very reasonable approximation of the ideal case.

11.2 FINITE VOLTAGE GAIN

The effect of the finite open-loop gain of an operational amplifier on circuit performance will be illustrated through the use of the basic noninverting and inverting configurations. Although this effect varies from one application to another, many of the observations made in conjunction with these circuits may be extended to other circuits as well.

11.2.1 Noninverting Amplifier

In the noninverting circuit of Fig. 11.1, the op amp is assumed to be ideal except for its finite, negative open-loop gain ($-A$) which, in this case, is independent of frequency. Because of the finite value of this gain, the input voltage V_a of the op amp may no longer be assumed to be zero. Instead its value is given by

$$V_a = \frac{V_o}{-A} \qquad (11.1)$$

Since

$$V_a = V_o \frac{R_1}{R_1 + R_2} - V_s$$

then

$$V_a = V_o F - V_s \qquad (11.2)$$

where the feedback factor F is defined by

$$F = \frac{R_1}{R_1 + R_2} \qquad (11.3)$$

The closed-loop gain of the configuration is found by equating (11.1) to (11.2):

$$G = \frac{V_o}{V_s} = \frac{A}{1 + AF} \qquad (11.4)$$

The form of (11.4) is similar to that of the closed-loop gain for the generalized feedback configuration, commonly found in basic control theory. In particular, the denominator of (11.4) may be expressed as

$$1 - (-AF) = 1 - LG \qquad (11.5)$$

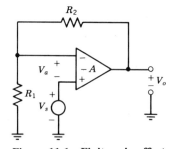

Figure 11.1 Finite gain effects–noninverting amplifier.

where $LG = -AF < 0$ is the loop gain of the configuration of Fig. 11.1. The loop gain, generally speaking, is obtained by finding the gain "around" the feedback loop in the absence of an input. The fact that the loop gain is negative indicates the presence of negative feedback, the benefits of which are numerous.

Assuming "large" magnitudes of loop gain, we may approximate (11.4) by

$$G \cong G' = \frac{A}{AF} \quad (|LG| \gg 1) \tag{11.6}$$

or,

$$\text{Closed-loop gain} \cong \frac{|\text{open-loop gain}|}{|\text{loop gain}|} \quad (|LG| \gg 1) \tag{11.7}$$

From (11.3) and (11.6),

$$G' = \frac{1}{F} = 1 + \frac{R_2}{R_1}$$

G' is the ideal case ($A = \infty$) closed-loop gain for the noninverting configuration of Fig. 11.1. An immediate conclusion is that increasing the magnitude of the loop gain results in a closed-loop gain that is independent of the characteristics of the operational amplifier. Of course, as seen by (11.7), as the magnitude of the loop gain is increased, the magnitude of the closed-loop gain of the configuration will have to be decreased assuming that the open-loop gain of the op amp remains constant. Otherwise, a higher-gain op amp should be sought.

The sensitivity $S(G, A)$ of the closed-loop gain to changes in the magnitude of the gain of the op amp is oftentimes used as a design criterion, both for qualitative and for quantitative purposes. In this case,

$$S(G, A) = \frac{A}{G}\frac{\partial G}{\partial A} = \frac{1}{1 + AF} \tag{11.8a}$$

or

$$S(G, A) = \frac{G}{A} \tag{11.8b}$$

making use of (11.4).

Approximating $S(G, A)$ by

$$S(G, A) \cong \frac{\Delta G/G}{\Delta A/A} \quad (\Delta G, \Delta A: \text{"small"})$$

in (11.8a) and (11.8b), we get

$$\frac{\Delta G}{G} = \frac{1}{1 + AF} \frac{\Delta A}{A} \qquad (\Delta G, \Delta A: \text{"small"}) \qquad (11.9a)$$

or

$$\frac{\Delta G}{G} = \frac{G}{A} \frac{\Delta A}{A} \qquad (\Delta G, \Delta A: \text{"small"}) \qquad (11.9b)$$

These last two expressions provide a quantitative measure of the expected fractional change in the nominal value of the closed-loop gain resulting from a small fractional change in the nominal value of the gain of the op amp. Here also the importance of the magnitude of the loop gain in reducing closed-loop gain sensitivity should be apparent.

To illustrate the use of (11.9b), let us assume nominal values of $A = 10^5$ and $G = 10^4$ in the circuit of Fig. 11.1. If the op amp gain changes by 10%, the expected change in the closed-loop gain will be

$$\frac{\Delta G}{G} = \frac{10^4}{10^5}(0.1) = 0.01$$

or 1%. On the other hand, if $G = 100$, its corresponding change is only 0.01%. This improvement in the gain stability is directly attributable to the increase in the magnitude of the loop gain from 9 to 999.

11.2.2 Inverting Amplifier

The operational amplifier in the inverting configuration of Fig. 11.2 is also assumed to be ideal except for its finite negative open-loop gain ($-A$). In this case, using superposition, we obtain

$$V_a = V_o \frac{R_1}{R_1 + R_2} + V_s \frac{R_2}{R_1 + R_2} \qquad (11.10a)$$

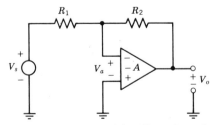

Figure 11.2 Finite gain effects–inverting amplifier.

or
$$V_a = V_oF + V_s(1 - F) \tag{11.10b}$$

where F is defined by (11.3) as
$$F = \frac{R_1}{R_1 + R_2}$$

Using the input-output relationship of the op amp,
$$V_a = \frac{V_o}{-A} \tag{11.11}$$

we can eliminate V_a between (11.10b) and (11.11) to obtain
$$G = \frac{V_o}{V_s} = -(1 - F)\frac{A}{1 + AF} \tag{11.12}$$

For large magnitudes of loop gain, (11.12) is approximated as
$$G \cong G' = -(1 - F)\frac{A}{AF} \quad (|LG| \gg 1) \tag{11.13}$$

or
$$\text{Closed-loop gain} \cong -(1 - F)\frac{|\text{open-loop gain}|}{|\text{loop gain}|} \quad (|LG| \gg 1) \tag{11.14}$$

In this case also, G' is the ideal-case ($A = \infty$) closed-loop gain for the inverting configuration of Fig. 11.2, since
$$G' = -\frac{1 - F}{F} = -\frac{R_2}{R_1} \tag{11.15}$$

making use of (11.3) and (11.13).

Finally, it can be shown that the closed-loop gain sensitivity to changes in the open-loop gain of the op amp for the inverting amplifier of Fig. 11.2 is also given by (11.8a),
$$S(G, A) = \frac{A}{G}\frac{\partial G}{\partial A} = \frac{1}{1 + AF} \tag{11.16a}$$

or, as a result of (11.12),
$$S(G, A) = -\frac{1}{1 - F}\frac{G}{A} \tag{11.16b}$$

Extending the results obtained in Section 11.2, we can conclude that a finite gain op amp may be approximated by an ideal one as far as open-loop gain is concerned provided that the magnitude of the loop gain of the configuration in which it is used is much larger than unity. This conclusion is also applicable in the case where the op amp gain varies with frequency, as will be shown in the following sections.

11.3 FINITE BANDWIDTH

In the previous section, the effects of finite op amp voltage gain on circuit performance were discussed, using the noninverting and inverting configurations to illustrate these effects. Although the treatment, strictly speaking, is appropriate for dc or low-frequency applications, the importance of the large magnitude of loop gain in making the circuit behavior independent of op amp behavior is still true. It is also true for higher-frequency applications where the gain of the op amp may no longer be assumed to be constant or real for that matter. This point will be illustrated using the two basic op amp configurations.

11.3.1 Noninverting Amplifier

The op amp in the noninverting circuit of Fig. 11.3 is assumed to be ideal except for its finite open-loop gain $-A(s)$, where

$$A(s) = \frac{A\omega_1}{s + \omega_1} \tag{11.17}$$

Even though a practical op amp may have a voltage-gain transfer function with several poles and zeros, the response of most general-purpose amplifiers is dominated by a single, real pole described by (11.17). The pole at $s = -\omega_1$ produces a -3 dB point for the magnitude response at $\omega = \omega_1$. For frequencies much smaller than ω_1, the open-loop gain is constant and real, approximated by $-A$. For frequencies much higher

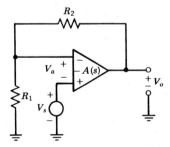

Figure 11.3 Finite bandwidth effects–noninverting amplifier.

than ω_1, the magnitude of the gain is inversely proportional to frequency since

$$|A(j\omega)| \cong \frac{A\omega_1}{\omega} \quad (\omega \gg \omega_1) \tag{11.18}$$

A magnitude response approximated by (11.18) provides unity gain when $\omega = A\omega_1$. Since this frequency is equal to the product of the passband gain and the -3 dB bandwidth of the low-pass response described by (11.17), it is defined as the gain-bandwidth product GB of the op amp,

$$GB = A\omega_1 \tag{11.19}$$

It should be noted that the numerator of (11.17) is also equal to GB.

To obtain the voltage-gain transfer function for the noninverting amplifier of Fig. 11.3, we only need to replace A by $A(s)$, as expressed by (11.17), in (11.4). After simplification, we have

$$G(s) = \frac{V_o}{V_s} = \frac{A\omega_1}{s + (1 + AF)\omega_1} \tag{11.20}$$

where

$$F = \frac{R_1}{R_1 + R_2}$$

This closed-loop gain transfer function is also a single-pole response, similar to the open-loop response of the op amp. Both responses have the same gain-bandwidth product. The corresponding expressions for the closed-loop passband gain G and the -3 dB frequency $\omega_{-3\,\text{dB}}$ are given by

$$G = \frac{A}{1 + AF} \tag{11.21a}$$

and

$$\omega_{-3\,\text{dB}} = (1 + AF)\omega_1 \tag{11.21b}$$

or, if we make use of (11.19) and (11.21a),

$$\omega_{-3\,\text{dB}} = \frac{GB}{G} \tag{11.21c}$$

Figure 11.4 shows the linearized amplitude Bode plots for the open-loop gain of the op amp and the closed-loop gain of the noninverting configuration of Fig. 11.3. The

250 Voltage Gain and Bandwidth

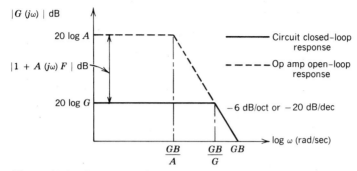

Figure 11.4 Closed-loop frequency response for the noninverting circuit.

negative feedback effectively reduces the passband open-loop gain by dividing it by $(1 + AF)$ and increases its cutoff frequency by multiplying it by the same factor. This factor, needless to say, is directly related to the dc loop gain of the configuration.

Because the open-loop gain of the op amp varies with frequency, the loop gain will also be a function of frequency. Noticing that

$$|G(j\omega)| = \frac{|A(j\omega)|}{|1 + A(j\omega)F|}$$

the corresponding expression in decibels is

$$|G(j\omega)|, \text{dB} = |A(j\omega)|, \text{dB} - |1 + A(j\omega)F|, \text{dB}$$

or

$$|A(j\omega)|, \text{dB} - |G(j\omega)|, \text{dB} = |1 + A(j\omega)F|, \text{dB} \quad (11.22)$$

As long as the magnitude of the loop gain is much larger than unity, the spacing between the open-loop gain and the closed-loop gain responses is approximately equal to the loop gain of the configuration. From Fig. 11.4, this spacing decreases with frequency starting at the cutoff frequency of the op amp and becomes zero at the cutoff frequency of the circuit. This gradual reduction in the magnitude of the loop gain is the cause for the dependence of the closed-loop gain on frequency.

To numerically illustrate the combined effects of the finite values of A and GB on the amplitude response of the noninverting circuit, let us assume $A = 10^5$ and $GB = 2\pi$ Mrad/sec. The resulting values for the closed-loop dc gain G and the corresponding -3 dB frequency ($f_{-3 \text{ dB}}$) for various pairs of values of R_1 and R_2 are tabulated below.

R_1	R_2	F	AF	G	$f_{-3 \text{ dB}}$
∞	0	1	10^5	1.00	1 MHz
1 kΩ	9 kΩ	0.1	10^4	10.0	100 kHz
1 kΩ	99 kΩ	0.01	10^3	99.9	10.01 kHz

11.3.2 Inverting Amplifier

In the inverting circuit of Fig. 11.5, the closed-loop response is

$$G(s) = \frac{V_o}{V_s} = -\frac{(1 - F)A\omega_1}{s + (1 + AF)\omega_1} \qquad (11.23)$$

where

$$F = \frac{R_1}{R_1 + R_2}$$

The above expression may be obtained by inspection, using (11.12) in conjunction with (11.20). The dc closed-loop gain is given by

$$G = -\frac{(1 - F)A}{1 + AF} \qquad (11.24)$$

Assuming a large magnitude of dc loop gain ($AF \gg 1$), we may approximate (11.24) by

$$G \cong -\frac{(1 - F)}{F} = -\frac{R_2}{R_1} \qquad (11.25)$$

The closed-loop cutoff frequency is provided by

$$\omega_{-3\,dB} = (1 + AF)\omega_1$$

or, making use of (11.24), by

$$\omega_{-3\,dB} = \frac{(1 - F)GB}{|G|} \qquad (11.26)$$

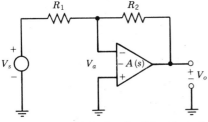

Figure 11.5 Finite bandwidth effects–inverting amplifier.

where $GB = A\omega_1$ is the open-loop gain-bandwidth product of the operational amplifier. This time, the magnitude of the closed-loop gain-bandwidth product [magnitude of numerator of (11.23)], is

$$\text{Gain-bandwidth product} = (1 - F)GB < GB \qquad (11.27)$$

For the large loop-gain case, (11.26) and (11.27) may be approximated by

$$\omega_{-3\text{ dB}} \cong \frac{GB}{1 + |G|} \qquad (AF \gg 1) \qquad (11.28)$$

and

$$\text{Gain-bandwidth product} \cong \frac{|G|GB}{1 + |G|} \qquad (AF \gg 1)$$

respectively, making use of (11.25). When we compare these last two expressions for the inverting circuit, with the corresponding expressions for the noninverting circuits of Fig. 11.3, it becomes apparent that, assuming equal closed-loop gain magnitudes in both cases, the noninverting circuit will provide a larger bandwidth. The difference between the two bandwidths is more pronounced as the magnitude of the closed-loop gain approaches unity, as can be seen from (11.28). To illustrate the point, let us assume $A = 10^5$ and $GB = 2\pi$ Mrad/sec in the circuit of Fig. 11.5. Using (11.3), (11.24), and (11.26), we obtain values for the closed-loop dc gain G and the corresponding -3 dB frequency $f_{-3\text{ dB}}$ for various pairs of values of R_1 and R_2.

R_1	R_2	F	AF	G	$f_{-3\text{ dB}}$
1 kΩ	1 kΩ	1/2	$5 \cdot 10^4$	-1.00	500 kHz
1 kΩ	10 kΩ	1/11	$9.09 \cdot 10^3$	-10.00	90.91 kHz
1 kΩ	100 kΩ	1/101	990	-99.90	9.91 kHz

A comparison of these results with those of the noninverting circuit shows that the closed-loop -3 dB bandwidth in the latter case is twice as large when the magnitude of the closed-loop gain of these circuits is unity. The difference between the corresponding corner frequencies becomes increasingly smaller as these gains become much larger than unity.

Figure 11.6 shows a linearized amplitude Bode plot for the inverting amplifier circuit together with the corresponding (open-loop) characteristic of the operational amplifier. The spacing between these plots is given by

$$|A(j\omega)|, \text{dB} - |G(j\omega)|, \text{dB} = \frac{|1 + A(j\omega)F|}{(1 - F)}, \text{dB}$$

where $F = R_1/(R_1 + R_2)$. For large values of dc loop gain, this expression may be

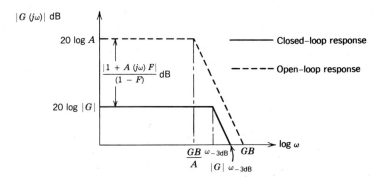

Figure 11.6 Closed-loop frequency response for the inverting amplifier.

approximated by

$$|1 + A(j\omega)|, \text{dB} - |G(j\omega)|, \text{dB}$$
$$\cong \frac{|1 + A(j\omega)F|(1 + |G|)}{|G|}, \text{dB}. \quad (AF \gg 1)$$

Thus, in the inverting case as well, the spacing between the open-loop gain of the op amp and the closed-loop gain of the circuit represents a measure of the magnitude of the loop-gain in decibels. A comparison of Figs. 11.4 and 11.5 verifies the previous conclusion that, for the same magnitude of closed-loop gain, the inverting configuration will have actually a smaller bandwidth, since rolloff in this case starts before the operational amplifier rolloff (intersection of the closed- and open-loop characteristics). This decrease in bandwidth becomes more pronounced as the feedback factor F approaches unity, that is, as more and more negative feedback is applied to the input of the operational amplifier.

11.4 FREQUENCY COMPENSATION SCHEMES

The op amp open-loop frequency response characteristic used in this chapter, as well as in the previous ones, is that of the 741 amplifier, or the 101A amplifier with an external 30 pF compensation capacitor. The frequency response of the 741 is fixed because its 30 pF compensation capacitor is connected internally. However, the frequency response of the 101A is externally controlled and may be modified. There are three compensation schemes for the 101A-type amplifier: one-pole, two-pole, and feedforward.

11.4.1 One-pole Compensation

The most common type of compensation, shown in Fig. 11.7, involves the use of a single capacitor C_c connected, internally or externally, between the base and collector of the Darlington pair in the op amp's second stage. The Miller version of this capacitor

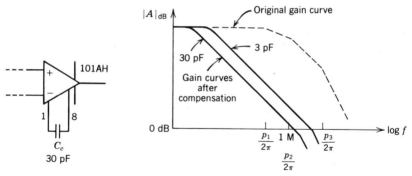

Figure 11.7 One-pole compensation.

and its associated resistance form the equivalent of a low-pass *RC* filter that has a single-pole response described by

$$A(s) = \frac{A_o \omega_c}{s + \omega_c}$$

where $-\omega_c$ is the location of low-frequency pole. The typical value of ω_c is $2\pi(5 \text{ Hz})t$ rad/sec. In the 741 and 101A, there are higher-order poles that cause additional phase shift, but they are above the unity-gain frequency of 1 MHz. The amplifier has a margin of stability commonly specified by phase margin. Phase margin equals 180° plus the phase of the loop gain where the magnitude of the loop gain is one. The 741 and 101A have a phase margin of about 60° and for a circuit consisting of resistors only, it is unconditionally stable. In externally compensated amplifiers, the compensation capacitor can be reduced to 3 pF for extended bandwidth but the tradeoff is less phase margin.

Without a compensation capacitor, the 741 and 101A type amplifiers would have a three-pole response as shown in Fig. 11.7. The three poles, p_1 through p_3, are the poles associated with the three circuits that make up the operational amplifier. The circuit that uses an amplifier with this type of response is potentially unstable. The addition of C_c introduces a dominant pole at a frequency such that, at the unity-gain frequency, the amplifier's phase margin is acceptable.

11.4.2 Two-pole Compensation

The one-pole compensation scheme is a conservative, general-purpose technique used in a variety of applications. There are, however, applications where a higher amplifier gain at intermediate frequencies is desired. If the magnitude of the amplifier's gain is increased in a scheme that has a dominant pole, the unity-gain frequency will increase, the phase margin will decrease, and the phase shift of the higher-order poles will be felt. A way to increase the amplifier's mid-frequency gain without increasing the unity-gain frequency is shown in Fig. 11.8.

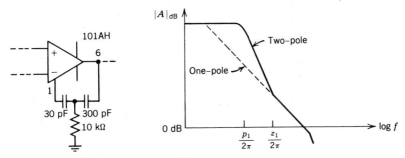

Figure 11.8 Two-pole compensation.

The gain expression for the two-pole scheme consists of two identical poles at $-p_1$ and a zero at $-z_1$. Beyond p_1 rad/sec, the gain decreases at a rate of 40 dB/decade until it reaches z_1 rad/sec where the rolloff returns to -20 dB/decade. The two-pole scheme, compared to the one-pole, increases the mid-frequency gain and slew rate while extending the full-power bandwidth to 12 kHz for the 101A manufacturer-recommended values shown in Fig. 11.8.

11.4.3 Feedforward Compensation

The feedforward compensation scheme shown in Fig. 11.9 involves capacitive coupling of the signal from the inverting input terminal of the operational amplifier to the input of the second stage. This scheme eliminates the frequency response of the first stage in the vicinity of the unity-gain frequency. The input stage uses lateral *pnp* transistors and represents a bandwidth bottleneck for the entire op amp. Since this stage is bypassed at high frequencies, the frequency response of the operational amplifier is essentially dependent on the last two stages. In this scheme, the unity-gain frequency is increased to 10 MHz, the slew rate is increased to 10 V/μsec, and the full power bandwidth is extended to about 250 kHz.

The feedforward approach has its limitations. Since only signals at the inverting input terminal are coupled to the output stage, the feedforward amplifier has a much lower bandwidth for signals applied to its noninverting input and generally cannot be

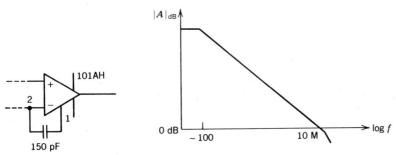

Figure 11.9 Feedforward compensation.

used in noninverting configurations. In addition, the phase shift between the output and the noninverting input approaches $-180°$ and makes feedforward amplifiers relatively intolerant to capacitive loading.

11.5 SUMMARY

The frequency dependence of the finite voltage gain of the nonideal operational amplifier is frequently the most important limitation placed on the application of the operational amplifier. The amplifier's voltage gain limits the maximum obtainable circuit gain, and the relative values of the open- and closed-loop gains establish the degree to which the circuit gain is a function of the discrete components in the feedback network.

A change in the amplifier's gain caused by time, temperature, etc. will be reflected to the circuit level. This change as expressed by the sensitivity function $S(G, A)$ is minimized by maintaining a high loop gain magnitude. Maintaining a high loop gain becomes the theme of the circuit designer because of the many positive circuit attributes derived from this condition.

The typical operational amplifier has a single-pole frequency response characteristic. The noninverting and inverting circuits, which contain resistors only, also display the same type of response. For the same closed-loop gain, the bandwidth of the noninverting circuit is slightly greater than for the inverting case.

The open-loop, frequency response of the 101A-type of amplifier may be externally modified by changing the effective value of the capacitor in the second stage of the op amp. There are two other compensation techniques: the two-pole and feedforward. Both of these schemes increase the gain, slew rate, and full-power bandwidth although at the expense of stability.

PROBLEMS

11.1 $A = 10^5$ in the otherwise ideal op amp of Fig. 11.1. What is the maximum ideal-case, closed-loop voltage gain of the circuit which ensures that the actual gain deviates at most by 1% from the ideal one. What is the corresponding value of the loop gain?

11.2 Repeat Problem 11.1 using the inverting amplifier of Fig. 11.2.

11.3 The magnitude of the ideal-case, closed-loop voltage gain of the noninverting amplifier of Fig. 11.1 is 100. What is the minimum value of A so that the actual voltage gain deviates a maximum of 1% from its ideal-case value?

11.4 Repeat Problem 11.3 using the inverting amplifier of Fig. 11.2.

11.5 The open-loop gain, for the otherwise ideal op amp in the circuit of Fig. 11.10, is finite.
(a) What is the loop gain of the circuit?
(b) What is the condition that must be met to ensure negative feedback in the circuit? Note that this is equivalent to having negative loop gain.
(c) What is the expression for the closed-loop gain of the circuit under negative feedback conditions?

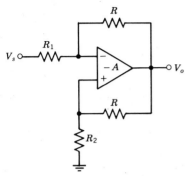

Figure 11.10 Circuit for Problems 11.5, 11.13, and 11.14.

11.6 The open-loop gain of the otherwise ideal op amp in Fig. 11.11 is finite. Derive an expression for its closed-loop gain. What is the ideal-case voltage gain of the circuit? What is the magnitude of the loop gain?

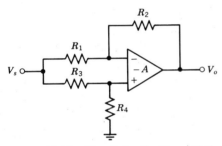

Figure 11.11 Circuit for Problems 11.6, 11.7, and 11.15.

11.7 Derive an expression for the closed-loop gain sensitivity $S(G, A)$ of the circuit of Fig. 11.11 to changes in the magnitude of the open-loop gain of its op amp.

11.8 The open-loop gain of the otherwise ideal op amp in Fig. 11.12 is finite. Derive an expression for its closed-loop gain. What is the ideal-case gain of the circuit? What is the magnitude of the loop gain?

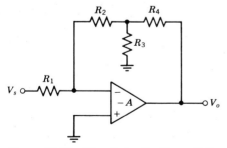

Figure 11.12 Circuit for Problems 11.8 and 11.16.

11.9 The otherwise ideal op amp in the noninverting circuit of Fig. 11.3 has a single-pole response given by (11.17). If its gain-bandwidth product is 1 MHz,

258 Voltage Gain and Bandwidth

$R_1 = 1 \text{ k}\Omega$, and $R_2 = 100 \text{ k}\Omega$, find the -3 dB point of the closed-loop response.

11.10 Repeat Problem 11.9 using the inverting circuit of Fig. 11.5.

11.11 The operational amplifier in the summing circuit of Fig. 11.13 may be assumed ideal except for its open-loop voltage gain that is given by (11.17). Derive an expression for the -3 dB frequency of the closed-loop response due to V_{s1} alone ($V_{s2} = 0$).

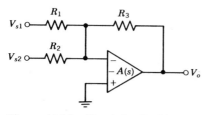

Figure 11.13 Circuit for Problems 11.11 and 11.12.

11.12 In Problem 11.11, $GB = 1 \text{ MHz}, R_1 = 1 \text{ k}\Omega, R_2 = 1 \text{ k}\Omega$, and $R_3 = 100 \text{ k}\Omega$. Find the -3 dB frequency (in Hz) for V_o/V_{s1} assuming $V_{s2} = 0$. What is the ideal-case value of V_o/V_{s1} assuming $V_{s2} = 0$? What is the magnitude of the feedback factor of the circuit?

11.13 The open-loop gain of the otherwise ideal op amp in the circuit of Fig. 11.10 is given by (11.17). Find the condition for stable operation (pole in the left-half of the s-plane).

11.14 In Problem 11.13, derive an expression for the -3 dB frequency of the closed-loop response.

11.15 The open-loop gain of the otherwise ideal op amp in the circuit of Fig. 11.11 is given by (11.17). Derive an expression for the -3 dB frequency of the closed-loop response.

11.16 Repeat Problem 11.15 using the circuit of Fig. 11.12.

Chapter 12

Input and Output Resistance

12.1 PROLOGUE

The differential input resistance and the output resistance of the *ideal* operational amplifier were assumed to be infinite and zero, respectively. These assumptions are no longer valid for the *nonideal* operational amplifier whose characteristics affect circuit performance. Finite input resistance R_i and nonzero output resistance R_o will affect the loop gain, introduce loading effects while cascading op amp stages, and cause additional phase shift that may result in instability. The effects of R_i and R_o are discussed in this chapter using the basic noninverting and inverting configurations as a vehicle.

12.2 FINITE INPUT RESISTANCE

The finite value of the differential input resistance in a nonideal operational amplifier affects circuit performance by reducing the loop gain of the configuration. This, in turn, limits the maximum values of the impedances of the circuit elements.

12.2.1 Noninverting Amplifier

In the noninverting configuration of Fig. 12.1, the finite differential input impedance of the op amp is modeled as an equivalent resistance R_i between its inverting and noninverting input terminals. The otherwise ideal op amp is assumed to have an open-

259

260 Input and Output Resistance

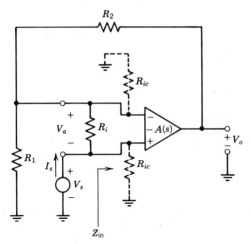

Figure 12.1 Finite input resistance effects–noninverting amplifier.

loop gain $-A(s)$ where

$$A(s) = \frac{A\omega_1}{s + \omega_1} \tag{12.1a}$$

or

$$A(s) = \frac{GB}{s + \frac{GB}{A}} \tag{12.1b}$$

To derive the closed-loop gain of the circuit, we use superposition to obtain the differential input voltage V_a seen by the op amp:

$$V_a = V_o \frac{R_1'}{R_1' + R_2} - V_s \frac{R_i}{R_i + R_{12}}$$

or

$$V_a = V_o F' - V_s B \tag{12.2}$$

where

$$R_1' = R_1 \| R_i \tag{12.2a}$$
$$R_{12} = R_1 \| R_2 \tag{12.2b}$$
$$F' = \frac{R_1'}{R_1' + R_2} < F = \frac{R_1}{R_1 + R_2} \tag{12.2c}$$

Finite Input Resistance

and

$$B = \frac{R_i}{R_i + R_{12}} < 1 \qquad (12.2d)$$

Because of the operational amplifier, we also have

$$V_a = \frac{V_o}{-A(s)} \qquad (12.3)$$

Eliminating V_a between (12.2) and (12.3), we obtain an expression for the closed-loop gain of the noninverting circuit when finite open-loop gain and differential input resistance effects are taken into account:

$$G(s) = \frac{V_o}{V_s} = \frac{B\,A(s)}{1 + A(s)F'} \qquad (12.4a)$$

or, using (12.1a), we get

$$G(s) = \frac{B\,A\omega_1}{s + (1 + AF')\omega_1} \qquad (12.4b)$$

Comparing (12.4b) with (11.20), we can see that the gain-bandwidth product of the circuit of Fig. 12.1 [numerator of (12.4b)] is less than the gain-bandwidth product of the op amp since

$$\text{Gain-bandwidth product} = B\,A\omega_1 = B\,GB < GB \qquad (12.5)$$

The -3 dB frequency $\omega_{-3\text{ dB}}$ of the closed-loop response is given by

$$\omega_{-3\text{ dB}} = (1 + AF')\omega_1 \qquad (12.6)$$

Its magnitude has been reduced, compared to the infinite R_i case, since the effective value of the feedback factor has also been reduced, that is, $F' < F$. Finally, the expression for the closed-loop, low-frequency passband gain G is

$$G = \frac{B\,A}{1 + AF'} \qquad (12.7)$$

In both (12.6) and (12.7), AF' represents the magnitude of the dc or low-frequency loop gain of the noninverting circuit of Fig. 12.1 where the op amp has a finite open-loop gain and a finite input resistance. If the magnitude of the dc loop gain is assumed to be large, the corresponding value of the passband gain G is approximated by

$$G \cong \frac{B}{F'} = \frac{1}{F} = \frac{R_1 + R_2}{R_1} \qquad (12.8)$$

262 Input and Output Resistance

The above expression confirms our previous observations regarding the importance of the magnitude of the loop gain in reducing the effects of op amp nonidealities.

The loop gain of the circuit must be much larger for the finite R_i case, compared to the infinite R_i case, to produce the same results in a given circuit. The effect of R_i in the circuit of Fig. 12.1 may be minimized by making R_1 much smaller than R_i. This makes F' [given by (12.2c)] and B [given by (12.2d)] approximately equal to F and unity, which would have been the case if R_i were assumed infinite in Fig. 12.1.

Closed-loop Input Impedance The finite magnitudes of R_i and $A(s)$ in the noninverting circuit of Fig. 12.1, besides affecting the frequency response of the circuit, cause the source V_s to draw a current I_s. This current was zero in the ideal case and resulted in an infinite equivalent input impedance seen by V_s.

In the nonideal case depicted by Fig. 12.1, Z_{in} defined by

$$Z_{in} = \frac{V_s}{I_s}$$

is given by

$$Z_{in} = \frac{V_s}{-\dfrac{V_a}{R_i}}$$

or

$$Z_{in} = \frac{R_i V_s}{\dfrac{V_o}{A(s)}} = R_i \frac{A(s)}{G(s)} \qquad (12.9)$$

where $G(s)$ is given by (12.4a) or (12.4b).

The significance of (12.9) lies in the fact that negative feedback, in this case, causes the open-loop input resistance of the op amp to be multiplied by the ratio of the open-to-closed-loop gain of the circuit. Specifically,

$$Z_{in} = R_i \frac{s + (1 + AF')\omega_1}{B(s + \omega_1)} \qquad (12.10)$$

after eliminating $A(s)$ and $G(s)$ in (12.9) with the use of (12.1a) and (12.4b). This last expression may be approximated at low frequencies by

$$Z_{in} \cong R_i \frac{1 + AF'}{B} > R_i \qquad (\omega \ll \omega_1) \qquad (12.11a)$$

and at high frequencies by

$$Z_{in} \cong \frac{R_i}{B} < R_i \qquad [\omega \gg \omega_1(1 + AF')] \qquad (12.11b)$$

At low frequencies where the magnitude of the loop gain in Fig. 12.1 is much higher than unity, the resulting input impedance will be real (purely resistive) and much larger than R_i. It is given as

$$Z_{in} \cong R_i \frac{AF'}{B} = R_i AF \qquad [\omega \ll \omega_1, \quad AF' \gg 1] \qquad (12.12)$$

This expression is obtained from (12.11a) using (12.8).

As an example, let the desired low-frequency closed-loop gain be 100 ($F = 0.01$) and the open-loop gain of the op amp be 10^5. The equivalent input resistance in the circuit of Fig. 12.1 is theoretically a thousand times the magnitude of R_i. In practice, this value of Z_{in} cannot be reached because of other considerations, notably the common-mode input resistances of the op amp. The common-mode input resistances R_{ic} of the op amp are modeled as resistances from the amplifier input terminals to ground and are shown in phantom in Fig. 12.1. The common-mode resistances are a function of the output resistance of the current source driving the emitters of the transistors in the differential amplifier stage.

12.2.2 Inverting Amplifier

To analyze the inverting amplifier circuit where the operational amplifier has a finite open-loop gain $-A(s)$ and a differential input resistance R_i, we will use Fig. 12.2. Using superposition, we obtain

$$V_a = V_o \frac{R_1'}{R_1' + R_2} + V_s \frac{R_2'}{R_2' + R_1} \qquad (12.13)$$

where R_1' was defined in (12.2a) while

$$R_2' = R_2 \| R_i \qquad (12.13a)$$

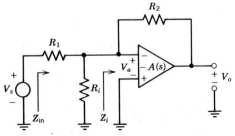

Figure 12.2 Finite input resistance effects–inverting amplifier.

264 Input and Output Resistance

The feedback factor

$$F' = \frac{R'_1}{R'_1 + R_2}$$

is the same as in the noninverting case. Using the above expression and (11.15), we obtain

$$\frac{R'_2}{R'_2 + R_1} = F'\frac{R_2}{R_1} = (1 - F)\frac{F'}{F} \qquad (12.14)$$

Thus, (12.13) may be expressed as

$$V_o = V_o F' + V_s(1 - F)\frac{F'}{F} \qquad (12.15)$$

Because of the operational amplifier,

$$V_a = -\frac{V_o}{A(s)}$$

we can eliminate V_a between (12.14) and (12.15) and obtain

$$G(s) = \frac{V_o}{V_s} = -(1 - F)\frac{F'}{F}\frac{A(s)}{1 + A(s)F'} \qquad (12.16a)$$

or, using (12.1a),

$$G(s) = -(1 - F)\frac{F'}{F}\frac{A\omega_1}{s + (1 + AF')\omega_1} \qquad (12.16b)$$

The closed-loop gain-bandwidth product of the inverting circuit [numerator of (12.16b)] is further reduced since

$$\text{Gain-bandwidth product} = (1 - F)\frac{F'}{F}A\omega_1 < (1 - F)A\omega_1 \qquad (12.17)$$

The -3 dB frequency of the closed-loop response

$$\omega_{-3\text{ dB}} = (1 + AF')\omega_1 \qquad (12.18)$$

is the same as in the noninverting case. Here also its magnitude has been reduced, compared to the infinite R_i case, because of the smaller value of the feedback factor, that is, $F' < F$.

Finally, the closed-loop, low-frequency passband gain may be expressed as

$$G = -(1 - F)\frac{F'}{F}\frac{A}{1 + AF'} \tag{12.19}$$

The -3 dB frequency and the low-frequency closed-loop gain are affected by the magnitude of the dc loop gain of the inverting circuit. If this magnitude is made much larger than unity, the corresponding value of the passband gain G is approximated by

$$G \cong -\frac{1 - F}{F} = -\frac{R_2}{R_1} \qquad (AF' \gg 1) \tag{12.20}$$

This last expression is obviously the ideal op amp closed-loop gain and again brings out the importance of maintaining a high-loop gain.

To minimize the effect of finite R_i in the inverting circuit, F' should always be made as close as possible to F. This amounts to making R_1 much smaller than R_i, the same as in the noninverting circuit.

Closed-loop Input Impedance The input impedance Z_{in} of the inverting amplifier circuit of Fig. 12.2 is equal to R_1 if the operational amplifier is ideal. When the finite values of the op amp open-loop gain and input resistance are taken into account, Z_{in} is equal to R_1 in series with the parallel combination of R_i and Z_i, the equivalent impedance between the inverting terminal and ground.

Z_i accounts for the current through R_2. Thus,

$$Z_i = \frac{V_a}{\dfrac{V_a - V_o}{R_2}}$$

or

$$Z_i = \frac{R_2}{1 - \dfrac{V_o}{V_a}} = \frac{R_2}{1 + A(s)}$$

Replacing $A(s)$ with the expression given by (12.1a), we finally get

$$Z_i = R_2 \frac{s + \omega_1}{s + (1 + A)\omega_1} \tag{12.21}$$

Z_i may be approximated at low frequencies by

$$Z_i \cong \frac{R_2}{1 + A} \ll R_2 \qquad (\omega \ll \omega_1)$$

Increasing the frequency causes Z_i to increase asymptotically toward R_2. Although, theoretically, this should be the case when $\omega \gg (1 + A)\omega_1 \cong GB$, (12.21) is no longer true at these frequencies since the open-loop response of the op amp as given by (12.1a) is not valid for $\omega > GB$. In any event, in the useful frequency range of operation of the inverting amplifier, Z_i will be a rather small number compared to the input resistance R_i of the operational amplifier.

12.3 NONZERO OUTPUT RESISTANCE

The open-loop output resistance R_o of an operational amplifier is relatively small, ranging from a few tens of ohms to several hundred ohms. In the presence of negative feedback, it can usually be neglected since the effective closed-loop output impedance is much smaller than R_o. R_o affects the loop gain and sets the lower limit of the impedance values of the passive components of the op amp configuration.

12.3.1 Noninverting Amplifier

In the noninverting amplifier of Fig. 12.3, the open-loop output resistance R_o of the op amp is modeled as a resistance in series with the amplifier's output. The op amp also has a finite input resistance R_i and an open-loop gain $-A(s)$, as discussed in previous sections.

The loop gain, in this case, is given by $-A(s)F''$, where

$$F'' = \frac{R_1'}{R_1' + R_2 + R_o} \qquad (12.22)$$

and

$$R_1' = R_1 \| R_i$$

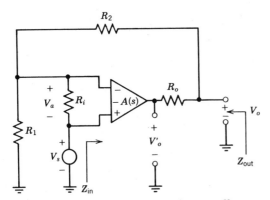

Figure 12.3 Nonzero output resistance effects–noninverting amplifier.

The presence of R_o reduces the magnitude of the effective feedback factor F'' and thus increases the op amp open-loop gain requirements to make the circuit relatively independent of the amplifier characteristics. To minimize the effect of R_o on F'', $(R_1' + R_2)$ or $(R_1 + R_2)$ should be made much larger than R_o. This places a lower limit on the magnitudes of R_1 and R_2, as contrasted to the finite R_i case that placed an upper limit on them.

The expression for the closed-loop gain of the noninverting amplifier of Fig. 12.3 may be found by making use of previously obtained results. They are repeated here for convenience:

$$V_a = V_o F' - V_s B \tag{12.2}$$

where

$$F' = \frac{R_1'}{R_1' + R_2} \tag{12.2c}$$

$$R_1' = R_1 \| R_i \tag{12.2a}$$

$$B = \frac{R_i}{R_i + R_{12}} \tag{12.2d}$$

and

$$R_{12} = R_1 \| R_2 \tag{12.2b}$$

Using superposition, we get

$$V_o = V_o' \frac{R_2}{R_2 + R_o} + (V_s + V_a) \frac{R_o}{R_2 + R_o}$$

or, since $V_o' = -V_a A(s)$,

$$V_o = \frac{1}{R_2 + R_o} [V_a(R_o - A(s)R_2) + V_s R_o] \tag{12.23}$$

If we insert (12.2) into (12.23) and simplify, we get

$$G(s) = \frac{V_o}{V_s} = \frac{R_o(1 - B) + BR_2 A(s)}{R_o(1 - F') + R_2[1 + A(s)F']} \tag{12.24}$$

The expression for the closed-loop gain of the noninverting amplifier in the presence of R_i, R_o, and $-A(s)$, although quite complicated, can be shown to reduce to

$$G(s) \cong \frac{A(s)}{1 + A(s)F} \tag{12.24a}$$

268 Input and Output Resistance

if $R_1 \ll R_i$ and $(R_1 + R_2) \gg R_o$. For large magnitudes of loop gain, the value of the passband (low-frequency) gain G is given by

$$G \cong \frac{B}{F'} = \frac{R_1 + R_2}{R_1} \qquad (AF' \gg 1)$$

Thus, again, large magnitudes of loop gain help eliminate the effects of the operational amplifier parameters such as R_i and R_o.

Closed-loop Input and Output Impedances The expression for the closed-loop input impedance Z_{in} of the noninverting configuration of Fig. 12.3 may be obtained directly from (12.10) after replacing F' by F'':

$$Z_{in} = R_i \cdot \frac{s + (1 + AF'')\omega_1}{B(s + \omega_1)}$$

where F'' and B were defined in (12.22) and (12.2d), respectively. It can be shown that R_o lowers the magnitude of Z_{in} and thus, in effect, increases the current flow from the voltage source V_s.

In the beginning of Section 12.3, it was stated that, generally speaking, negative feedback results in a closed-loop output impedance Z_{out} having a magnitude smaller than its open-loop value. Z_{out} in this case is defined as the impedance of the circuit seen at its output looking back into the circuit with V_s set to zero, as shown in Fig. 12.4. Z_{out} is the parallel combination of $(R_2 + R_1')$ where $R_1' = R_1 \| R_i$, and the effective impedance Z_o. Z_o accounts for the current that would flow through R_o if a voltage source V_s' were applied to the output. Thus,

$$Z_{out} = (R_2 + R_1') \| Z_o \qquad (12.25)$$

where

$$Z_o = \frac{V_s'}{\dfrac{V_s' - V_o'}{R_o}}$$

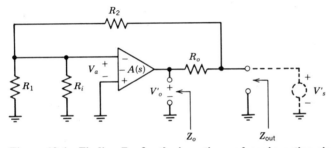

Figure 12.4 Finding Z_{out} for the inverting and noninverting circuits.

or

$$Z_o = \frac{R_o}{1 - \dfrac{V'_o}{V'_s}}$$

Noticing that

$$\frac{V'_o}{V'_s} = -A(s)F'$$

Z_o may be expressed as

$$Z_o = \frac{R_o}{1 + A(s)F'} \quad (12.26a)$$

or

$$Z_o = R_o \frac{s + \omega_1}{s + (1 + AF')\omega_1} \quad (12.26b)$$

by making use of (12.1a).

The last expression shows that R_o at low frequencies is effectively divided by the magnitude of the loop gain of the circuit resulting, typically, in a very low impedance. For this reason, Z_{out} may be approximated by Z_o:

$$Z_{\text{out}} \cong R_o \frac{s + \omega_1}{s + (1 + AF')\omega_1} \quad (12.27)$$

At low frequencies, Z_{out} is real (purely resistive) and approaches $R_o/(1 + AF')$, a value typically less than one ohm. As the frequency is increased, Z_{out} appears to be approaching R_o. The model though used for the operational amplifier in Fig. 12.4, and thus (12.27), is no longer valid at frequencies higher than the gain-bandwidth product $GB = A\omega_1$ of the op amp.

12.3.2 Inverting Amplifier

Figure 12.5 is a model of the inverting amplifier configuration in the presence of finite R_i and $-A(s)$, and nonzero R_o. Following an approach similar to that used in deriving the corresponding expression for the noninverting case in Section 12.3.1, we can show that

$$G(s) = \frac{V_o}{V_s} = -(1 - F)\frac{F'}{F} \frac{R_2 A(s) - R_o}{R_o(1 - F') + R_2[1 + A(s)F']} \quad (12.28)$$

270 Input and Output Resistance

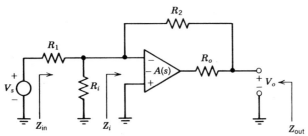

Figure 12.5 Nonzero output resistance effects–inverting amplifier.

where

$$F' = \frac{R_1'}{R_1' + R_2}$$

and

$$R_1' = R_1 \| R_i$$

The expression (12.28), complicated as it may look, reduces to

$$G(s) \cong -(1 - F)\frac{A(s)}{1 + A(s)F}$$

if $R_1 \ll R_i$ and $R_2 \gg R_o$. The conditions set on R_1 and R_2 represent limits on the maximum and minimum values needed to minimize the effects of R_i and R_o in this case.

At low frequencies and for large magnitudes of loop gain, the value of the passband gain G is given by

$$G \cong -\frac{1 - F}{F} = -\frac{R_2}{R_1} \quad (AF' \gg 1)$$

Closed-loop Input and Output Impedances The input impedance Z_{in} for the inverting configuration of Fig. 12.5 is equal to R_1 in series with the parallel combination of R_i and Z_i. Z_i may be found directly from (12.21) after replacing R_2 by $(R_2 + R_o)$. Here also Z_{in} is approximately equal to R_1 for all practical purposes.

Shorting V_s in Fig. 12.5 to find its closed-loop output impedance Z_{out} results in Fig. 12.4. Thus, the resulting expressions for Z_{out} is identical to the ones for the noninverting case given by (12.25) and (12.26b) or, approximately, by (12.27).

12.4 LOAD CONSIDERATIONS

In an operational amplifier with zero or negligible output resistance R_o, a load impedance Z_L connected across the output of the noninverting and inverting circuits will not have any effect on the output voltage V_o or the closed-loop input impedance Z_{in}. However,

Load Considerations 271

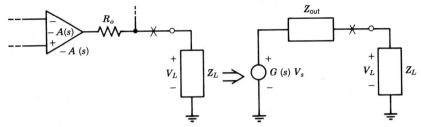

Figure 12.6 Finding the effect of Z_L on V_o and $G(s)$.

the minimum allowable magnitude of Z_L will be determined by the ability of the op amp to supply the current demanded by the load.

In the presence of nonzero R_o, the output voltage of the circuit may vary somewhat with Z_L depending on the relative magnitudes of the closed-loop output impedance Z_{out} and Z_L. The load impedance will also affect the loop gain of the circuit, the closed-loop input impedance Z_{in} and, more important, the stability of the circuit.

The no-load ($Z_L = \infty$) expressions for the closed-loop gain $G(s)$ and output impedance Z_{out} derived for the nonzero, open-loop output resistance case can be used to take into account the effect of finite Z_L on the closed-loop gain. Noticing that the no-load expressions for $G(s)$ and Z_{out} constitute the Thevenin equivalent of the noninverting or inverting amplifier circuit as shown in Fig. 12.6, we have

$$V_L = G(s)V_s \frac{Z_L}{Z_L + Z_{out}} \quad (12.29\text{a})$$

or

$$G_L(s) = \frac{V_L}{V_s} = G(s) \frac{Z_L}{Z_L + Z_{out}} \quad (12.29\text{b})$$

where $G_L(s)$ is the "loaded" gain of the circuit, while $G(s)$ and Z_{out} are given by the appropriate expressions in Section 12.3.1 or 12.3.2.

The expressions for the loop gain and the closed-loop input impedance Z_{in} of the noninverting and inverting circuits of Figs. 12.3 and 12.5 under finite load conditions may be easily derived by including the load Z_L in the Thevenin equivalent circuit of the op amp, as shown in Fig. 12.7. Noticing that the expressions for the effective

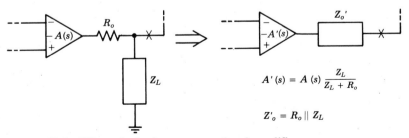

Figure 12.7 Effect of a load on an operational amplifier.

Input and Output Resistance

open-loop gain $-A(s)$ and output impedance Z'_o are now given by

$$A'(s) = A(s) \frac{Z_L}{Z_L + R_o} \qquad (12.30a)$$

and

$$Z'_o = R_o \parallel Z_L \qquad (12.30b)$$

the loop gain with a finite Z_L is $-A(s)F''$, where F'' is defined by (12.22) with R_o replaced by Z'_o. The load impedance will affect both the magnitude and phase of the loop gain since Z_L may include reactive (L and/or C) elements.

Finally, the closed-loop input impedance Z_{in} in the presence of Z_L may be obtained from the material in Sections 12.3.1 and 12.3.2 after replacing $A(s)$ and R_o with (12.30a) and (12.30b), respectively.

12.5 STABILITY CONSIDERATIONS

A circuit will oscillate if its actual transfer function produces poles in the right-half of the s plane. Typically, this is due to effectively positive feedback that, in the presence of sufficient loop gain, can sustain unwanted oscillations.

The poles of feedback systems are the roots of the characteristic equation

$$1 - \text{loop gain} = 0 \qquad (12.31)$$

If the characteristic equation has any roots with positive real parts, then the circuit is unstable and thus subject to oscillations.

We will use the noninverting amplifier of Fig. 12.8 to illustrate these ideas. For simplification purposes, we have assumed $R_i \gg R_1$ and $R_o \ll (R_1 + R_2)$. Lumping the effect of the capacitive load in $A(s)$ and R_o, we find that the loop gain is given by

$$\text{Loop gain} = -\frac{A(s)F''}{1 + sR_oC_L} \qquad (12.32a)$$

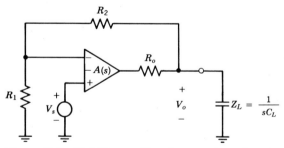

Figure 12.8 Stability considerations—noninverting circuit.

where

$$F'' \cong F = \frac{R_1}{R_1 + R_2} \qquad (12.32b)$$

In addition, if we assume an operational amplifier with a two-pole frequency response given by

$$-A(s) = -\frac{A\omega_1\omega_2}{(s + \omega_1)(s + \omega_2)} \qquad (A, \omega_1, \omega_2 > 0) \qquad (12.33)$$

the characteristic equation for the circuit of Fig. 12.8 is approximated by

$$1 + \frac{AF\omega_1\omega_2\omega_3}{(s + \omega_1)(s + \omega_2)(s + \omega_3)} = 0 \qquad (12.34a)$$

or

$$(s + \omega_1)(s + \omega_2)(s + \omega_3) + AF\omega_1\omega_2\omega_3 = 0 \qquad (12.34b)$$

where

$$\omega_3 = \frac{1}{R_o C_L}$$

making use of (12.31) through (12.33).

As seen from (12.34a), capacitive loading in conjunction with nonzero R_o contributes an additional pole to the open-loop response of the operational amplifier.

If we expand (12.34b) and simplify, the characteristic equation becomes a cubic equation of the form

$$a_3 s^3 + a_2 s^2 + a_1 s + a_0 = 0$$

where

$$a_3 = 1$$
$$a_2 = \omega_1 + \omega_2 + \omega_3 > 0$$
$$a_1 = \omega_1\omega_2 + \omega_1\omega_3 + \omega_2\omega_3 > 0$$
$$a_0 = \omega_1\omega_2\omega_3(1 + AF) > 0$$

Using Routh's criterion, we find that the characteristic equation will have roots with positive real parts if

$$a_0 > \frac{a_1 a_2}{a_3}$$

or, using the expressions for a_0 and a_3,

$$AF > \frac{a_1 a_2}{\omega_1 \omega_2 \omega_3} - 1 \qquad (12.35)$$

Noticing that AF is the approximate value of the dc loop gain of the circuit of Fig. 12.8, a rather startling observation is now in order based on (12.35). Even though, generally speaking, the larger the loop gain of a given, negative feedback circuit, the closer it approximates an ideal operation, it now becomes apparent that too much loop gain could produce instability. The worst-case condition for (12.35) occurs when $F = 1$, a case corresponding to the unity-gain noninverting circuit.

12.6 SUMMARY

The finite and nonzero values of the open-loop input resistance and output resistance of the operational amplifier effectively reduce the loop gain of the negative feedback configuration. The reduction in loop gain, in turn, affects the deviation between the ideal and actual values of the low-frequency closed-loop gain, the frequency response, and the closed-loop input and output impedances of the circuit. In addition, the nonzero output resistance of the op amp in conjunction with capacitive loading may cause sufficient phase shift to make the circuit unstable.

PROBLEMS

12.1 An operational amplifier with $R_i = 2$ MΩ and $A = 100{,}000$ is used in the noninverting circuit of Fig. 12.1 to provide a low-frequency gain of 10. Assuming $R_1 \leq R_i/1000$, find
  (a) The low-frequency loop gain of the circuit.
  (b) The upper limits of R_1 and R_2 to minimize the effect of R_i on circuit performance.
  (c) The low-frequency value of Z_{in}.

12.2 An operational amplifier with $R_i = 2$ MΩ and $A = 100{,}000$ is used in the inverting circuit of Fig. 12.2 to provide a low-frequency closed-loop gain of -10. Assuming $R_1 \leq R_i/1000$, find
  (a) The low-frequency loop gain of the circuit.
  (b) The upper limits of R_1 and R_2 to minimize the effect of R_i on circuit performance.
  (c) The low-frequency value of Z_i under the conditions specified in part b.

12.3 The otherwise ideal op amp used in the circuit of Fig. 12.9 has an open-loop gain $-A(s)$ and a finite differential input resistance R_i. Find
  (a) The expression for the loop gain of the circuit.
  (b) The expression for the closed-loop transfer function $T(s) = V_o/I_s$.
  (c) The condition that R must satisfy to minimize the effect of R_1 on circuit performance.

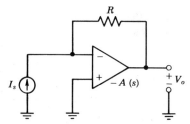

Figure 12.9 Circuit for Problems 12.3 to 12.6.

12.4 Repeat Problem 12.3 assuming, in addition, nonzero op amp output resistance R_o. For part c, find the conditions that R must meet to minimize the effects of R_i and R_o on circuit performance.

12.5 The operational amplifier used in the circuit of Fig. 12.9 has $R_i = 1 \text{ M}\Omega$, $R_o = 100 \text{ }\Omega$, and a single-pole response described by (12.1a) with $A = 100{,}000$ and $\omega_1 = 10$ rad/sec. If $R = 100 \text{ k}\Omega$, find
 (a) The value of the low-frequency impedance seen by the current source I_s.
 (b) The -3 dB frequency of the frequency response for V_o/I_s.

12.6 Find the low-frequency value of the closed-loop output impedance for the circuit of Problem 12.5.

12.7 In the noninverting circuit of Fig. 12.3, assume $R_1 \leq R_i/k$ and $(R_1 + R_2) \geq kR_o$ to minimize the effects of R_i and R_o on circuit performance. The number k is an arbitrarily large positive number, and an ideal-case, closed-loop gain G' is desired. What is the expression for the *maximum* value of k in terms of R_i, R_o, and G'?

12.8 In Problem 12.7, find the values of R_1 and R_2 that will minimize the effects of R_i and R_o on circuit performance if $R_i = 1 \text{ M}\Omega$, $R_o = 100 \text{ }\Omega$, and $G' = 100$.

12.9 In the inverting circuit of Fig. 12.5, let $R_1 \leq R_i/k$ and $R_2 \geq kR_o$ to minimize the effects of R_i and R_o on circuit performance. The number k is an arbitrarily large positive number, and an ideal-case, closed-loop low-frequency gain G' is desired. What is the expression for the maximum value of k in terms of R_i, R_o, and G'?

12.10 In Problem 12.9, find the values of R_1 and R_2 that will minimize the effects of R_i and R_o on circuit performance if $R_i = 1 \text{ M}\Omega$, $R_o = 100 \text{ }\Omega$, and $G' = -100$.

12.11 Derive expression (12.28) for the inverting circuit of Fig. 12.5.

12.12 A three-pole operational amplifier with an open-loop gain $-A(s)$ given by

$$A(s) = \frac{A\omega_1\omega_2\omega_3}{(s + \omega_1)(s + \omega_2)(s + \omega_3)} \qquad (A, \omega_1, \omega_2, \omega_3 > 0)$$

is used in the inverting circuit of Fig. 12.5. Neglecting the effects of R_i and R_o, find the condition that the dc loop gain must meet in order to avoid oscillations.

12.13 In Problem 12.12, $A = 10^5$, $\omega_2 = 10^6$ rad/sec, and $\omega_3 = 2 \cdot 10^6$ rad/sec. Find the condition for ω_1 to ensure stability for a closed-loop gain of -1.

12.14 The operational amplifier in the noninverting circuit of Fig. 12.8 has a two-pole open-loop gain described by (12.33), where $A = 5 \cdot 10^4$ and $\omega_1 = \omega_2 = 10^5$ rad/sec. Neglecting the effect of R_i, find the maximum allowable value of C_L before oscillations occur. Let $R_o = 20\,\Omega$, $R_1 = \infty$, $R_2 = 0$, and $C_L \leq 100$ pF.

Chapter 13

Secondary Parameters

13.1 PROLOGUE

The real operational amplifier deviates from its ideal behavior because of the finite and nonzero values associated with the primary characteristics of voltage gain, input and output impedance, and bandwidth. These values introduce error terms in the input-output relationship for the circuit in which the amplifier is used, and they limit the performance of the nonideal amplifier. *Secondary parameters* also introduce *error terms* in the circuit, frequently greater in magnitude than those caused by the primary characteristics. These parameters are specified for a given set of test conditions (hence the name parameters instead of characteristics) and can be found in the manufacturer's list of specifications or data sheets. They are a result of the manufacturing process and the components and devices used in the design of the amplifier circuit. A composite data sheet for the 101A integrated circuit operational amplifier, shown in Fig. 13.1, contains most of the primary characteristics and the key secondary parameters. Complete data sheets for the 741 and 101A are provided in the appendix.

The secondary parameters can be divided into two categories, dc and ac. The dc parameters can be further divided between those associated with the input terminals, the output terminal, and the dc supply pins. The key input dc parameters are input offset voltage (V_{OS}) and input bias current (I_B). The key ac parameter is slew rate (S_R).

Figure 13.1 Data Sheet for 101A.

Single pole compensation

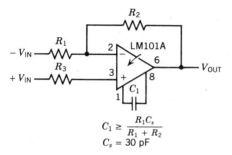

Two pole compensation

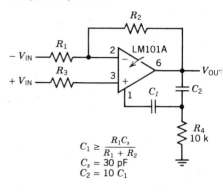

Feedforward compensation

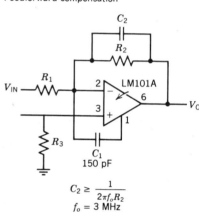

Absolute Maximum Ratings LM101A/LM201A

		Section
Supply voltage	±22 V	13.4.1
Power dissipation	500 mW	13.4.4
Differential input voltage	±30 V	13.2.3
Input voltage	±15 V	13.2.2
Output short-circuit duration	Indefinite	
Operating temperature range LM101A	−55 to 125°C	
LM201A	−25 to 85°C	
Storage temperature range	−65 to 150°C	
Lead temperature (soldering 10 sec)	300°C	

Electrical Characteristics

Parameter	Conditions	Min	Typ	Max	Units	Section
Input offset voltage	$T_A = 25°C$, $R_s \leq 50$ kΩ		0.7	2.0	mV	13.2.1
Input offset current	$T_A = 25°C$		1.5	10	nA	13.2.5
Input bias current	$T_A = 25°C$		30	75	nA	13.2.4
Input resistance	$T_A = 25°C$	1.5	4		MΩ	12.2
Supply current	$T_A = 25°C$, $V_s = \pm 20$ V		1.8	3.0	mA	13.4.2
Large-signal voltage gain	$T_A = 25°C$, $V_s = \pm 15$ V, $V_{out} = \pm 10$ V, $R_L \geq 2$ kΩ	50	160		V/mV	11.2
Input offset voltage	$R_s \leq 50$ kΩ			3.0	mV	
Average temperature coefficient of input offset voltage			3.0	15	μV/°C	13.2.1
Input offset current				20	nA	
Average temperature coefficient of input offset current	$25°C \leq T_A \leq 125°C$ $-55°C \leq T_A \leq 25°C$		0.01 0.02	0.1 0.2	nA/°C nA/°C	13.2.5
Input bias current				100	nA	
Supply current	$T_A = +125°C$, $V_s = \pm 20$ V		1.2	2.5	mA	
Large-signal voltage gain	$V_s = \pm 15$ V, $V_{out} = \pm 10$ V, $R_L \geq 2$ kΩ	25			V/mV	
Output voltage swing	$V_s = \pm 15$ V, $R_L = 10$ kΩ $R_L = 2$ kΩ	±12 ±10	±14 ±13		V V	13.3.1
Input voltage range	$V_s = \pm 20$ V	±15			V	
Common mode rejection ratio	$R_s \leq 50$ kΩ	80	96		dB	13.2.5
Supply voltage rejection ratio	$R_s \leq 50$ kΩ	80	96		dB	13.4.3

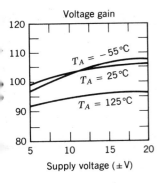
(Voltage gain)

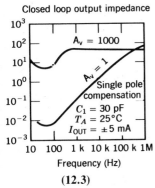
(12.3) Closed loop output impedance

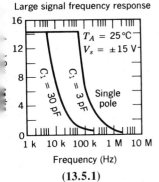
(13.5.1) Large signal frequency response

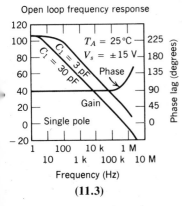
(11.3) Open loop frequency response

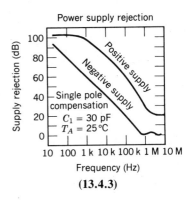
(13.4.3) Power supply rejection

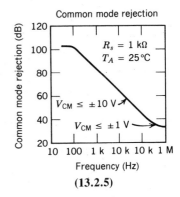
(13.2.5) Common mode rejection

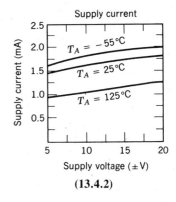
(13.4.2) Supply current

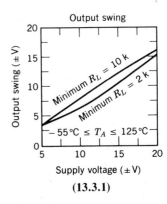
(13.3.1) Output swing

13.2 INPUT (dc) PARAMETERS

13.2.1 Input Offset Voltage (V_{OS})

The (initial) input offset voltage V_{OS} is defined as the difference in potential between the amplifier input terminals when the output is at zero volts. Physically, this potential difference represents a mismatch in the input voltages of the transistors in the differential amplifier of the input stage. The input offset voltage can be further related to the mismatch in the transistor parameters and the mismatch in the loading of the two halves of the differential amplifier circuit.

The effect of V_{OS} can be graphically illustrated utilizing the transfer characteristic of the amplifier. In the ideal amplifier case, Fig. 13.2a, the straight line representing the linear relationship between the input and output goes through zero, that is, the output voltage is zero when the differential input voltage is zero. In a real amplifier, where V_{OS} is nonzero, the straight line is offset from the origin (Fig. 13.2b). This means that some, typically small, input voltage is required to cause the output to go to zero. The typical range of values for V_{OS} is from 0.5 to 15 mV, and represents a parameter that manufacturers have sought to improve in new IC amplifier designs.

The effect of V_{OS} can be modeled as a dc voltage source in series with either of the amplifier inputs and can be of either polarity. In a circuit, V_{OS} is reflected to the output and sums, as an error, with the reflected input signal. In the inverting amplifier circuit in Fig. 13.3, V_o will be a function of the signal source V_s and the error voltage V_{OS} if we assume ideal, otherwise, operation. The output voltage is found using superposition,

$$V_o = -\frac{R_2}{R_1} V_s + \left(1 + \frac{R_2}{R_1}\right) V_{OS} \qquad (13.1)$$

The offset error voltage is reflected to the output with a magnitude of $[1 + (R_2/R_1)]V_{OS}$.

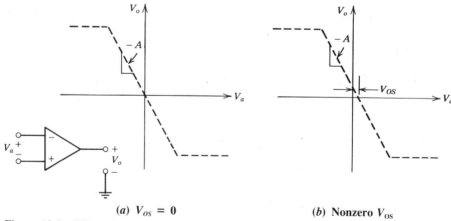

(a) $V_{OS} = 0$ (b) Nonzero V_{OS}

Figure 13.2 Effect of V_{OS} on the transfer characteristic.

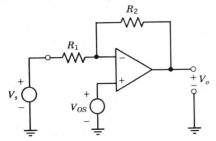

Figure 13.3 Inverting amplifier circuit with nonzero V_{OS}.

For example, if the closed-loop gain in Fig. 13.3 is -10 and V_{OS} is 1 mV, an error of 11 mV will be summed with the amplified input signal at the output. To minimize the percentage of error caused by this term, the magnitude of the signal source voltage should be kept significantly larger than V_{OS}, or a device with a lower specified value of V_{OS} should be used. Oftentimes, V_{OS} can be adjusted to zero using an external network. Manufacturer's data sheets specify the proper offset null circuit for each type of amplifier. The offset null circuits for the 741 and 101A op amps are shown in Fig. 13.4. The offset voltage is set to zero, for a given temperature, by setting the input source to zero and adjusting the potentiometer until the output voltage is zero. The effort involved in this procedure is significant for high-volume, low-cost products and is usually avoided, or V_{OS} is accounted for in the design. The offset null circuitry unbalances the emitter (source) currents in the transistors of the differential amplifier stage, thus forcing the matching of the base-to-emitter (gate-to-source) voltages. Manufacturing techniques are currently available that guarantee offset voltages to less than 0.5 mV.

The input offset voltage is temperature dependent. The temperature dependence of the offset voltage originates from the temperature dependence of the base-to-emitter voltage of bipolar transistors. The resolution in a circuit determines how small a signal that system can process and is heavily dependent on the temperature coefficient of V_{OS}, $\Delta V_{OS}/\Delta T$. This parameter also plays a key role in establishing the performance of the circuit over a broad temperature range. A typical range of values for this parameter is from 5 μV/°C (108A) to 15 μV/°C (101A and 741A).

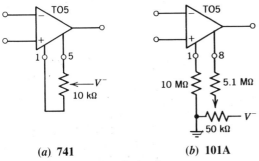

(a) 741 (b) 101A

Figure 13.4 Offset null circuits.

If an external offset null circuit is used to zero out the initial offset voltage, the drift parameter or temperature sensitivity of V_{OS} still applies and, in fact, can get worse because of the compensating circuit. A bipolar amplifier with a 1 mV offset voltage and a 3 µV/°C drift will acquire an additional 3 µV/°C drift component when its offset is nulled. This additional drift term may have the same or opposite direction as the original amplifier drift, so that the net offset drift can either increase or decrease. Since ΔV_{OS} is directly related to the ac or small-signal emitter resistance r_e of the transistors involved, which in turn is highly temperature dependent, the nulled-out offset voltage is reduced to zero *only* at one temperature.

13.2.2 Input Voltage (V_i)

This parameter does not introduce an error term in the circuit but identifies the maximum voltage (with respect to ground) that can be applied to any one of the input terminals. For the 101A, the maximum input voltage is ±15 V.

13.2.3 Differential Input Voltage (V_{id})

In normal linear circuit operation, the differential input voltage of the amplifier will be of the order of a few millivolts. However, under fault conditions or in nonlinear applications, a large differential voltage may appear between the amplifier input terminals. The result of such a condition is the breakdown of the emitter-base (or gate-source) junctions of the transistors in the differential amplifier stage. The differential input voltage restricts the maximum applied differential voltage to typically ±5 V (709) to ±30 V (101A and 741).

13.2.4 Input Bias Current (I_B)

The input bias current is defined as the magnitude of the average of the dc biasing currents at the inputs of an operational amplifier with the output at zero volts. Mathematically, I_B is defined as follows:

$$I_B = \frac{|I_B^+ + I_B^-|}{2} \qquad (13.2)$$

where I_B^+ and I_B^- represent the currents entering or leaving the noninverting (+) and inverting (−) input terminals of the op amp. Physically, I_B represents the average of the base (or gate) currents of the differential amplifier's transistors. For monolithic op amps with *npn* transistors in the differential amplifier stage, the direction of the current will be into the amplifier. Newer IC designs use superbeta transistors and Darlington or complementary pairs to reduce the value of the base or bias current. Amplifiers that employ JFET transistors in the differential amplifier stage have bias current ratings one or two orders of magnitude smaller than their bipolar counterparts. The typical range of values for I_B for amplifier's with bipolar input transistors is from 100 pA to

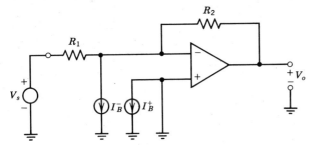

Figure 13.5 Inverting amplifier circuit with nonzero I_B^- and I_B^+.

1.5 μA. I_B is theoretically related to the differential input resistance R_{in} by

$$R_{in} = \frac{V_T}{I_B}$$

where $V_T = kT/q = 26$ mV at ambient temperature ($T = 300°K$).

The input bias currents are modeled by two dc current sources from the input terminals of the op amp to ground. In an actual circuit, I_B is reflected to the output and sums, as an error, with the reflected input signal. If we assume ideal conditions, except for I_B, in the inverting amplifier circuit in Fig. 13.5, V_o is a function of the signal source V_s and the bias current and can be found using superposition,

$$V_o = -\frac{R_2}{R_1} V_s + (I_B^-)R_2 \qquad (13.3)$$

13.2.5 Input Offset Current (I_{OS})

The input offset current I_{OS} is defined as the magnitude of the difference between the amplifier input currents with the output at zero volts,

$$I_{OS} = |I_B^+ - I_B^-| \qquad (13.4)$$

I_{OS} is typically much smaller than the bias current ($\approx 0.1 I_B$) and is in the 10 pA to 0.5 μA range. I_{OS} is also temperature dependent with a typical value of temperature coefficient of 0.1 nA/°C (101A).

In amplifier circuits, a resistor is added in series with an input terminal of the op amp to shift the dependence of the error at the output from I_B to the smaller I_{OS}. The following example illustrates the use of this technique in the inverting amplifier circuit.

Example

The typical values of V_{OS}, I_B, and I_{OS} for the 741 are 2 mV, 70 nA, and 3 nA, respectively. Design an inverting amplifier with a gain of -10 so that the minimum error at the output from the above parameters is less than or equal to 25 mV. Assume that all other parameters are ideal.

Solution

The circuit in Fig. 13.6a is an inverting amplifier with resistor R_3 added to reduce the effect of the bias on the output voltage. Using superposition, we obtain

$$V_o = -\frac{R_2}{R_1} V_s + V_{os}\left(1 + \frac{R_2}{R_1}\right) + (I_B^-)R_2 - (I_B^+)R_3\left(1 + \frac{R_2}{R_1}\right)$$

If R_3 is made equal to the parallel combination of R_1 and R_2,

$$R_3 = (R_1 \| R_2)$$

then,

$$V_o = -\frac{R_2}{R_1} V_s + V_{os}\left(1 + \frac{R_2}{R_1}\right) + (I_B^- - I_B^+)R_2$$

or

$$V_o = -\frac{R_2}{R_1} V_s + V_{os}\left(1 + \frac{R_2}{R_1}\right) \pm I_{os}R_2$$

If V_{os} equals 2 mV and since $R_2/R_1 = 10$, then the error at the output due to V_{os} alone is

$$V_o = 11(2 \text{ mV}) = 22 \text{ mV}$$

The magnitude of the error due to I_{os} must be

$$I_{os}R_2 \leq (25 \text{ mV} - 22 \text{ mV}) = 3 \text{ mV}$$

or

$$R_2 \leq \frac{3 \text{ mV}}{3 \text{ nA}} = 1 \text{ M}\Omega$$

We select

$$R_2 = 100 \text{ k}\Omega$$

which, to satisfy the gain conditions, makes

$$R_1 = 10 \text{ k}\Omega$$

Resistor R_3 is the parallel combination of the two resistors,

$$R_3 = 9.1 \text{ k}\Omega$$

It is a good engineering practice to design a circuit to guarantee its performance against worst-case conditions. In terms of this example, the guaranteed *maximum* values of V_{OS}, I_B, and I_{OS} should be used.

Example

Find the relationship between V_o, V_s, and the parameters V_{OS} and I_{OS} for the *noninverting* amplifier circuit in Fig. 13.6b. Let $R_3 = R_1 \parallel R_2$.

Solution

Resistor R_3 performs the same function in the noninverting circuit as in the inverting circuit; it reduces the error at the output from an I_B error to an I_{OS} error. The relationship between V_o, V_s, V_{OS}, and I_{OS} can be found using superposition or using KVL, KCL, and the amplifier's input constraints. Using the latter approach, we find the voltage at the amplifier's noninverting input V_{ni} by applying KVL to the loop formed by V_s, R_3,

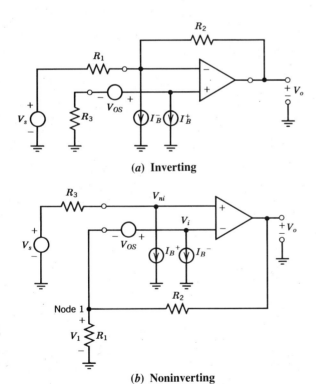

Figure 13.6 Practical amplifier circuits.

and I_B^+. This results in

$$V_{ni} = V_s - (I_B^+)R_3$$

From the input voltage constraint of the amplifier,

$$V_i = V_{ni} = V_s - (I_B^+)R_3$$

Again applying KVL to the loop formed by R_1, V_{OS}, and I_B^-, we get

$$V_1 = -V_{OS} + V_i$$
$$= -V_{OS} + V_s - (I_B^+)R_3$$

Applying KCL to node 1 produces

$$\frac{V_o - V_1}{R_2} = \frac{V_1}{R_1} + I_B^-$$

If we eliminate V_1 from the last expression, substitute $R_3 = (R_1 R_2)/(R_1 + R_2)$ and simplify, we arrive at

$$V_o = \left(1 + \frac{R_2}{R_1}\right) V_s - \left(1 + \frac{R_2}{R_1}\right) V_{OS} \pm I_{OS} R_2 \qquad ■$$

The range of values for both offset voltage and offset current can be traced back to the amplifier components and their parameters. However, the specific value of an op amp's secondary parameter is related to the manufacturing processes and is often statistically determined.

13.2.6 Common-Mode Rejection Ratio (CMRR)

In many applications, the voltage at both inputs of the op amp is raised above ground potential. For example, the amplifier's inputs will be near the input signal voltage when the operational amplifier is used in a unity-gain, noninverting voltage amplifier circuit.

The average of the two input voltages v_1 and v_2, shown in Fig. 13.7a, is called the common-mode input voltage v_c and is defined as

$$v_c = \frac{v_1 + v_2}{2} \qquad (13.6a)$$

On the other hand, the differential-mode input voltage v_d is defined as

$$v_d = v_1 - v_2 \qquad (13.6b)$$

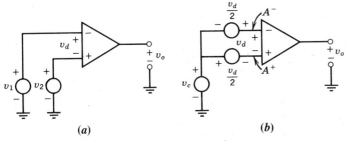

Figure 13.7 Modeling a common-mode voltage.

The input voltages v_1 and v_2, in terms of v_c and v_d, are

$$v_1 = v_c + \frac{v_d}{2}$$

and

$$v_2 = v_c - \frac{v_d}{2}$$

and are shown in Fig. 13.7b.

In an actual op amp, the output is not zero volts when the two input voltages are identical, even if dc errors are eliminated. The nonzero output is the result of the difference in the voltage gains from the individual input terminals to the output. The individual voltage gains are designated A^+ and A^- where

A^+ = the gain from the noninverting input terminal to the output with the inverting terminal grounded ($A^+ > 0$)

A^- = the gain from the inverting input terminal to the output with the noninverting terminal grounded ($A^- < 0$)

A common-mode error, the effect of which can be simulated by an equivalent voltage source in series with either of the input terminals of the op amp, is present when $v_1 \cong v_2 \neq 0$. Using superposition in the equivalent circuit in Fig. 13.7b, we get

$$v_o = v_c(A^- + A^+) + \frac{v_d}{2}(A^- - A^+)$$

or

$$v_o = v_c A_c + v_d A_d \tag{13.7}$$

where A_c is the common-mode gain defined as

$$A_c = A^- + A^+ \tag{13.8}$$

288 Secondary Parameters

and A_d is the differential-mode voltage gain defined as

$$A_d = \frac{(A^- - A^+)}{2} \quad (13.9)$$

A_d is the open-loop voltage gain of the op amp discussed in Chapter 11.
Common-mode rejection ratio CMRR is defined as

$$CMRR = \frac{A_d}{A_c} \quad (13.10)$$

or, if expressed in decibels, as

$$CMRR, \ dB = 20 \log CMRR$$

After eliminating A_c between (13.7) and (13.10) and rearranging terms, we obtain

$$v_o = A_d \left(v_d + \frac{v_c}{CMRR} \right) \quad (13.11a)$$

If the two input voltages v_1 and v_2 are nearly equal, then

$$v_o \cong A_d \left(v_d + \frac{v_1}{CMRR} \right) \quad (v_1 \cong v_2) \quad (13.11b)$$

The term $(v_c/CMMRR)$ is the error, reflected back to the input, due to the common-mode voltage and can be represented as an equivalent input voltage V'_{OS} as shown in Fig. 13.8. CMRR is in the 90 dB range (101A, 741, and 108A) and decreases with frequency.

The equivalent circuit of the operational amplifier, which includes the key secondary parameters, is shown in Fig. 13.9. V'_{OS} represents a "change" in input offset voltage because of parameters like CMRR and PSRR.

Example

An operational amplifier in a unity-gain, noninverting circuit has an offset voltage V_{OS} of 0.5 mV and a CMRR of 80 dB. The op amp is otherwise ideal. If the input voltage is 10 V, which parameter contributes the larger error?

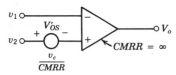

Figure 13.8 Modeling the effect of CMRR.

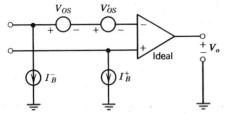

Figure 13.9 Op amp equivalent circuit reflecting effects of I_B, V_{os}, and V'_{os}.

Solution

The magnitude of the error at the input, due to V_{OS}, is

$$V_{OS} = 0.5 \text{ mV}$$

The magnitude of the error at the input, due to CMRR, is determined using (13.11a),

$$V'_{OS} = \frac{v_c}{CMRR} = \frac{10 \text{ V}}{10,000} = 1 \text{ mV}$$

where

$$v_c = 10 \text{ V}$$

The error due to CMRR is twice that due to the offset voltage. ∎

13.3 OUTPUT (dc) PARAMETERS

13.3.1 Output Voltage Swing [V_O(swing)]

The output voltage swing is the peak output voltage, referred to zero, that can be obtained with the amplifier operating in its linear region. Its value is primarily established by the dc supply voltages and the output current. The manufacturer guarantees this parameter using a graph of output voltage versus supply voltage for some value of load resistance (typically, 2 and 10 kΩ). The output of the 101A op amp will swing between +10 and −10 V if it has ±15 V supplies and a 2 kΩ load resistor.

13.3.2 Output Current (I_O)

The output current, like the output voltage swing, is determined through a graph. This is not a "guaranteed" parameter and is listed under the typical performance characteristics shown in the appendix. The 101A and 741 op amps have a current limiting circuit in the output stage and will limit the maximum output current, at ambient temperature, to about 25 mA. The amplifier (101A or 741) output, typically, can swing ±10 V (±15 V supplies) and source, or sink, 25 mA.

13.4 SUPPLY AND PACKAGE RELATED PARAMETERS

13.4.1 Supply Voltage Range

The typical or nominal supply voltages, V^+ and V^-, used in IC op amp circuits are ± 15 V. These supplies can be increased to ± 22 V for aerospace-grade devices (101A and 741A) or decreased to ± 5 V (101A) or ± 2 V (108A). Specialized integrated amplifiers are available with supply voltage ranges up to ± 40 V.

13.4.2 Supply Current (I_S)

Supply current I_S is the power supply current required to operate the amplifier at given supply voltages, with no load, and the output at zero volts. It varies from 0.60 mA (108A) to 3.0 mA (101A). The low supply current and voltage values for the 108A op amp make it an attractive choice in battery-operated applications.

13.4.3 Power Supply Rejection Ratio (PSRR)

When the power supply voltages change, the equivalent input offset voltage also changes. The ratio of the change in this voltage to the change in power supply voltage producing it is called the power-supply rejection ratio (*PSRR*) and is defined as

$$PSRR = \frac{\Delta V_{OS}}{\Delta V_{PS}} \quad (13.12a)$$

where $V'_{OS} = \Delta V_{OS}$ is the apparent change in input offset voltage due to the change in power supply voltage. *PSRR* is expressed in μV/V, or in decibels where

$$PSRR, \quad dB = 20 \log \frac{\Delta V_{OS}}{\Delta V_{PS}} \quad (13.12b)$$

PSRR is in the 80 dB range (101A and 741). The advent of low-cost, high-performance IC regulators has minimized the error contributed by the *PSRR* parameter in op amp applications.

13.4.4 Power Dissipation (P_D)

This parameter identifies the maximum amount of power that a particular device is capable of dissipating safely on a continuous basis while operating within its specified temperature range. Power dissipation is package dependent; ceramic packages have the highest rating with metal and plastic packages following in that order. The typical power dissipation capability for an IC op amp is 500 mW.

13.4.5 Operating Temperature Range (T_{OR})

This parameter specifies the range of temperature over which the device will perform within its rated specifications. Military-grade devices operate from -55 to $+125°C$; industrial-grade devices from -25 to $+85°C$; and commercial-grade devices from 0 to $+70°C$.

13.5 AC PARAMETERS

13.5.1 Slew Rate (S_R)

Slew rate S_R is the maximum rate of change of the amplifier's output voltage. This specification represents, primarily, the ability of the amplifier to charge and discharge its compensation capacitor, within the period of the input signal, because of the finite amount of current available from the differential amplifier stage. S_R governs the large-signal behavior of the op amp as opposed to its small-signal performance, which is limited by frequency response or GB considerations. A current source charging a capacitor models the frequency behavior under large-signal conditions, while a low-pass RC filter circuit models the frequency behavior of the op amp under small-signal conditions. S_R is measured in V/μsec and is typically 0.4 V/μsec (101A) and 0.8 V/μsec (741).

A sinusoidal output signal will cease being a small signal when its maximum rate of change exceeds the slew rate of the amplifier. The maximum rate of change for a sinusoidal signal occurs at the zero crossings. Assuming that

$$v_o = V_{op} \sin 2\pi f t \qquad (13.13)$$

the derivative of v_o with respect to time gives us its rate of change. If we evaluate this derivative at $t = 0$ when the output signal goes through zero, we obtain

$$\left.\frac{dv_o}{dt}\right|_{t=0} = 2\pi f V_{op} \cos(2\pi f t)\Big|_{t=0}$$

$$= 2\pi f V_{op} \leq S_R \qquad (13.14)$$

As long as the maximum value of the rate of change of the output voltage with respect to time is less than the amplifier's slew rate, the output voltage will be sinusoidal and the performance will be small signal. However, if the product $2\pi f V_{op}$ is greater than the slew rate limit, the output waveform will be slew rate limited and distorted. The maximum undistorted ($<3\%$) frequency $f(\max)$ is given by

$$f(\max) = \frac{S_R}{2\pi V_{op}} \qquad (13.15)$$

To increase $f(\max)$, we must either increase S_R or decrease V_{op}.

292 Secondary Parameters

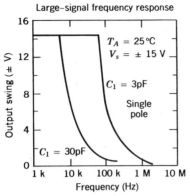

Figure 13.10 Large-signal frequency response characteristic (101A).

The value of the slew rate for an amplifier is not always given explicitly. It can be determined from the large-signal frequency response curve, Fig. 13.10, which is a plot of output swing (V_p) versus frequency. Slew rate defines the device's full power bandwidth (large-signal response), or the maximum frequency at which the amplifier will deliver the full-rated output voltage with less than 3% distortion.

For a given type of operational amplifier, the slew rate of the amplifier can be proportionally increased if the compensation capacitor is reduced. This, of course, does not hold true for internally compensated amplifiers.

13.5.2 Transient Response (T_r)

A manufacturer frequently provides transient response specifications of an op amp used as unity-gain amplifier. The circuit is shown in Fig. 13.11. If the input is a voltage step, the output rises exponentially toward the peak value of this step provided that there is no slew rate limiting. The rise time T_r of the output response, defined as the time it takes v_o to rise from 10 to 90% of its peak-to-peak value, is

$$T_r \cong 2.2\tau \qquad (13.16a)$$

where τ is the time constant of the equivalent low-pass RC circuit modeling the open-

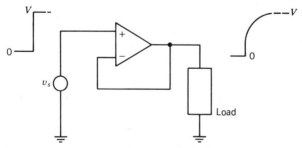

Figure 13.11 Transient response test circuit.

loop frequency response of the op amp. This time constant is related to the -3 dB cutoff frequency of the closed-loop response of the unity-gain amplifier circuit of Fig. 13.11, under sinusoidal input conditions, by

$$2\pi f_c = 2\pi GB_f = \frac{1}{\tau} \qquad (13.16\text{b})$$

where the cutoff frequency f_c is equal to the gain-bandwidth product GB_f (in hertz) of the op amp. Eliminating τ between (13.16a) and (13.16b), we have

$$T_r \cong \frac{2.2}{2\pi GB_f} \cong \frac{0.35}{GB_f} \qquad (13.17)$$

Equation (13.17) relates the small-signal step response of the circuit of Fig. 13.11 to its small-signal frequency response. The parameter T_r may be used to calculate the gain-bandwidth product of the op amp.

13.5.3 Noise

Noise is defined as any signal appearing at the op amp's output that could not have been predicted by a dc and ac input error analysis. Noise can be random or repetitive, internally or externally generated, current or voltage type, narrow band or wideband, or high frequency or low frequency. Whatever its nature or source, it places a limit on a circuit's resolution, gain, and signal quality and has to be minimized in many applications.

The noise, which is present at the output of an operational amplifier, is a combination of the externally and internally generated amplifier noise. Noise produced internally by an op amp is modeled by a noiseless amplifier with noise voltage and noise current generators, shown in Fig. 13.12, at its input terminals. Noise generated by external resistors is similarly represented by equivalent noise generators connected to the amplifier's input terminals. Several noise sources can be combined into a single equivalent source, and in a closed-loop circuit, this total input noise appears at the output multiplied by the closed-loop noise gain.

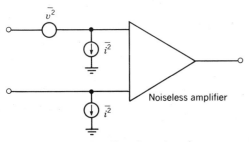

Figure 13.12 Modeling the noise of an op amp.

294 Secondary Parameters

The main contributors to the internal noise in a 741 type amplifier are the active loads in the differential amplifier stage. Using active loads in op amps allows the realization of a very high voltage gain in a relatively few stages. However, the active loads also amplify their own internal noise. Since the input stage has both current and voltage gain, the noise in the following stages is generally not significant because the noise is not amplified by the full gain of the op amp. IC amplifiers designed for low noise operation generally use a simple common-emitter or differential input circuit with resistive loads.

Externally generated noise covers the frequency spectrum with mechanical vibrations, pc (printed circuit) board contamination, line frequency pickup, and power supply ripple on the low end and a radar's prf (pulse recurrence frequency), radio station pickup, and SCR (silicon controlled rectifier) and relay switching and arcing on the high end. These external sources of noise can be minimized by proper shielding, single-point grounding, decoupling the bias supplies, securing cables and connectors, and sometimes even reorienting the locations of the components.

Types of Noise The three major types of noise found in operational amplifiers are $1/f$ or flicker, shot or Schottky, and thermal or Johnson noise.

Internally generated noise is dominated by $1/f$ or flicker noise on the low end of the frequency spectrum and thermal or Johnson noise on the high end. Flicker noise, the most critical noise in limited bandwidth applications, is found in all active devices and is caused mainly by contamination and crystal defects in the emitter-base depletion layer of bipolar devices. Flicker noise is always associated with a dc current.

Shot (Schottky) noise, like $1/f$ noise, is associated with a direct current and is present in diodes and bipolar transistors. In these devices, the passage of each carrier across a *pn* junction is a purely random event and is dependent on the carrier having sufficient energy and velocity directed toward the junction. Thus the external dc current, which appears to be a steady current, is in fact composed of a large number of random current pulses. The fluctuations in a dc current I_{DC} are termed shot noise, which is generally specified in terms of its mean-square variation about the average value or

$$\overline{i^2} = 2qI_{DC}\,\Delta f \quad\quad (13.18)$$

where q is the electronic charge and Δf is the bandwidth, in hertz, for which the noise is defined.

Since the noise current has a mean-square value that is directly proportional to bandwidth, a noise current spectral density $\overline{i^2}/\Delta f$ (A²/Hz) can be defined that is constant as a function of frequency. Noise with such a spectrum is called white noise.

Thermal or Johnson noise, also classified as white noise, is generated in resistances by the random thermal motion of the electrons. It is dependent on temperature and can be represented by a voltage or a current noise generator whose magnitudes are

$$\overline{v^2} = 4kTR\,\Delta f \quad\quad (13.19a)$$

and

$$\overline{i^2} = \frac{4kT\,\Delta f}{R} \tag{13.19b}$$

where k is Boltzmann's constant and T is the temperature (°K). Thermal noise is present in any linear passive resistor; the higher the resistance, the higher the noise voltage. The noise voltage of a resistor may also be represented as a spectral density measured in V^2/Hz. The amplitude distribution of thermal and shot noise is Gaussian and, since each has a flat frequency spectrum, they are indistinguishable once they are introduced in the circuit.

The equivalent mean-square value of a number of independent voltage noise generators in series, or the parallel combination of current noise generators, is the sum of the corresponding individual mean-square values:

$$\overline{v_t^2} = \overline{v_1^2} + \overline{v_2^2} + \cdots \tag{13.20a}$$

and

$$\overline{i_t^2} = \overline{i_1^2} + \overline{i_2^2} + \cdots \tag{13.20b}$$

respectively.

Noise Specifications Noise voltage, or more properly equivalent short-circuit input RMS noise voltage, is noise voltage that appears to originate at the input of a noiseless amplifier if the input terminals were shorted. It is typically expressed in nV per root Hz (nV/$\sqrt{Hz}$) at a specified frequency, or in μV in a given frequency band. It is measured by shorting the input terminals, measuring the output RMS noise voltage, dividing by the amplifier's gain, and referring it to the input. An output bandpass filter of known characteristic is used in the measurement, and the measured voltage is divided by the square root of the filter's bandwidth if the noise is expressed in root Hz. The noise in an op amp is not constant with frequency; it increases at the lower frequencies because of the $1/f$ noise.

Similarly, the equivalent open-circuit RMS noise current is expressed in pA/$\sqrt{Hz}$ at a specified frequency or in nA for a given frequency band. The typical performance curves for noise voltage and noise current for the 741 are shown in Fig. 13.13.

To minimize the noise or its effect in a circuit:

(A) Minimize the externally generated noise.
(B) Limit the circuit bandwidth to the input-signal bandwidth.
(C) Choose a low noise amplifier.
(D) Keep the input current levels low.
(E) Eliminate excessive resistances in the input portion of the external circuit.

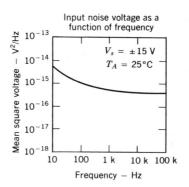
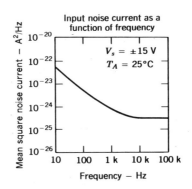
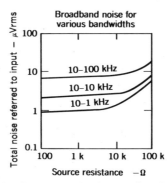

Figure 13.13 Current and voltage noise of a 741 op amp.

Example

Calculate (a) the shot noise associated with a 100 nA dc semiconductor current and (b) the thermal current noise associated with a 10 kΩ metal film resistor. The bandwidth for each case is 10 kHz and the temperature is 27°C. What is the sum of the two noise currents if they are in parallel?

Solution

The shot noise is calculated using (13.18),

$$i = \sqrt{(3.2 \cdot 10^{-19})(100 \cdot 10^{-9})(1 \cdot 10^4)}$$
$$= 17.9 \text{ pA RMS}$$

The thermal noise, given by (13.19b), is

$$i = \sqrt{4(1.38 \cdot 10^{-23})(300)(10^{-4})(10^4)}$$
$$= 128.7 \text{ pA RMS}$$

The thermal noise of wirewound and metal film resistors may be computed using (13.19). Another term is added to (13.19) for discrete carbon resistors because they also display flicker noise.

The total noise current is the square root of the sum of the mean-square values of the individual noise currents,

$$i_t = \sqrt{(17.9)^2 + (128.7)^2} \quad \text{pA RMS}$$
$$= 130.0 \text{ pA RMS} \quad \blacksquare$$

13.6 MEASURING DATA SHEET PARAMETERS

13.6.1 V_{OS}, I_{OS}, and I_B

Generally speaking, operational amplifier parameters are not measured directly. Rather, they are measured, as shown in Fig. 13.14a, by connecting the amplifier-under-test (AUT) in series with a buffer amplifier A_2 in a closed-loop configuration. The AUT input parameters, multiplied by the closed-loop gain of the test circuit, are reflected to the output of the buffer amplifier A_2. The closed-loop gain of the test circuit from the potential V_{IN} to the output of A_2 is given by $(1 + R_2/R_1)$, which is 1000 in this case. Thus,

$$V_O = 1000 V_{IN} \quad (13.21)$$

The voltage V_{IN} can be a function of the amplifier's offset voltage, an amplifier parameter reflected back to the input, or a voltage drop caused by inserting a resistor in series with an amplifier input.

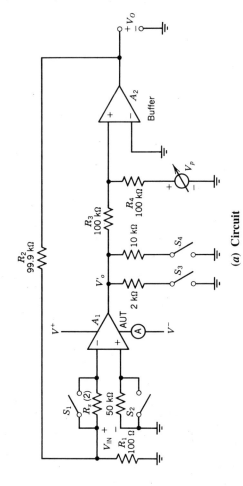

Figure 13.14 Measuring op amp parameters.

(*a*) Circuit.

298

Parameter	S_1	S_2	V_P	V^+/V^-	R_L	V_o'	Measure
V_{OS}	Close	Close	0 V	+15 V/−15 V		0 V	$V_o = 1000\ V_{OS}$
I_B-	Open	Close	0 V	+15 V/−15 V		0 V	$V_o = 1000\ \{V_{OS} \pm I_B^- R_S\}$
I_B+	Close	Open	0 V	+15 V/−15 V		0 V	$V_o = 1000\ \{V_{OS} \mp I_B^+ R_S\}$
I_{OS}	Open	Open	0 V	+15 V/−15 V		0 V	$V_o = 1000\ \{V_{OS} \pm I_{OS} R_S\}$
I_S	Close	Close	0 V	+15 V/−15 V		0 V	ammeter
V_o(Swing)	Close	Close	±10 V	+15 V/−15 V	2 kΩ	∓10 V	$V_o' = \mp 10$ V
V_o(Swing)	Close	Close	±12 V	+15 V/−15 V	10 kΩ	∓12 V	$V_o' = \mp 12$ V
Supply voltage	Close	Close	0 V	+5 V/−5 V		0 V	$V_o = 1000\ V_{OS}$
Supply voltage	Close	Close	0 V	+22 V/−22 V	2 kΩ	0 V	$V_o = 1000\ V_{OS}$
Dc gain (T_1)	Close	Close	0 V	+15 V/−15 V	2 kΩ	0 V	$V_o = 1000\ V_{OS}$
(T_2)	Close	Close	−10 V	+15 V/−15 V	2 kΩ	+10 V	$V_o = 1000\ \{V_{OS} \pm V_d\}$
CMRR (T_1)	Close	Close	0 V	+22 V/−5 V		0 V	$V_o = 1000\ \{V_{OS} \pm V_{OS}'\}$
(T_2)	Close	Close	0 V	+5 V/−22 V		0 V	$V_o = 1000\ \{V_{OS} \pm V_{OS}'\}$
PSRR (T_1)	Close	Close	0 V	+5 V/−5 V		0 V	$V_o = 1000\ \{V_{OS} \pm V_{OS}'\}$
(T_2)	Close	Close	0 V	+22 V/−22 V		0 V	$V_o = 1000\ \{V_{OS} \pm V_{OS}'\}$

T_1: Test 1 T_2: Test 2

(*b*) Test circuit conditions (typical for 101A-like amplifier).

Figure 13.14   (*Continued*)

The output voltage of the AUT V'_O is controlled by the externally programmable source V_P, and resistors R_3 and R_4. Noticing that the noninverting input of A_2 is at 0 V (virtual ground), the output of the AUT in the circuit of Fig. 13.14a will be

$$V'_O = -V_P \tag{13.22}$$

since $R_3 = R_4$.

Resistors in series with the input terminals convert the input bias currents to voltages that are reflected to the output along with the offset voltage V_{OS}. Normally, V_{OS} is the first parameter measured, and its value is stored in memory and subtracted from subsequent measurements. To measure V_{OS}, switches S_1 and S_2 are closed and V_P is set to zero volts. In this case, $V_{IN} = V_{OS}$ and

$$V_O = \left(1 + \frac{R_2}{R_1}\right) V_{IN} = 1000(\pm V_{OS})$$

or

$$|V_{OS}| = \frac{V_o}{1000}$$

If both input resistors (R_S) are not shorted, then the voltage V_{IN} is

$$V_{IN} = \pm(I_B^- - I_B^+)R_S \pm V_{OS}$$

and

$$V_O = 1000 V_{IN}$$
$$= \pm 1000 V_{OS} \pm 1000(I_B^- - I_B^+)R_S \tag{13.23}$$

Thus,

$$I_{OS} = |I_B^+ - I_B^-|$$
$$= \frac{\left|\frac{\pm V_o}{1000} \pm V_{OS}\right|}{R_S} \tag{13.24}$$

Similarly, the individual bias currents can be measured by connecting R_S in series with the appropriate input terminal. The purpose of inserting the resistors R_S in series with the amplifier inputs is to convert the op amp's input currents to voltages that can be summed with the offset voltage. The current in the feedback loop must be much greater than the input currents of the operational amplifier.

Switches S_3 and S_4 are used to connect a 2 kΩ or 10 kΩ load resistor to the amplifier's output in those cases where a parameter is measured with a specified load resistor. An example will help to illustrate some of the above ideas.

Example

Let the AUT have the following known specifications (see Fig. 13.9):

$$V_{OS} = 1.5 \text{ mV}$$
$$I_B^+ = 70 \text{ nA}$$
$$I_B^- = 75 \text{ nA}$$

To measure V_{OS}, switches S_1 and S_2 are closed and V_P is programmed to 0 V. The output voltage of the AUT V_O' will also be 0 V, but the output of the test circuit will be

$$V_O = \left(1 + \frac{R_2}{R_1}\right) V_{OS} = 1000(1.5 \text{ mV})$$

or

$$V_O = 1.50 \text{ V}$$

To measure I_B^-, switch S_1 is opened, S_2 is closed, and V_P is again programmed to 0 V. The output voltage of the test circuit V_O, this time, will be a function of both V_{OS} and the voltage drop across R_S caused by I_B^-,

$$V_O = 1000 V_{IN} = 1000(I_B^- R_S + V_{OS})$$
$$= 1000 \ (3.75 \text{mV} + 1.5 \text{ mV})$$

or

$$V_O = 5.250 \text{ V}$$

Since V_{OS} is known at this point, the contribution of the current I_B^- on the output voltage can be calculated from the measured value of V_O by subtracting 1.5 V.

To measure I_B^+, switch S_1 is closed, S_2 is opened, and V_P is programmed to zero, causing $V_O' = 0$ V. The voltage drop across R_S, which is in series with the noninverting amplifier input, will be negative with respect to ground and will be subtracted from the offset voltage,

$$V_{IN} = V_{OS} - I_B^+ R_S$$

At the output of the test circuit,

$$V_O = 1000 V_{IN} = 1000(V_{OS} - I_B^+ R_S)$$
$$= 1000(1.5 \text{ mV} - 3.5 \text{ mV})$$

or

$$V_O = -2.00 \text{ V}$$

Once the values of I_B^+ and I_B^- are known, its average value I_B can be easily computed. ∎

The supply terminals of the AUT in the test circuit of Fig. 13.14 are connected to programmable voltage sources. The equivalent of an ammeter in series with one of the bias supplies is used to measure the supply current parameter I_S. PSRR is determined, along with the supply voltage range, by measuring the change in input offset voltage when the supplies are programmed from their minimum to their maximum values.

CMRR is measured by making the supplies unsymmetric, which causes the inputs to see a common-mode voltage. This results in a change of input offset voltage that is reflected to the output of the test circuit. For example, suppose V^+ and V^- were programmed to $+10$ and -20 V, respectively. The reference for the amplifier's internal circuitry would be -5 V with respect to ground. The internal reference point for the op amp is at the average value of the two supply voltages. Since the amplifier input terminals are near ground potential, the inputs with respect to the internal reference are at $+5$ V. Hence, the inputs see a 5 V common-mode voltage. This common-mode voltage will cause an equivalent offset voltage (V'_{OS}) that will sum with the initial offset voltage. The sum of the two will be reflected to the output with a gain of 1000 where the computer can calculate the CMRR of the device.

13.6.2 DC Gain

The measurement of the dc gain $A_O(A)$ requires two tests. For the first test, $V_P = 0$ V and

$$V_O = 1000 V_{IN}$$

For the second test, $V_P = -10$ V. This forces the large-signal output voltage of the AUT to go to $+10$ V. Because of its finite gain, the input voltage will change by ΔV_a. This voltage along with V_{OS} will be reflected to the output of A_2. Since V_{OS}, the closed-loop gain, and ΔV_O are known, the dc gain of the op amp can be computed,

$$A_O = \frac{\Delta V'_O}{\Delta V_a} = \frac{\Delta V'_O}{\Delta V_{IN}}$$

$$= \frac{10}{\frac{\Delta V_O}{1000}} = \frac{10^4}{\Delta V_O}$$

The test conditions to measure the basic input parameters are presented in Fig. 13.14b. The test circuit discussed in this section is used commercially to measure op amp parameters. With a few modifications, the circuit can be used to measure all dc and ac parameters.

13.7 SUMMARY

Secondary parameters introduce error terms in the input-output relationship of op amp circuits. The dc parameters V_{OS} and I_B represent the offset voltage and input currents of the transistors in the amplifier's input stage. These parameters are modeled using voltage and current sources and they cause dc errors at the circuit's output. The parameter *CMRR* describes the difference in gain between the inverting and noninverting input terminals and is important in noninverting circuits where the average of the two input voltages is typically nonzero. The effect of *CMRR* is modeled using a voltage source in series with one of the amplifier's input terminals. Other parameters limit the input voltage, output voltage, output current, supply voltage, power dissipation, and operating temperature range of the op amp.

The parameter S_R describes the frequency limitation of the op amp under large-signal conditions. S_R is a measure of the maximum rate of change of the output voltage with respect to time. The noise in the op amp is of the $1/f$, shot, and thermal types. Voltage and current noise generators connected to the amplifier's input terminals model the cumulative noise effect.

A two-amplifier test circuit may be used to measure the primary and secondary parameters of the op amp. The input parameters of the op amp are reflected to the output of the circuit at a higher, measurable level.

PROBLEMS

13.1 The circuit in Fig. 13.15 may be used to measure the approximate value of the input offset voltage V_{OS} of an otherwise ideal operational amplifier. If $|V_{OS}|/R \gg I_B$, prove that

$$V_{OS} \cong \frac{\pm V_o}{k}$$

Why is the measured value of V_{OS} only an approximation?

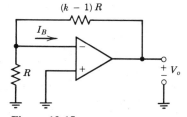

Figure 13.15

13.2 The offset voltage of the amplifier in Fig. 13.15 is nulled to zero volts at room temperature (25°C). Calculate the output voltage of the circuit if the temperature is increased to 75°C, $\Delta V_{OS}/\Delta T = 25\ \mu V/°C$, and $k = 500$.

13.3 The circuit in Fig. 13.16 may be used to measure the approximate value of the input bias currents, I_B^+ and I_B^-, and the offset current, I_{OS}, of an otherwise ideal operational amplifier. Show the positions of switches S_1 and S_2 (open or closed) to implement these measurements. Express V_O in terms of these input bias currents for each of the three cases.

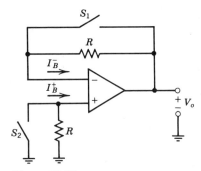

Figure 13.16

13.4 The circuit in Fig. 13.17 may be used to measure the approximate value of the input currents, I_B^+ and I_B^-, as well as the input offset current, I_{OS}, of an otherwise ideal operational amplifier. Show the positions of switches S_1 and S_2 (open or closed) that will allow implementation of these measurements. Express $v_o(t)$ in terms of these input currents for each of the three cases. Assume that the initial voltage for the capacitors is zero.

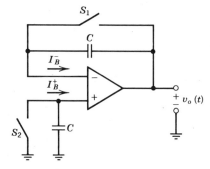

Figure 13.17

13.5 Sketch $v_o(t)$ for each of the three cases in Problem 13.4 if $I_B^+ = 100$ nA, $I_B^- = 90$ nA, and $C = 0.01\ \mu F$. Assume zero initial voltages for the capacitors.

13.6 Draw the output voltage for the circuit in Fig. 13.18, for the switch waveform shown, if (1) the op amp is ideal and (2) if the bias current I_B^- of an otherwise ideal op amp is 100 nA.

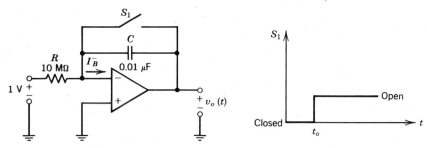

Figure 13.18

13.7 The operational amplifier in the summing amplifier circuit of Fig. 13.19 has the following parameters: $V_{OS} = 5$ mV, $I_B^+ = 100$ nA, and $I_B^- = 120$ nA. The voltage at the noninverting input terminal of the amplifier is positive with respect to the inverting terminal ($V_{ni} > V_i$), and the transistors in the input stage are *npn*. The amplifier is ideal except for these parameters. Calculate the exact value of V_O.

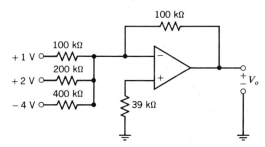

Figure 13.19

13.8 The operational amplifier in the difference amplifier circuit of Fig. 13.20 has the following parameters: $V_{OS} = 1$ mV ($V_{ni} > V_i$), $I_B^+ = 50$ nA, and $I_B^- = 35$ nA (*npn* input transistors). The amplifier is ideal except for these parameters. Calculate the exact value of the output voltage.

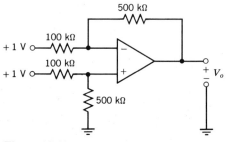

Figure 13.20

13.9 The circuit of Fig. 13.21 may be used to measure the approximate value of CMRR of an operational amplifier (otherwise assumed ideal). Prove that

$$CMRR \cong \frac{R_2 V_s}{R_1 V_o}$$

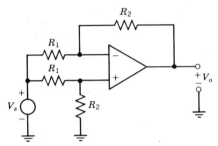

Figure 13.21

13.10 The same 101A operational amplifier is used in the configurations of Figs. 13.22a and 13.22b. If the op amp is ideal except for $CMRR = 60$ dB, calculate the exact value of the output voltage for each circuit.

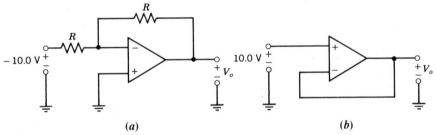

Figure 13.22

13.11 The operational amplifier in the circuit of Fig. 13.23 has a slew rate of 0.5 V/μsec. Determine whether the output of the circuit will be distorted or not?

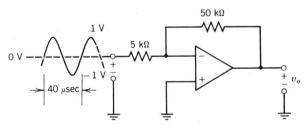

Figure 13.23

13.12 Sketch the output voltage waveform for the circuit in Fig. 13.24 if the slew rate of the otherwise ideal op amp is 1 V/μsec.

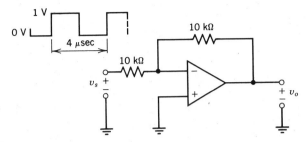

Figure 13.24

13.13 Estimate the maximum peak-to-peak input signal for the circuit in Fig. 13.25 if the peak output swing [V_o(swing)] is determined by the data sheet in Fig. 13.1.

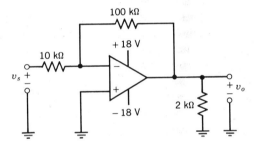

Figure 13.25

13.14 Estimate the life (in hours) of a 12 V battery with a rating of 1 ampere-hour if it is used to bias the circuit in Fig. 13.26. Assume in the first case that the operational amplifier is a 101A and in the second case that it is a 108A. The circuit is operating under no input signal conditions.

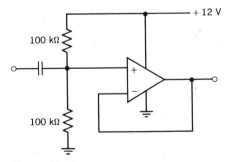

Figure 13.26

13.15 Using Fig. 13.13, model the noise of the 741 op amp at 100 Hz, 1 kHz, and 10 kHz in the equivalent circuit of Fig. 13.12.

13.16 Using Fig. 13.14, calculate the output voltage for each appropriate test configuration (V_{OS}, I_B^-, I_B^+, and I_{OS}) if the AUT has the following parameters (see Fig. 13.9):
(a) $V_{OS} = 5$ mV
(b) $I_B^+ = 40$ nA
(c) $I_B^- = 35$ nA

13.17 Calculate the values of the parameters V_{OS}, I_B, and I_{OS} of an amplifier tested in the circuit of Fig. 13.14 if
(a) $V_O = -5$ V (V_{OS} Test)
(b) $V_O = +4$ V (I_B^- Test)
(c) $V_O = -10$ V (I_B^+ Test)

PART IV

PRACTICAL IMPLEMENTATIONS

CHAPTER 14

Design Examples

14.1 PROLOGUE

Chapters 1, 2, and 15 of this book dealt with the operational amplifier as an active device. Chapters 3 through 10 looked at the extensive use of the op amp in various configurations. To focus on the particular application of the op amp, the device was treated as ideal in these chapters. The practical limitations of the real amplifier were examined in Chapters 11 through 13, where we saw how the finite and nonzero values of the primary characteristics and the secondary parameters modified the ideal relationships of the two most basic amplifier circuits. The objective of this chapter is to help the reader make the transition from understanding the amplifier, its applications, and its limitations to the design of actual amplifier circuits.

The design of a circuit or a system to a set of criteria involves not only an analysis of the basic circuit but also a number of practical considerations that are often referred to as higher-order effects. Many times these higher-order effects dominate the performance of a circuit, but cannot be described mathematically. The practical considerations in a circuit vary from the hardware, to assembly, to economy, to the limitations of the devices themselves.

A select number of circuits are presented in this chapter. By no means will they cover every circuit permutation or all design criteria. They should, however, provide the reader with a better understanding and greater insight of what a "working" circuit requires and help make the transition from analysis to design.

312 Design Examples

The design examples reflect the summation of information presented thus far. For the design engineer, design is the creative application of the knowledge derived from the understanding of the devices and the analysis of the basic circuits.

The considerations that an engineer has in designing a particular circuit do not always produce singular answers. As an example, there are a number of types of general-purpose op amps that have similar performance specifications and are in the same price range. In many designs, any one of the amplifiers would satisfy the requirements. The choosing of a particular one may be made based on availability, delivery time, or past experience with the device. This type of considerations cannot be quantified nor can it be adequately described in words. Let us just say they are the result of experience and the particular employer and circumstances.

In the initial example presented in this chapter, the design will be quite detailed. In the second example, some of the details will not be shown, but their presence will be implied. For example, it is good design practice to always bypass the dc supplies at the amplifier. This important feature is covered in the first circuit, but not in the subsequent ones. The physical location of the amplifiers with respect to each other and other components, economy, and types of components are other examples. An experienced analog design engineer will be aware of most of these design essentials.

It is common in industrial amplifier applications to use a number of basic integrated circuits along with discrete and digital devices. Usually, the basic op amp circuits are modified to fit the particular requirements and then connected to other types of circuits to form the end product.

Rather than use a step-by-step cookbook procedure, we present the material in a conversational-like manner. This approach somewhat simulates the designer's thoughts as he or she proceeds through the design process.

14.2 RAMP GENERATOR

The design of a programmable ramp generator is presented in two parts; (1) an analysis of the basic circuit so that the reader is familiar with the operation of the circuit, and (2) the evolution of the circuit, showing how the limitations are compensated for in satisfying the design requirements.

14.2.1 Analysis

The circuit in Fig. 14.1a is a simplified version of a programmable ramp generator. The circuit provides an output voltage varying linearly with time, that is, a voltage ramp, and is a functional block in a computer-based system. The ramp's initial voltage, termination voltage, and slope are computer controlled and programmable using digital-to-analog convertors.

The generator consists of three basic amplifier configurations. The A_1 circuit is an integrator, the A_2 circuit is a comparator, and the A_3 circuit is a summer. Voltages V_2 and V_4 control the generator's characteristics, and switches S_1 and S_2 control the time response of the integrator. The initial voltage of the generator output is called the

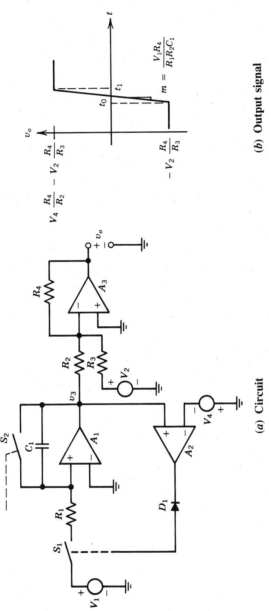

Figure 14.1 Simplified schematic of the ramp generator.

"start" voltage and the final voltage is called the "stop" voltage. The output will go from the start voltage to the stop voltage at a defined slope and on command from the computer. The summing amplifier A_3 sums the dc voltage V_2, which sets the start voltage of the ramp, and the integrator output v_3. Since the input to the integrator V_1 is a dc voltage, the outputs of the integrator and the generator will be linearly varying voltages. The sum of V_2 and the integrator output v_3 provides a linear output voltage that is offset from zero.

The voltage v_o at the output of the ramp generator is initiated, say at $t = t_0$, with the opening of switch S_2. Up until that time, v_o was at its start value $-V_2(R_4/R_3)$, as shown in Fig. 14.1b. The positive output of comparator A_2 had kept the diode D_1 reverse-biased and the solid-state switch S_1 closed.

When, at $t = t_0$, S_2 is opened by a command from the computer, the constant voltage V_1 will start charging capacitor C_1 at a constant rate and will generate a negative voltage, negative-slope ramp at v_3, the output of the integrator circuit A_1. As long as v_3 is smaller in magnitude that V_4, comparator A_2 will remain at its positive output state and keep switch S_1 closed. When v_3 becomes slightly larger in magnitude than V_4 (at $t = t_1$), A_2 goes to its negative output state, forward-biasing D_1, and opening S_1. With S_1 open, the charging of C_1 will be interrupted, causing v_3 to reach its most negative level, $-V_4$. The output of the ramp generator is the negative weighted sum of v_3, and the constant reference voltage V_2 and is given by

$$v_o(t) = -\frac{R_4}{R_2} v_3(t) - \frac{R_4}{R_3} V_2 \tag{14.1}$$

While C_1 in the integrator circuit is being charged,

$$v_3(t) = v_3(t_0) - \frac{1}{R_1 C_1} \int_{t_0}^{t_1} V_1 \, dt$$

$$= -\frac{V_1}{R_1 C_1} t \Big|_{t_0}^{t_1}$$

or

$$v_3(t) = -\frac{V_1}{R_1 C_1} (t - t_0) \qquad (t_0 \leq t \leq t_1) \tag{14.2}$$

Combining (14.1) and (14.2), we obtain the expression for the output of the ramp generator,

$$v_o(t) = \frac{V_1 R_4}{R_1 R_2 C_1} (t - t_0) - V_2 \frac{R_4}{R_3} \qquad (t_0 \leq t \leq t_1) \tag{14.3}$$

Since

$$v_3(t_1) = -\frac{V_1}{R_1 C_1} (t_1 - t_0) = -V_4$$

then (14.3) may be used to obtain the maximum value of v_o (stop voltage),

$$V_{stop} = v_o(t_1) = \frac{V_4 R_4}{R_2} - \frac{V_2 R_4}{R_3} \quad (S_1, S_2 \text{ open}) \quad (14.4)$$

The output of the ramp generator will remain at the value given by (14.4) until S_2 is closed again. With S_2 closed, the output goes to its minimum value or start voltage,

$$V_{start} = -\frac{V_2 R_4}{R_3} \quad (S_2 \text{ closed}) \quad (14.5)$$

From (14.4) and (14.5), we may obtain the amplitude of the output voltage ramp,

$$V_{stop} - V_{start} = \frac{V_4 R_4}{R_2} \quad (14.6)$$

Finally, using (14.3), we see that the slope of the output ramp is given by

$$m = \frac{V_{stop} - V_{start}}{t_1 - t_0} = \frac{V_1 R_4}{R_1 R_2 C_1} \quad (14.7)$$

At the systems level, the voltages V_1, V_2, and V_4 are the outputs of digital-to-analog convertors and thus are computer-controlled and programmable. The switch S_2 is also computer-controlled where a bit in the computer's control word either closes or opens the switch.

The feedback loop formed by A_1 and A_2 acts like a regulator and forces the maximum voltage across C_1 to be equal in magnitude to V_4. If C_1 discharges slightly, the output of A_2 will go positive and turn S_1 on briefly. With S_1 on, the capacitor C_1 will again charge to the magnitude determined by V_4.

14.2.2 Design

The design of any circuit is influenced by its specifications. The specifications must include not only the technical performance requirements but also other factors such as cost, development time, and environment. These nontechnical requirements are beyond the scope of this book. To illustrate some of the specifics of an amplifier circuit design procedure, the ramp generator will be designed to meet the following criteria only:

Start voltage: -5 V, $\pm 2\%$
Stop voltage: $+5$ V, $\pm 2\%$
Ramp slope: 1000 V/sec, $\pm 2\%$
Environment: commercial ($25° \pm 5°C$)
End product: test equipment

Specific values were assigned to the ramp start, stop, and slope magnitudes to simplify the design procedure. In actuality, the performance and requirements of the digital-to-analog convertors that establish these voltages would have to be reflected to the generator or circuit level.

Supplies and Ground The system, part of which is the ramp generator, has ± 15 V dc supplies for the analog electronics and a $+2$ V precision reference supply. These dc supplies are used to bias the amplifiers in the ramp generator, but they must be bypassed physically near the devices. Manufacturers of general-purpose amplifiers recommend that each supply is bypassed with a 0.1 µF ceramic disc in parallel with a 2.2 µF tantalum capacitor. The specific value of the capacitors is not crucial, but the combination of the capacitor values and type collectively provide a frequency response characteristic that minimizes external interference and oscillations. The large value capacitor is used to cancel out the effects of the distributed lead inductance of the power supply leads, and the small value capacitor is used to improve the bypass-circuit's high-frequency response.

It is extremely important in any analog circuit or system to have a proper grounding system. One physical point should be selected and designated as "the ground," which is typically located near the supply's regulator output terminals. The common connection between the two dc supplies used to bias the analog circuits is usually designated as ground. All terminals or nodes that are to be grounded in the circuit should be brought individually to "the ground." This is done to avoid ground loops. The technique is illustrated in Fig. 14.2.

Summer Circuit Initially, we will focus on components that establish the input-output relationship, and then add components that compensate for the amplifier non-idealities. The choosing of a summer circuit, shown in Fig. 14.3, should be obvious.

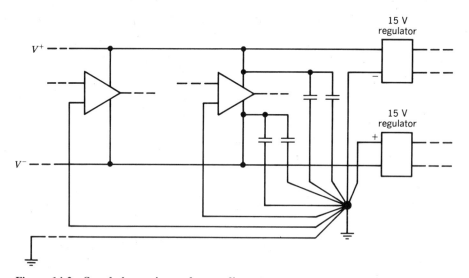

Figure 14.2 Supply bypassing and grounding.

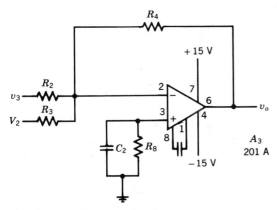

Figure 14.3 Summer circuit.

The specific amplifier chosen for the A_3 circuit is the industrial-grade 201A. It has better electrical performance specifications than the commercial-grade 301A, but it does not have to operate in a stringent environment that would require the expensive, aerospace-grade 101A. The important characteristics and parameters of the 201A that apply to the design of the summer are:

$$A = 50{,}000 \text{ (min)}$$
$$R_i = 1.5 \text{ M}\Omega \text{ (min)}$$
$$R_o = 75 \text{ }\Omega \text{ (typ)}$$
$$V_{OS} = 2 \text{ mV (max)}$$
$$I_B = 75 \text{ nA (max)}$$
$$I_{OS} = 10 \text{ nA (max)}$$
$$S_R = 0.5 \text{ V}/\mu\text{sec (typ)}$$
$$I_O \text{ (max)} = 18 \text{ mA (typ)}$$

The summer is designed for unity gain, which means

$$R_2 = R_3 = R_4$$

and, using (14.1),

$$v_o(t) = -[v_3(t) + V_2]$$

Under these conditions, the start voltage is

$$V_{\text{start}} = -V_2 = -5 \text{ V}$$

or

$$V_2 = 5 \text{ V}$$

From Chapter 12, we found that the parallel combination of R_2 and R_3 must be much smaller than R_i by some factor k. In our case,

$$R_2 \parallel R_3 = \frac{R_4}{2} \leq \frac{R_i}{k}$$

or

$$R_4 \leq \frac{2R_i}{k}$$

The higher the value of k, the smaller the error at the output of the summer because of the finite input resistance R_i of the op amp. We also found that R_4 must be greater than R_o by the same factor k, that is,

$$R_4 \geq kR_o$$

As k is increased to minimize the errors due to R_i and R_o, the upper and lower limits for R_4 converge. The maximum value for k, k_{max}, is obtained when these limits become equal, resulting in

$$k_{max}R_o = \frac{2R_i}{k_{max}}$$

or solving for k_{max},

$$k_{max} = \sqrt{\frac{2R_i}{R_o}}$$

Using the specifications for the 201A gives us

$$k_{max} = \sqrt{\frac{2(1.5 \cdot 10^6)}{75}}$$

$$= 200$$

This value of k means that in the worst case R_i and R_o will affect the accuracy of the closed-loop gain (compared to the ideal-case value) by at most 0.5%. For $k = k_{max} = 200$,

$$R_2 = R_3 = R_4 = \frac{2(1.5 \cdot 10^6)\Omega}{200} = 15 \text{ k}\Omega$$

The finite open-loop gain has a minimal effect on the closed-loop gain of the summer circuit of Fig. 14.3. This is due to the large magnitude of the loop gain of the circuit, which at dc is equal to (worst case)

$$50{,}000 \; \frac{15 \text{ k}\Omega \; \| \; 15 \text{ k}\Omega}{15 \text{ k}\Omega \; + \; (15 \text{ k}\Omega \; \| \; 15 \text{ k}\Omega)} = 16{,}667$$

This amounts to an error corresponding to 0.006%.

The offset voltage (V_{OS}) and bias current (I_B) of the op amp will introduce errors at the output. V_{OS}, shown in Fig. 14.4a, has a maximum value of 2 mV and, when reflected to the output, equals

$$V_o = \left(1 + \frac{R_4}{R_2 \; \| \; R_3}\right) = 6 \text{ mV (max)}$$

The equivalent circuit used to determine the magnitude of the error due to I_B is shown in Fig. 14.4b. In the absence of R_8, the input bias current will introduce an error at the output with a magnitude of

$$V_o = (I_B^-)R_4 \cong 1.13 \text{ mV (max)} \qquad (R_8 = 0)$$

The error at the output due to I_B, although small, may be further reduced from $I_B R_4$ to $I_{OS} R_4$ by adding a resistor R_8, shown in phantom in Fig. 14.3, in series with the noninverting input. Setting

$$R_8 = R_2 \; \| \; R_3 \; \| \; R_4 \cong 5.1 \text{ k}\Omega$$

results in an output voltage caused by I_{OS},

$$V_o = I_{OS} R_4 = 0.15 \text{ mV (max)} \qquad (R_8 = 5.1 \text{ k}\Omega)$$

In this case, the effect of V_{OS} and I_{OS} are minimal. Of course, the percentage of error contributed by these parameters is greater for smaller values of v_o.

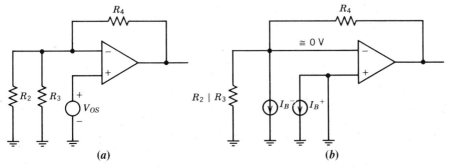

Figure 14.4 Equivalent circuits used to determine V_{OS} and I_B errors.

In Fig. 14.3, the resistor R_8 is shunted with a capacitor C_2 to force the noninverting input to zero volts under ac conditions. Its value is chosen so that the cutoff frequency of the $R_8 C_2$ filter is much less than the line frequency or 60 Hz,

$$C_2 = \frac{1}{2\pi R_8 f_c} \geq 5.2 \ \mu\text{F}$$

for $f_c \leq 6$ Hz.

Starting at $t = t_0$ when S_2 is opened, the summer output must slew at a rate of 1000 V/sec. This rate is 0.2% of the typical value of the slew rate S_R for the 201A, which means that its slew rate limit will introduce minimum distortion at the summer's output. The S_R rating of the 201A applies when the amplifier's compensation capacitor is 30 pF. Although the value of the capacitor can be increased based on small-signal requirements, it should not be made larger because of the slew rate considerations.

The operational amplifier has a limited output current I_O. We must check to see if the amplifier can satisfy the feedback current requirements. The maximum output voltage of the amplifier is ±5 V, which means that the maximum current that is needed to satisfy the feedback current requirement is ±5 V/15 kΩ or ±0.333 mA. This is well within the maximum I_O specification of 18 mA. If a load was specified, it would have to be added to the feedback current. Two other important data sheet parameters are common-mode and power-supply rejection ratios. The *CMRR* parameter does not apply to the summing circuit because the input terminals of the amplifier are always near zero volts and never see a common-mode voltage.

The *PSRR* parameter introduces a negligible error at the output if the dc bias supplies are reasonably well regulated. The voltage regulation for a typical system dc supply or local IC regulator is 2% and the guaranteed *PSRR* for the 201A is 80 dB. This causes the error at the output to be much less than 1 mV.

The components that are used in establishing the input-output relationship should have a 1% tolerance. The calculated values of the resistors are standard and available in low-cost, 1% film resistors.

Integrator Circuit The operational amplifier selected for the integrator circuit in Fig. 14.5 is the 208A. It was selected because of its low values of V_{OS} and I_B. The important parameters of the 208A that apply to the design of the integrator are:

$V_{OS} = 0.5$ mV (max)

$I_B = 2$ nA (max)

$S_R = 0.5$ V/μsec

The voltage gain (80,000 min), input resistance (30 MΩ min), and output resistance (75 Ω typ) have a minimal impact on the circuit's performance.

Since a system 2 V reference is available,

$$V_1 = 2.0 \text{ V}$$

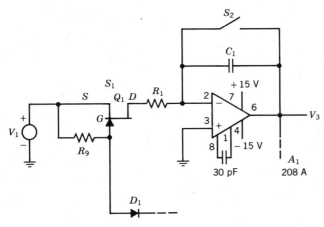

Figure 14.5 Integrator circuit.

Assuming S_2 open and S_1 closed, we see that the integrator output will decrease linearly as a function of time until it reaches $-V_4$. From the generator's specifications and (14.7),

$$m = \frac{V_1 R_4}{R_1 R_2 C_1} = 1000 \text{ V/sec}$$

or

$$R_1 C_1 = 2 \text{ msec}$$

for $V_1 = 2.0$ V and $R_2 = R_4$.

The values of R_1 and C_1 will be selected to minimize the effects of the operational amplifier and switches S_1 and S_2.

The charging current for C_1 will be a function of V_1 and R_1, although V_{OS} and I_B may affect its magnitude and thus the slope of the ramp. With the 208A op amp, V_{OS} may be assumed negligible compared to 2 V, the magnitude of V_1. To minimize the I_B effects, we arbitrarily let the charging current be 0.2 mA,

$$\frac{V_1}{R_1} = \frac{2 \text{ V}}{R_1} = 0.2 \text{ mA}$$

resulting in

$$R_1 = 10 \text{ k}\Omega$$

The value of C_1 is

$$C_1 = \frac{2 \cdot 10^{-3} \text{ sec}}{10 \text{ k}\Omega} = 0.2 \text{ }\mu\text{F}$$

The capacitor must be a high quality mylar, mica, or polystyrene type because the integrator operates in a sample-and-hold type of mode. The higher the quality of the dielectric, the higher the shunt resistance of the practical capacitor, and the less effect the capacitor has on the integrator's decay rate in the hold mode.

We must now consider the r_{ds}(on) rating of the JFET transistor switch S_1. This is the effective resistance of S_1 when closed. A 2N4391 n-type JFET was selected because of its low r_{ds}(on) rating of 30 Ω (max). The maximum error caused by r_{ds}(on) in series with R_1 is 0.3%, an acceptable value. The r_{ds}(on) value of S_2 is not as critical because this JFET is used to discharge C_1.

Finally, the S_R rating for the 208A is 0.5 V/μsec and, like the 201A in the A_3 circuit, will cause minimum distortion at the output.

Comparator Circuit   A 201A operational amplifier is also selected for the A_2 comparator circuit although, in this application, almost any general-purpose amplifier or comparator would suffice. The amplifier in the comparator circuit shown in Fig. 14.6 is operated open loop and its output goes from V_o(min) to V_o(max) depending on the relative values of v_3 and V_4. For

$$v_3 > -V_4$$

$V_o = V_o(\text{max}) \cong +13$ V, and for

$$v_3 < -V_4$$

$V_o = V_o(\text{min}) \cong -13$ V.

When $V_o = V_o(\text{max})$, D_1 is reverse-biased and the gate-to-source voltage V_{gs} of the JFET in switch S_1 is 0 V. Resistor R_9 provides a dc path from gate to source. Since the gate current is zero, the voltage drop across R_9 is zero and the gate and source are at the same potential. A $V_{gs} = 0$ V will turn the transistor switch S_1 on, and the A_1 output will integrate the input voltage V_1. When $V_o = V_o(\text{min})$, D_1 is forward-biased and V_{gs} of the S_1 JFET is approximately -14.5 V, assuming a 0.5 V drop across D_1. Since the V_{gs}(off) rating of the 2N4391 is -8 V, S_1 will be open.

The required performance of the A_2 circuit puts minimum demands on the components. One important consideration in selecting the amplifier or a comparator is the

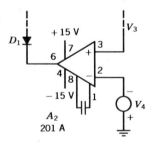

Figure 14.6 Comparator circuit.

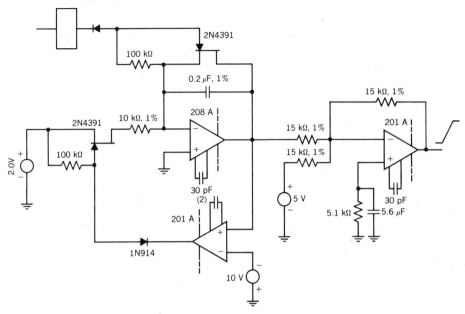

Figure 14.7 Schematic of the ramp generator.

input differential voltage (V_{id}) rating. The inputs of A_2 will see a maximum differential voltage of 10 V (V_4), but this value will not exceed the rating of the 201A, which is ±30 V.

The complete schematic of the ramp generator is shown in Fig. 14.7. It may appear at this point that there may be simpler ways to implement the design criteria. Based on the given criteria only, there are easier ways. However, the example was taken from a more complex project, and its only purpose is to illustrate some basic ideas. The thought of always looking for a better way to implement a design is valid because an important design objective is to minimize the component count and, thus, the cost.

The design reflects two levels of considerations:

1. Input-output relationships.
2. Compensation of component nonidealities.

A third level of consideration is possible depending on the environment of the generator.

Device Protection If a circuit is operated in an environment that could cause damage to the amplifiers, appropriate protection schemes must be employed. The amount and type of protection used is dependent on the extent and location of the hazardous conditions. Techniques employed to protect the amplifier from adverse supply, input, and output conditions are shown in Fig. 14.8.

The diodes D_1 and D_2 in Fig. 14.8a protect the amplifier if the supply voltages are accidentally reversed. The scheme is extended in Fig. 14.8b to include limiting the

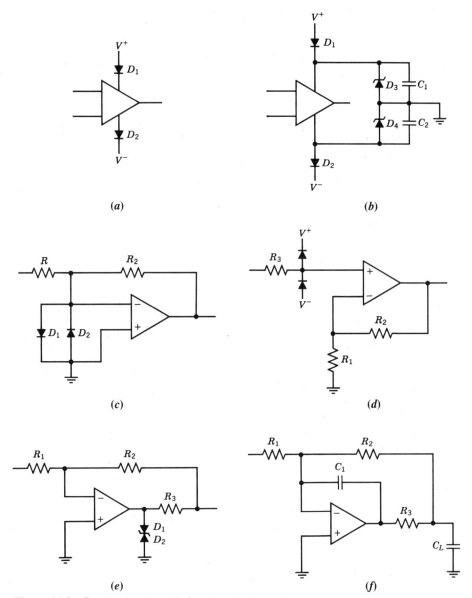

Figure 14.8 Supply, input, and output protection schemes.

maximum value of the supply voltages. The zener diodes limit the supply voltages and the parallel combination of the zeners and capacitors filters out transients on the power supply lines. In Fig. 14.8c, the two diodes insure that the inverting amplifier's differential input voltage is about 0.6 V in magnitude. The input to the noninverting amplifier in Fig. 14.8d is bounded to no more than 0.6 V more than the supply. This particular technique is frequently used to protect the internal circuitry of an IC. The back-to-back zeners D_1 and D_2, and the current-limiting resistor R_3, in Fig. 14.8e limit the amplifier's output voltage when the circuit output is inadvertently connected to a high voltage.

When an amplifier circuit drives a capacitive load, the phase margin is reduced and, because of the additional phase shift, the likelihood of oscillations is strong. The operational amplifier's output resistance R_o and the load capacitance C_L introduce an additional pole at $-1/R_o C_L$. The addition of R_3 and C_1 as shown in Fig. 14.8f decouples the amplifier from C_L and adds a zero to the frequency response characteristic to minimize the effect of the pole due to C_L.

The hardware plays a vital role in many amplifier applications. In low-current, low-voltage, high-frequency, and high-resistance circuits, the circuit construction technique will affect the performance as much or more than the tolerance of the components.

In high-gain and wide-bandwidth applications, the circuit should be assembled using a printed circuit board with short copper traces and a ground plane. Unintentional feedback paths through distributed capacitance should be avoided. In low-current and high-resistance applications, again a gound plane should be used and shielding employed to prevent interference from radiant noise and light sources. In addition, the amplifier input terminals should be isolated from adjacent terminals and traces using the ground plane.

14.3 ACTIVE FILTER

This design example illustrates the design and application of a higher-order, active filter. The design specifications are:

1. Maximally flat magnitude response.
2. Gain at dc = 20 dB ±2%.
3. Gain at 60 Hz = −40 dB ±2%.
4. −3 dB (cutoff) frequency = 15 Hz.

The filter is specified for a low-frequency control system and must attenuate any 60 Hz component that is present in the input signal.

14.3.1 Analysis

Since the filter must have a maximally flat magnitude response, a Butterworth filter must be used. The order of the Butterworth filter is determined from items 3 and 4 of the specifications. The magnitude of the transfer function of the Butterworth filter is given as

$$|T(j\omega)| = \frac{A_o}{\sqrt{1 + \left(\dfrac{f}{f_c}\right)^{2n}}} \qquad (14.8)$$

where A_o is the low-frequency passband gain and n is the filter order. For $A_o = 10$ (20 dB), $|T(j\omega)| = 0.01$ (-40 dB) at $f = 60$ Hz, and $f_c = 15$ Hz, (14.8) evaluates to $n = 4.98$.

The lowest order Butterworth filter that satisfies the requirements is 5 and can be implemented using a first-order, and two second-order filter stages or sections.

The normalized transfer function (-3 dB cutoff at $\omega = 1$ rad/sec) for a fifth-order Butterworth filter has the form of

$$G(s) = \frac{A_o}{(s' + 1)\,[(s')^2 + 0.618034s' + 1]\,[(s')^2 + 1.618034s' + 1]}$$

The normalized transfer functions of the individual sections are

$$T_1(s') = \frac{A_{o1}}{s' + 1} \tag{14.9a}$$

$$T_2(s') = \frac{A_{o2}}{(s')^2 + 0.618034s' + 1} \tag{14.9b}$$

and

$$T_3(s') = \frac{A_{o3}}{(s')^2 + 1.618034s' + 1} \tag{14.9c}$$

where A_{o1}, A_{o2}, and A_{o3} are the passband gains of the individual filter sections.

To maximize the ratio between the open- and closed-loop gain of each section, we set

$$|A_{o1}| = 1$$
$$|A_{o2}| = |A_{o3}| = \sqrt{10} \cong 3.162$$

To denormalize the transfer functions (-3 dB cutoff for overall response at $\omega = 30\pi$ rad/sec), we substitute

$$s' = \frac{s}{30\pi}$$

in (14.9a), (14.9b), and (14.9c) to obtain

$$T_1(s) = \frac{\pm 30\pi}{s + 30\pi} = \frac{A_{o1}\omega_{c1}}{s + \omega_{c1}}$$

$$T_2(s) = \frac{\pm 3.162(900\pi^2)}{s^2 + (0.618034)30\pi s + 900\pi^2}$$

$$= \frac{A_{o2}\omega_{o2}^2}{s^2 + \dfrac{\omega_{o2}}{Q_2} s + \omega_{o2}^2}$$

and
$$T_3(s) = \frac{\pm 3.162(900\pi^2)}{s^2 + (1.618034)30\pi s + 900\pi^2}$$
$$= \frac{A_{o3}\omega_{o3}^2}{s^2 + \dfrac{\omega_{o3}}{Q_3}s + \omega_{o3}^2}$$

By matching coefficients, we get

$\omega_{c1} = 30\pi$ rad/sec

$A_{o1} = \pm 1$

$\omega_{o2} = 30\pi$ rad/sec

$A_{o2} = \pm 3.162$

$Q_2 = 1.618$

$\omega_{o3} = 30\pi$ rad/sec

$A_{o3} = \pm 3.162$

$Q_3 = 0.618$

14.3.2 Design

First Section The passband gain and cutoff frequency requirements for the low-pass filter are very low and can be satisfied using almost any general-purpose operational amplifier. The 101A op amp was chosen because its compensation capacitor is externally added. This capacitor will be increased from the standard 30 pF value to reduce the amplifier's bandwidth and increase the phase margin and, hence, reduce the chances for instability.

There are two first-order, low-pass circuits: one inverting and the other noninverting. Somewhat arbitrarily, the latter circuit, shown in Fig. 14.9, was chosen to implement the first section. Components R_1 and C_1 establish ω_{c1}, and R_2 is used to reduce the I_B error at the output to an I_{OS} error.

For the cutoff frequency,

$$\omega_{c1} = \frac{1}{R_1 C_1} = 30\pi \text{ rad/sec}$$

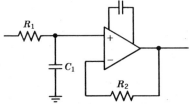

Figure 14.9 First filter section of active filter design example.

A combination of component values that satisfies this relationship is

$$R_1 = 156 \text{ k}\Omega$$

and

$$C_1 = 0.068 \text{ }\mu\text{F}$$

Resistor R_2 is made equal to R_1,

$$R_2 = 156 \text{ k}\Omega$$

to balance the resistance that each amplifier input terminal sees.

Second and Third Sections An IGMF filter was chosen to implement the two second-order filter sections. This particular configuration has a minimum number of components and has stable characteristics (see Fig. 9.18). The circuit, shown in Fig. 14.10, has a transfer function (see Section 9.2.2)

$$T(s) = -\frac{\dfrac{1}{R_3 R_4 C_2 C_3}}{s^2 + \left(\dfrac{1}{C_2}\right)\left(\dfrac{1}{R_3} + \dfrac{1}{R_4} + \dfrac{1}{R_5}\right)s + \dfrac{1}{R_4 R_5 C_2 C_3}}$$

$$= \frac{A_o \omega_o^2}{s^2 + \dfrac{\omega_o}{Q} s + \omega_o^2}$$

where the filter characteristics, in terms of the components of the second section, are

$$\omega_{o2} = \frac{1}{\sqrt{R_4 R_5 C_2 C_3}}$$

$$Q_2 = \sqrt{\frac{C_2}{R_4 R_5 C_3}} \frac{R_3 R_4 R_5}{R_4 R_5 + R_3 R_4 + R_3 R_5}$$

and

$$A_{o2} = -\frac{R_5}{R_3} < 0$$

The above set of equations may be solved simultaneously by arbitrarily specifying the values of any two of the circuit components. Most publications that specialize in

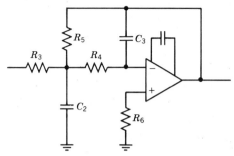

Figure 14.10 Second section.

the theory and design of active filters provide the appropriate expressions for the circuit components in terms of the filter characteristics.

In this particular case, we can arbitrarily set the values of C_2 and C_3 provided that the resulting values of R_3, R_4, and R_5 are real and within the constraints imposed by the open-loop input and output resistances of the operational amplifier. The selection of capacitors is preferred because of the limited choice of standard capacitor values.

If we solve the above set of three simultaneous equations for R_3 and R_4, we obtain

$$R_3 = -\frac{R_5}{A_{o2}}$$

and

$$R_4 = \frac{1}{(\omega_{o2})^2 \, R_5 C_2 C_3}$$

It can be shown that using these two relationships in the expression for Q_2 results in a second-order equation for R_5 in terms of C_2, C_3, and the filter characteristics. If the discriminant of this quadratic equation is made equal to zero by letting

$$C_2 = 4Q_2^2(1 - A_{o2})C_3$$

then R_5 will have a single value given by

$$R_5 = \frac{1}{2Q_2 \omega_{o2} C_3}$$

The step-by-step procedure for finding the values of the circuit elements in Fig. 14.10 is as follows:

1. Arbitrarily select an appropriate value for C_3. Let

   $C_3 = 0.03 \ \mu F$.

2. Use $C_2 = 4Q_2^2(1 - A_{o2})C_3$ to find C_2:

   $C_2 = 1.31\ \mu F$.

3. Use $R_5 = 1/(2Q_2\omega_{o2}C_3)$ to find R_5:

   $R_5 = 109.3\ k\Omega$.

4. Use $R_3 = -R_5/A_{o2}$ to find R_3:

   $R_3 = 34.57\ k\Omega$.

5. Use $R_4 = 1/(\omega_{o2}^2 R_5 C_2 C_3)$ to find R_4:

   $R_4 = 26.21\ k\Omega$.

6. To reduce the I_B error at the circuit output to the smaller I_{OS} error, make $R_6 = R_4 + (R_3 \parallel R_5)$:

   $R_6 = 52.47\ k\Omega$

If we repeat this procedure for the third section shown in Fig. 14.11 and let $C_5 = 0.047\ \mu F$, we get

$$C_4 = 0.30\ \mu F$$
$$R_9 = 182.7\ k\Omega$$
$$R_7 = 57.76\ k\Omega$$
$$R_8 = 43.71\ k\Omega$$

and

$$R_{10} = 87.59\ k\Omega$$

The complete schematic of the active filter is shown in Fig. 14.12.

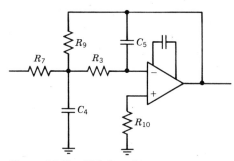

Figure 14.11 Third section.

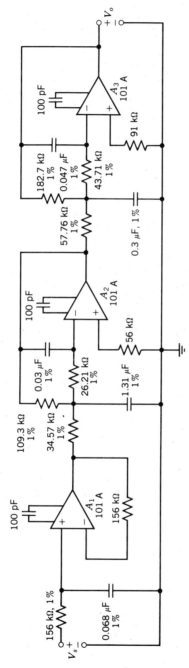

Figure 14.12 Schematic of a fifth-order low-pass active filter.

The resistors and capacitors have reasonable values. The calculations procedure may be repeated to obtain a different set of values, if we assume a different value for C_3 or C_5 in Step 1. Using CAD, a complete set of component values can be quickly calculated for different standard values of capacitors. The components should be at least 1% to meet the 2% accuracy requirement unless a trimming procedure is to be used. High quality resistors and capacitors should be used.

The compensation capacitor for the 101A op amp has been increased to 100 pF. For a 30 pF capacitor, the GB_f of the op amp is approximately 900 kHz (see Section 15.4.7). If we triple the value of the capacitor, we reduce the bandwidth by one-third and increase the phase margin. GB_f is, in general, not a guaranteed parameter or precise. The reduced bandwidth is still sufficient to provide a difference of approximately 80 dB of gain between the open- and closed-loop values. The effect of distributed capacitance on stability is minimal in this example because of the low frequencies. However, it is always a good design practice to avoid any type of coupling between the outputs and inputs when laying out the printed circuit board.

Of course, all of the good design practices discussed in the last example should also be employed here. The dc supplies must be adequately bypassed, lead lengths should be kept to a minimum, heavy grounds with no loops should be employed, device protection schemes should be used if the environment warrants them, and so on.

High-Pass and Band-Pass Design Procedures The step-by-step procedure for finding the values of the circuit elements of the IGMF high-pass filter in Fig. 14.13 (see Section 9.2) is as follows:

1. Given: ω_o, Q, and A_o.
2. Arbitrarily select an appropriate value for C_1.
3. Let $C_1 = C_3$.
4. Use

$$R_2 = \frac{\frac{1}{Q}}{\omega_o C_1 \left(2 + \frac{1}{|A_o|}\right)}$$

to find R_2.

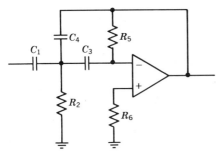

Figure 14.13 IGMF high-pass filter.

5. Use

$$R_5 = \frac{|A_o|\left(2 + \dfrac{1}{|A_o|}\right)}{\dfrac{\omega_o C_1}{Q}}$$

to find R_5.

6. Use $R_6 = R_5$ to find R_6.

The step-by-step procedure for finding the values of the circuit elements of the IGMF band-pass filter in Fig. 14.14 (see Section 9.2) is as follows:

1. Given: ω_o, Q, and A_o.
2. Arbitrarily select an appropriate value for C_3.
3. Let $C_4 = C_3$.
4. Use

$$R_1 = \frac{1}{|A_o|\omega_o C_3}$$

to find R_1.

5. Use

$$R_2 = \frac{1}{(2Q - |A_o|)\omega_o C_3}$$

to find R_2.

6. Use

$$R_5 = \frac{2Q}{\omega_o C_3}$$

to find R_5.

7. Use $R_6 = R_5$ to find R_6.

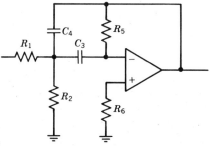

Figure 14.14 IGMF bandpass filter.

14.4 SUMMARY

The circuits chosen in the design examples of this chapter represent a cross section of the more important applications areas of the operational amplifier. The first design example used two linear circuits that performed mathematical operations, the summing amplifier and the integrator. Included in this design example was a nonlinear application, the comparator. The second design example illustrated how basic, second-order active filter circuits are cascaded to implement a higher-order response that must meet a given set of specifications.

The intent of this chapter was not to provide the analog designer with a compendium of design examples to apply when needed. Rather, the intent was to illustrate how the body of knowledge encompassed in the first 13 chapters is translated to sound engineering practices.

PART V

Integrated Circuit Amplifiers

Chapter 15

Circuit Description of the 741

15.1 PROLOGUE

The 741-type amplifier represents the second generation of monolithic operational amplifiers and incorporates design techniques that are significantly different from the conventional ones that make use of discrete components. These new techniques are a direct result of the limitations of integrated-circuit technology that restrict the choice of the types of devices and magnitudes of the passive elements that may be efficiently used in a given design, while also imposing large tolerances on most parameter values. These drawbacks force the IC designer to make extensive use of the advantages provided by the IC technology by replacing passive elements with active ones and parameter tolerance by parameter matching at little or no extra cost.

Our coverage of the 741-type amplifier contains three parts: (1) a qualitative discussion of the operation of the block diagram of the amplifier, (2) a dc analysis to determine the dc currents and voltages in the device, and (3) a small-signal analysis to determine the voltage gain, input and output resistances, and bandwidth. Along the way, those circuit characteristics that *cause* the key secondary parameters will be identified along with their typical values.

In order to simplify the coverage of the design of an operational amplifier, certain assumptions are made (unless specified otherwise): (1) all transistors are operating in their active region, (2) the output resistance of the transistor is extremely high, (3) base currents are negligible (4) β is high, especially for the *npn* transistors, (5) and the output of the op amp is at 0 V in the absence of a differential input voltage.

The numbers used in this chapter represent nominal values for the characteristics of the amplifier and are intended to provide its user with a background on the derivation and quantitative evaluation of the nonidealities of the device.

15.2 CIRCUIT OPERATION

The general-purpose bipolar operational amplifier consists of three basic blocks: the differential amplifier, the common-emitter amplifier, and the emitter-follower. A simplified, discrete version of each of these circuits was shown in Fig. 2.1*b*.

The complete schematic for the 741 operational amplifier is shown in Fig. 15.1. This rather complex circuit is reduced to the simplified, conceptual version in Fig. 15.2, which is used to identify the key building blocks. In Fig. 15.2, transistor pairs Q_1–Q_3 and Q_2–Q_4 are the equivalent of a pair of *pnp* transistors connected as a differential amplifier. Q_1 and Q_2 are emitter-followers that drive the differential pair of common-base *pnp* transistors Q_3 and Q_4. Transistors Q_5 and Q_6, represented by collector resistors in the discrete equivalent circuit shown in Fig. 2.2*a*, are active loads. To avoid the use of large resistors, which use an excessive die area, transistors are used to provide the required large voltage drops at the relatively low values of current through them.

The characteristics provided by the differential-input first stage include moderate (about 500) voltage gain, high input resistance, low input (bias) current, high common-mode rejection, and low offset voltage. In addition, the first stage shifts the differential-input signal and converts it to a single-ended voltage. The level shifting is accomplished by *pnp* transistors Q_3 and Q_4 that allow the input signal to be shifted in the negative direction. The differential to single-ended conversion is accomplished by taking only one of the outputs of the emitter-coupled pair Q_5–Q_6 and feeding it directly into the single-ended common-emitter circuit.

Transistors Q_{16} and Q_{17} form an *npn* Darlington pair, with its inherent high-β property, and implement the common-emitter second stage. The high input impedance of the Darlington pair in the second stage reduces the loading of the input stage. The second stage uses an active load (Q_{13B}) to increase the voltage gain and is frequency-compensated by adding a capacitor C_c, internal or external, between the base and collector of the Darlington pair. This stage, along with the differential amplifier circuit, produces the overall voltage gain of the op amp.

The output stage is a class *AB*, push-pull emitter-follower formed by transistors Q_{14} and Q_{20}. The *pnp* transistor Q_{20} is buffered by Q_{23} because of its relatively low current gain compared to the *npn* Q_{14}. Transistors Q_{20} and Q_{23} form the equivalent of a Darlington pair whose β more closely matches the β of the *npn* Q_{14}. Key properties of an emitter-follower are current gain, low output resistance, and wide bandwidth. The push-pull action of the output circuit allows the op amp to swing to positive and negative output voltages and, thus, to source or sink output current.

Transistor Q_{13} is a multicollector, lateral *pnp* device. It has two collectors with the collector of Q_{13A} exposed to $\frac{1}{4}$ of the emitter area and that of Q_{13B} exposed to the

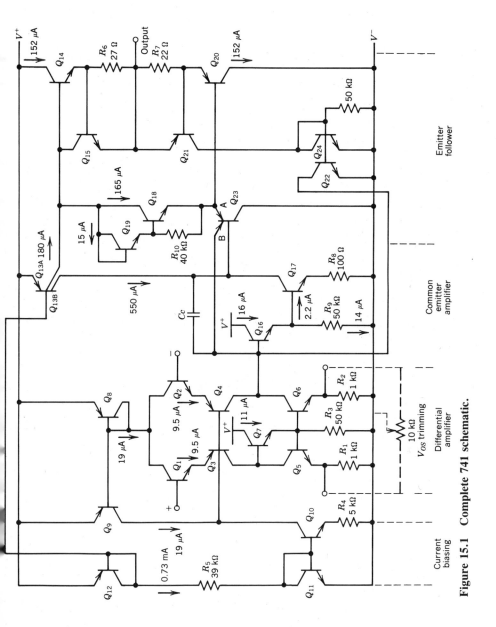

Figure 15.1 Complete 741 schematic.

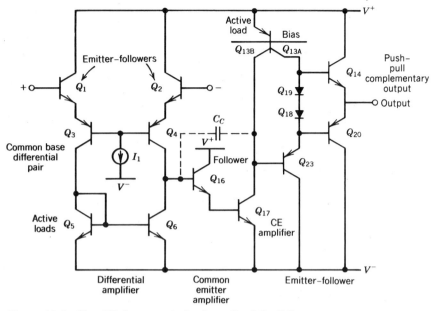

Figure 15.2 Simplified, conceptual schematic of the 741.

remaining $\frac{3}{4}$. This means that Q_{13A} will conduct $\frac{1}{4}$ of the total current in the transistor and, of course, Q_{13B} will conduct the remaining $\frac{3}{4}$. As we will see shortly, the total current through Q_{13} is determined by the base-emitter voltage of Q_{12} (Fig. 15.1).

15.3 DC CURRENTS AND VOLTAGES

The dc analysis of an integrated circuit relies heavily upon transistor parameters. For example, it is commonplace in IC designs to have the base-emitter voltage *directly* establish the value of the collector current, or vice versa. For these cases, we make use of relationships

$$I_C = I_S \exp \frac{V_{BE}}{V_T} \tag{15.1a}$$

or

$$V_{BE} = V_T \ln \frac{I_C}{I_S} \tag{15.1b}$$

These relationships are derived from the Ebers–Moll equations assuming $V_{BE} \gg V_T$. Voltage $V_T = kT/q$ at room temperature is about 26 mV, while I_S is the saturation current. The saturation current along with the many other transistor parameters will differ for the *npn* and *pnp* transistors and will also vary for the various forms of IC transistor structures. A typical set of device parameters, many of which we will use in our calculations, is shown in Fig. 15.3. Parameters such as V_A, τ_F, C_{je}, and C_μ will be explained when their application is required.

In the discussion that follows, we often use a shorthand notation to represent the currents, voltages, resistances, and conductance of the various devices and circuits involved. Generally speaking, currents and voltages are identified by $I_{\text{subscript}}$ (device) or $V_{\text{subscript}}$ (device), while their respective directions and polarities are assumed to be those causing these currents and voltages to be positive. To illustrate this notation, $I_C(Q_2)$ in Fig. 15.1 is the current entering the collector of Q_2, while $V_{BE}(Q_5)$ is the forward base-to-emitter voltage for Q_5. On the other hand, $I(R_5)$ is the current assumed to be entering R_5 from the collector of Q_{12}. Similarly, $r_o(Q_4)$ and $R_o(Q_4)$ represent the output resistances of the transistor Q_4 and the Q_4 circuit, respectively.

Integrated circuits, compared to their discrete counterparts, are frequently current-biased instead of voltage-biased. The current-biasing subcircuit composed of Q_{12}, Q_{11}, Q_{10}, Q_9, Q_8, and Q_{13} directly establishes the quiescent collector currents of Q_1, Q_2, Q_3, Q_4, Q_5, Q_6, Q_{14}, Q_{17}, Q_{18}, Q_{19}, Q_{20}, and Q_{23} in Fig. 15.1. The series combination of Q_{11} and Q_{12}, connected as diodes, and R_5 establishes the reference current $I(R_5)$. Assuming supply voltages of ± 15 V, we have

$$I(R_5) = \frac{V^+ + V^- - V_{BE}(Q_{11}) - V_{BE}(Q_{12})}{39 \text{ k}\Omega} = 730 \text{ }\mu\text{A} \quad (15.2)$$

where the base-to-emitter voltages are assumed to be 0.7 V.

Matched transistors Q_{10} and Q_{11} together with R_4 form a mirrorlike Widlar current source. The relationship between $I(R_5)$ and $I_C(Q_{10})$ is found by summing the voltages around the loop formed by Q_{10}, Q_{11}, and R_4,

$$V_{BE}(Q_{11}) - V_{BE}(Q_{10}) - I_C(Q_{10})R_4 = 0 \quad (15.3)$$

Using (15.1b), we can express (15.3) in the form of

$$V_T \ln \frac{I_C(Q_{11})}{I_S(Q_{11})} - V_T \ln \frac{I_C(Q_{10})}{I_S(Q_{10})} - I_C(Q_{10})R_4 = 0 \quad (15.4)$$

	npn		*pnp*
β	250	β	50
V_A	130 V	V_A	50 V
I_S	10^{-14} A	I_S	$2 \cdot 10^{-15}$ A
τ_F	0.35 nsec	τ_F	30 nsec
C_{je}	1 pF	C_{je}	0.3 pF
C_μ	0.3 pF	C_μ	1 pF

Figure 15.3 Typical 741 device parameters.

For identical transistors, $I_S(Q_{10})$ and $I_S(Q_{11})$ are equal, reducing (15.4) to

$$V_T \ln \frac{I_C(Q_{11})}{I_C(Q_{10})} = I_C(Q_{10})R_4 \tag{15.5}$$

This is a transcendental equation that may be solved for $I_C(Q_{10})$ by trial-and-error. For $I_C(Q_{11}) = 0.73$ mA, $R_4 = 5$ kΩ, and $V_T = 26$ mV, the collector current of Q_{10} is

$$I_C(Q_{10}) = 19 \ \mu\text{A}$$

This current flows from V^+ to V^- through the emitter and collector of Q_9. Transistor Q_9 will have a V_{BE} determined by (15.1b) and corresponding to a collector current of 19 μA. This voltage is impressed across the base-emitter junction of Q_8. Since Q_8 and Q_9 are also matched, the collector current of Q_8 will also be 19 μA. Transistor Q_8 is the current source that drives the equivalent *pnp* differential-pair formed by Q_1 through Q_4. Hence each of the input transistors conducts a quiescent current of $I_C(Q_{10})/2$ or

$$I_C(Q_{1-4}) = 9.5 \ \mu\text{A}$$

Since the collectors of Q_5 and Q_6, the active load transistors, are in series with Q_3 and Q_4, their collector currents are also 9.5 μA:

$$I_C(Q_{5,6}) = 9.5 \ \mu\text{A}$$

At this point we have sufficient information to calculate the nominal values of the input bias (I_B) and offset (I_{OS}) currents. Assuming that the input *npn* transistors have betas of 250, we find the nominal value of the base (bias) currents using

$$I_B = \frac{9.5 \ \mu\text{A}}{250} = 38 \text{ nA} \tag{15.6}$$

Offset current I_{OS}, on the other hand, is caused by the mismatches in the input devices and is, typically, 10% of the bias current or

$$I_{OS} = 3.8 \text{ nA} \tag{15.7}$$

Input offset voltage V_{OS} is a random parameter with a mean value near zero. V_{OS} arises from randomly occurring mismatches between the components making up the input stage. Mismatches are caused by the *difference* in values of R_1 and R_2, transistor saturation currents $I_S(Q_{1-6})$, and transistor betas of Q_5 and Q_6. V_{OS} is determined using statistics and the standard deviation of the distribution of the above mismatches, assuming that $\Delta R/R = 1\%$, $\Delta I_S/I_S = 5\%$, and $\Delta \beta = 50$ equals

$$\sigma(V_{OS}) = 2.6 \text{ mV} \tag{15.8}$$

The largest single offset contribution is the mismatch of I_S and β in the active-load devices Q_5 and Q_6.

The offset voltage can be nulled out by unbalancing the emitter currents of Q_5 and Q_6 and, therefore, of Q_1 and Q_2, thus forcing a matching of the base-to-emitter voltages of the input transistors. This is accomplished for the 741 by using a 10 kΩ potentiometer in a manner shown in Fig. 15.1.

We now return to the discussion of the dc biasing of the 741. From (15.2), the collector current of Q_{12} is 730 μA, which means that the transistor will have a V_{BE} corresponding to the value given by (15.1). This value of V_{BE} establishes the base-to-emitter voltage of Q_{13} and, hence, its total collector current of 730 μA. Transistor Q_{13} has two collectors and, because of the size and location of these collectors, Q_{13B} will conduct three-fourths of the total current of 730 μA with Q_{13A} conducting the remaining one-quarter. Hence,

$$I_C(Q_{13B}) = 550 \ \mu A$$

and

$$I_C(Q_{13A}) = 180 \ \mu A$$

The emitter current of Q_7 in the input stage consists of the base currents of Q_5 and Q_6, which are neglected, and the current flowing in R_3. The voltage across this resistor is the sum of the base-to-emitter drop of Q_5 (or Q_6) and the drop across the 1 kΩ emitter resistor R_1 (or R_2). The base-to-emitter voltage, assuming an I_S of 10^{-14} A and a collector current of 9.5 μA in (15.1b), is 537 mV. The voltage drop across R_1, being 9.5 mV, provides a potential difference across R_3 of about 547 mV with a resulting collector current for Q_7 of

$$I_C(Q_7) = 11 \ \mu A$$

The voltage drop across R_9 in the common-emitter stage is the sum of the voltage drop across R_8 (100 Ω) and the base-to-emitter voltage of Q_{17}. The collector current of Q_{17}

$$I_C(Q_{17}) = I_C(Q_{13B}) = 550 \ \mu A$$

causes a voltage drop across R_8,

$$V(R_8) = (550 \ \mu A) \cdot (100 \ \Omega) = 55 \ mV$$

Again assuming an I_S of 10^{-14} A, we can use (15.1b) to calculate the base-to-emitter voltage of Q_{17}:

$$V_{BE}(Q_{17}) = V_T \ln \frac{550 \ \mu A}{10^{-14} \ A} = 642 \ mV$$

which summed with the voltage drop across R_8 results in the voltage developed across R_9,

$$V(R_9) = V(R_8) + V_{BE}(Q_{17}) = 697 \text{ mV}$$

Since the collector current of Q_{17} in the common-emitter stage is relatively high, we must estimate its base current to determine $I_C(Q_{16})$. Assuming a β of 250, we have

$$I_B(Q_{17}) = \frac{550 \text{ μA}}{250} = 2.2 \text{ μA}$$

The current through R_9 added to this base current gives us the collector current in Q_{16},

$$I_C(Q_{16}) = \frac{697 \text{ mV}}{50 \text{ kΩ}} + 2.2 \text{ μA} = 16 \text{ μA}$$

We now shift our attention to the output stage. Neglecting the base current of Q_{14} and noticing that Q_{15} is normally off, we see that the collector current of Q_{13A} (180 μA) splits into two parts: $I_C(Q_{18})$ and $I_C(Q_{19})$. The collector currents of Q_{18} and Q_{19} are

$$I_C(Q_{19}) = \frac{V_{BE}(Q_{18})}{R_{10}} = 15 \text{ μA}$$

and

$$I_C(Q_{18}) = I_C(Q_{13A}) - I_C(Q_{19}) = 165 \text{ μA}$$

The currents through these transistors, with their corresponding values of base-to-emitter voltages, establish the quiescent current of Q_{14} and Q_{20}. The voltage $V_{BE}(Q_{18})$ is found by using KCL at the node where the collectors of Q_{13A}, Q_{18}, and Q_{19} are connected. The expression for this voltage is a transcendental equation that produces a value of $V_{BE}(Q_{18})$ of 0.600 V.

The output stage is a class AB, push-pull emitter-follower composed of Q_{14} and Q_{20}. Neglecting the voltage drops across R_6 and R_7, we have

$$V_{BE}(Q_{19}) + V_{BE}(Q_{18}) = V_{BE}(Q_{14}) + V_{BE}(Q_{20})$$

or, using (15.1b),

$$V_T \ln \frac{I_C(Q_{19})I_C(Q_{18})}{I_S(Q_{19})I_S(Q_{18})} = V_T \ln \frac{I_C(Q_{14})I_C(Q_{20})}{I_S(Q_{14})I_S(Q_{20})}$$

Since there is no output (load) current, $I_C(Q_{14})$ and $I_C(Q_{20})$ must be equal. Solving the last expression for either of these currents, we get

$$I_C(Q_{14}) = I_C(Q_{20}) = \sqrt{I_C(Q_{19})I_C(Q_{18})}\sqrt{\frac{I_S(Q_{14})I_S(Q_{20})}{I_S(Q_{18})I_S(Q_{19})}} \qquad (15.9)$$

The collector current that flows in the output transistors thus depends on their saturation current, which in turn, depends on their physical geometry. The output transistors must carry heavier currents than the other transistors in the circuit and hence their geometry is larger, typically by a factor of three. Thus, using (15.9), we get

$$I_C(Q_{14}) = I_C(Q_{20}) = 3\sqrt{(15 \ \mu A)(165 \ \mu A)} = 149 \ \mu A$$

Transistors Q_{15}, Q_{21}, Q_{22}, and Q_{24} implement the current-limiting feature of the op amp. Transistor Q_{15} protects the amplifier when it is sourcing current, turning on only when the voltage across R_6 exceeds about 550 mV, or when the output current is 550 mV/27 Ω or 20 mA. Transistors Q_{21}, Q_{22}, and Q_{24} perform a similar function when the amplifier is sinking current. Transistor Q_{21} turns on when the current through R_7 exceeds about 550 mV/22 Ω or 25 mA. When Q_{21} turns on, it initiates a chain reaction and causes Q_{24} and then Q_{22} to also turn on. Transistor Q_{22} will sink some of the base current of Q_{16}, which will force the amplifier's output to go into a current-limiting mode.

The extra emitter of Q_{23} protects Q_{16} from dissipating an excessive amount of power. Under nonlinear operating conditions, the collector current of Q_{16} could rise to nearly 20 mA. The large value of V_{CE} for this transistor, nearly 30 V, and the high current could cause the device to dissipate as much as 600 mW. Q_{23B} prevents Q_{17} from saturating and limits the power dissipation of Q_{16} by shunting the base drive away from it.

The specific values of the voltages at the various portions of the circuit can be calculated since the various transistor parameters and the collector currents are known. As an example, let us determine the quiescent collector voltage of Q_3 (to ground). This voltage is

$$V_C(Q_3) = V^- + V(R_3) + V_{BE}(Q_7)$$

The potential drop across R_3 is

$$V(R_3) = (11 \ \mu A) (50 \ k\Omega) = 0.55 \ V$$

and because the base-to-emitter voltage of Q_7 is

$$V_{BE}(Q_7) = V_T \ln \frac{11 \ \mu A}{10^{-14} \ A} = 0.52 \ V$$

then

$$V_C(Q_3) = -15 \ V + 0.55 \ V + 0.52 \ V = -13.93 \ V$$

15.4 SMALL-SIGNAL ANALYSIS

15.4.1 Transistor Hybrid-π Model

Under small-signal conditions, analog circuits operate with signal levels that are "small" compared to the bias currents and voltages in the circuit. To calculate the magnitudes of voltage gain and terminal resistances of the 741 op amp, the small-signal model of the bipolar transistor is used. The first-order, hybrid-π, small-signal model of the transistor, shown in Fig. 15.4, has only those parameters necessary to provide the required information for our purposes.

Transconductance (g_m) In simple terms, transconductance relates the collector current of the transistor to its base-to-emitter voltage. Since we are talking about the small-signal or incremental changes of these variables, the differentials define their relationship,

$$g_m = \frac{\partial I_C}{\partial V_{BE}} = \frac{i_c}{v_{be}}$$

Since

$$I_C = I_S \exp \frac{V_{BE}}{V_T}$$

then

$$g_m = \frac{I_c}{V_T} \qquad (15.10)$$

Transconductance is dependent on the specific value of the collector bias current and is given as $I_C/26$ mV at room temperature. For $I_C = 1$ mA, $g_m = 38$ mS or 38 mA/V.

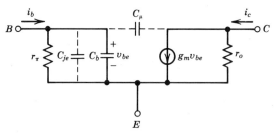

Figure 15.4 Basic hybrid-π model.

Input Resistance (r_π) The input voltage of the transistor (v_{be}) divided by the input current (i_b) defines the low-frequency input resistance or r_π,

$$r_\pi = \frac{v_{be}}{i_b} \tag{15.11}$$

$$= \beta \frac{v_{be}}{i_c} = \frac{\beta}{g_m}$$

Occasionally, it is necessary to convert the input resistance of the hybrid-π model to an equivalent resistance in series with the emitter. An example is the transistor T model where r_e is the emitter resistance in this small-signal, equivalent circuit. Since the emitter current is approximately beta times larger than the base current, the equivalent resistance effectively in series with the emitter will be

$$r_e \cong \frac{r_\pi}{\beta} = \frac{1}{g_m} = \frac{I_C}{V_T} \quad (\beta \gg 1) \tag{15.12}$$

At $I_C = 1$ mA ($g_m = 38$ mS) and for an *npn* transistor with a β of 250, r_π and r_e are 6.6 kΩ and 26 Ω at room temperature.

Output Resistance (r_o) A small-signal change in V_{CE} produces a corresponding change in I_C. The ratio defines the small-signal collector-to-emitter resistance of the bipolar transistor,

$$r_o = \frac{\partial V_{CE}}{\partial I_C} = \frac{v_{ce}}{i_c}$$

It can be shown that, while in the linear region of the transistor's characteristics,

$$r_o = \frac{V_A}{I_C} \tag{15.13}$$

The variation of I_C with V_{CE} is described by the Early effect and V_A, the Early voltage, is a common parameter for circuit analysis. Typical values of V_A for integrated-circuit transistors are from 50 to 130 V. Thus, for $I_C = 1$ mA, the range of r_o is in the order of 50 to 130 kΩ.

Input Capacitances A change in the base-emitter voltage causes a change in the charge in the base region. The ratio of these changes defines the equivalent input capacitance C_b which, like r_π, is dependent on I_C. C_b is computed using g_m and the base transit time τ_F through

$$C_b = \tau_F g_m \tag{15.14}$$

where τ_F is from 0.1 to 1 nsec for integrated *npn* transistors and 20 to 40 nsec for certain *pnp* transistors.

The capacitance C_π, used in an expanded version of the hybrid-π model, is the sum of the base charging capacitance C_b and the emitter-base depletion layer capacitance C_{je},

$$C_\pi = C_b + C_{je} \qquad (15.15)$$

If we add C_π to the base-collector depletion layer capacitance C_μ, we can relate C_π and C_μ to the familiar current gain-bandwidth product f_T through

$$f_T = \frac{g_m}{2\pi(C_\pi + C_\mu)} \qquad (15.16)$$

C_{je} and C_μ, the two depletion layer capacitances, are typically less than 1 pF each, and the input capacitance is frequently approximated as

$$C_b \cong C_\pi \cong \frac{g_m}{2\pi f_T}$$

15.4.2 Basic Circuits

As mentioned previously, the three basic stages of a typical, bipolar IC operational amplifier are the differential amplifier, the common-emitter amplifier, and the emitter-follower output stage. However, the complete analysis reveals a common-base circuit, a Widlar current source, several Darlington pairs, among other types of circuit. An understanding of the basic characteristics and relationships of a common-emitter (*CE*), common-base (*CB*), common-collector (*CC*) or emitter-follower, and differential amplifier configurations is assumed as a necessary prerequisite for the reader to follow the analysis of the operational amplifier circuitry. The equivalent of a *CE* stage with a resistance in series with the emitter is frequently used in the analysis of integrated circuits, and a brief summary of its characteristics is presented here because of its relative importance.

The emitter resistance in the *CE* circuit shown in Fig. 15.5 has several noteworthy effects, including the reduction of the circuit's voltage gain and equivalent transconductance, and the increase of its input and output resistances. The emitter resistor actually causes negative feedback to occur since the voltage across it, which is proportional to the output current, subtracts directly from the input voltage.

The input resistance, like that of the emitter-follower, is the sum of r_π and the equivalent value of R_E reflected back to the base. To reflect R_E back to the base, we must multiply it by $\beta + 1$ because of the different current levels in the base and emitter. For the circuit of Fig. 5.15,

$$R_i = r_\pi + R_E(\beta + 1)$$

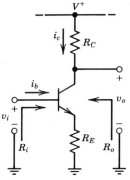

Figure 15.5 CE amplifier with emitter degeneration.

or, if $\beta \gg 1$,

$$R_i = r_\pi(1 + g_m R_E) \qquad (15.18)$$

using (15.11).

The equivalent transconductance G_m of the configuration of Fig. 15.5 relates the input voltage v_i and the output current i_c,

$$G_m = \frac{i_c}{v_i}$$

The input voltage equals

$$v_i = i_b r_\pi + (i_b + i_c)R_E$$
$$= i_c \left(\frac{r_\pi}{\beta} + R_E\right)$$

Solving for i_c/v_i produces

$$G_m = \frac{i_c}{v_i} = \frac{g_m}{1 + g_m R_E} \qquad (\beta \gg 1) \qquad (15.19)$$

An analogous analysis for the output resistance yields

$$R_o = r_o(1 + g_m R_E) \qquad (15.20)$$

The voltage gain of this circuit is the product of the circuit's transconductance and the parallel combination of the transistor's output resistance and the collector's load resistance R_L,

$$A_V = G_m (R_o \| R_L)$$

15.4.3 Differential Amplifier Circuit

The 741-type amplifier, under small-signal conditions, can be ultimately reduced to the equivalent circuit shown in Fig. 15.6. The equivalent circuit, with component values, provides sufficient information to calculate the voltage gain and the input and output resistances of the operational amplifier.

The voltage gain for the amplifier is established by the product of the voltage gains of the first two stages. Each of these gains is determined by the product of the equivalent circuit conductance and the load resistance seen by it. For the first stage shown in Fig. 15.7,

$$A_{v1} = G_{m1}(R_{o1} \| R_{i2})$$

The equivalent transconductance G_{m1} of the first stage is equal to the series combination of the transconductance of the Q_1–Q_3 or Q_2–Q_4 pairs. For one of these transistors,

$$g_m = \frac{I_C}{V_T} = \frac{9.5\ \mu A}{26\ mV} = 0.365\ mS$$

and for the differential amplifier circuit,

$$G_{m1} = \frac{g_m}{2} = 0.183\ mS \tag{15.21}$$

The output resistance R_{o1} of the first stage is the parallel combination of the output resistances of the Q_4 and Q_6 circuits,

$$R_{o1} = R_o(Q_4) \| R_o(Q_6) \tag{15.22}$$

For this parameter, we cannot idealize the output resistance of the transistor. Each transistor in the output circuit has a resistance in its emitter; R_2 in the Q_6 circuit and the emitter resistance of Q_2 ($r_e = 1/g_m$) in the Q_4 circuit. These emitter resistances make the output resistance of the transistor r_o appear higher by the amount $1 + g_m R_E$, where R_E is the resistance in series with the emitter.

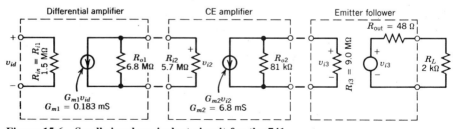

Figure 15.6 Small-signal equivalent circuit for the 741.

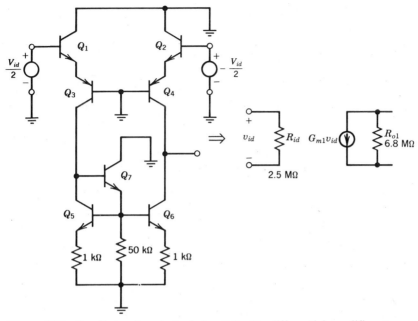

Figure 15.7 Small-signal equivalent circuit for the differential amplifier stage.

The typical value for the Early voltage V_A for the *npn* transistors in the 741 is 130 V and for the *pnp* transistors is 50 V. Thus, using (15.20) and (15.22) in conjunction with Fig. 15.7, we get

$$R_{o1} = r_o(Q_4)\left[1 + \frac{g_m(Q_4)}{g_m(Q_2)}\right] \| r_o(Q_2)\left[1 + g_m(Q_6)R_2\right] \quad (15.23)$$

where

$$r_o(Q_2) = r_o(Q_6) = \frac{130\text{ V}}{9.5\ \mu\text{A}} = 13.7\text{ M}\Omega$$

$$r_o(Q_4) = \frac{50\text{ V}}{9.5\ \mu\text{A}} = 5.3\text{ M}\Omega$$

$$1 + \frac{g_m(Q_4)}{g_m(Q_2)} = 2$$

and

$$1 + g_m(Q_6)R_2 = 1 + \left(\frac{9.5\ \mu\text{A}}{26\text{ mV}}\right)1\text{ k}\Omega = 1.365$$

Thus,

$$R_{o1} = (10.6 \| 18.7) \text{ M}\Omega = 6.8 \text{ M}\Omega$$

while the input resistance R_{i2} of the common-emitter stage will be found in the following section.

The differential input resistance R_{id} of the first stage and the op amp is the sum of the input resistances r_π of the four transistors Q_1 through Q_4 where

$$r_\pi(Q_3) = r_\pi(Q_4) = \frac{\beta}{g_m} = \frac{250}{0.4 \text{ mS}}$$
$$= 625 \text{ k}\Omega$$

and

$$r_\pi(Q_1) = r_\pi(Q_2) = \frac{50}{0.4 \text{ mS}} = 125 \text{ k}\Omega$$

For the operational amplifier,

$$R_{id} = 2(625 \text{ k}\Omega) + 2(125 \text{ k}\Omega) \quad (15.24)$$
$$= 1.5 \text{ M}\Omega$$

15.4.4 Common-Emitter Circuit

The small-signal configuration and equivalent circuits for the second stage are shown in Fig. 15.8. From them, we will again calculate the equivalent conductance and the input and output resistances of the stage. Since the voltage gain of Q_{16} is essentially unity, the equivalent transconductance G_{m2} of this stage is determined by Q_{17} or

$$G_{m2} = \frac{g_m(Q_{17})}{1 + g_m(Q_{17})R_E}$$
$$= \frac{21 \text{ mS}}{1 + 21 \text{ mS}(100 \text{ }\Omega)} \quad (15.25)$$
$$= 6.8 \text{ mS}$$

The input resistance R_{i2} of the second stage is equal to r_π of Q_{16} plus β times the resistance in its emitter. The resistance in the emitter, shown in Fig. 15.8b, is the parallel combination of R_9 and the input resistance of Q_{17} designated as R_{eq1}. Resistance R_{eq1} is given by

$$R_{eq1} = r_\pi(Q_{17}) + \beta R_8$$

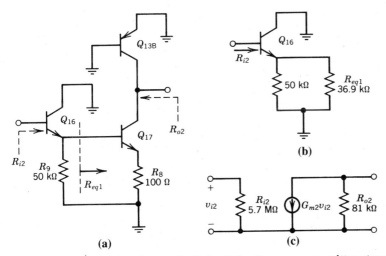

Figure 15.8 Small-signal equivalent circuit for the common-emitter stage

yielding

$$R_{eq1} = 11.8 \text{ k}\Omega + 250(100 \text{ }\Omega)$$
$$= 37 \text{ k}\Omega$$

The parallel combination of R_9 and R_{eq1} is about 21 kΩ. To reflect this resistance back to the base of Q_{16}, we must multiply it by its beta. The sum of this reflected resistance and $r_\pi(Q_{16})$ gives

$$R_{i2} = r_\pi(Q_{16}) + \beta(R_{eq1} \parallel R_9)$$
$$= 406 \text{ k}\Omega + 250(21 \text{ k}\Omega) \quad (15.26)$$
$$= 5.7 \text{ M}\Omega$$

The output resistance of the second stage is the parallel combination of the output resistances of the Q_{17} and Q_{13B} circuits,

$$R_{o2} = r_o(Q_{13B}) \parallel R_o(Q_{17})$$

Since Q_{17} has R_8 in its emitter, its output resistance $r_o(Q_{17})$ [$V_A = 130$ V, $I_C(Q_{17}) = 0.55$ mA] is multiplied by $1 + g_m(Q_{17})R_8$,

$$R_o(Q_{17}) = r_o(Q_{17})[1 + g_m(Q_{17})R_8]$$
$$= 236 \text{ k}\Omega (3.1)$$
$$= 736 \text{ k}\Omega$$

For the output resistance of Q_{13B} [$V_A = 50$ V, $I_C(Q_{13B}) = 0.55$ mA]

$$r_o(Q_{13B}) = 91 \text{ k}\Omega$$

which produces an equivalent output resistance of

$$R_{o2} = R_o(Q_{17}) \parallel r_o(Q_{13B}) \qquad (15.27)$$
$$= 81 \text{ k}\Omega$$

The equivalent circuit for the second stage shown in Fig. 15.8c reflects the key parameters.

15.4.5 Emitter-Follower Stage

The output stage, shown in a simplified form in Fig. 15.9a, is a push-pull emitter-follower with a voltage gain near unity. Since it is a push-pull circuit and drives a variable load, we will look at only one-half of this circuit using a load of 2 kΩ at 2 mA for calculation purposes. The results will vary, of course, for other loads. The equivalent circuits used to calculate the input and output resistance are shown in Figs. 15.9b and 15.9c.

The input resistance R_{i3} of the circuit shown in Fig. 15.9b is equal to the sum of $r_\pi(Q_{23A})$ and the equivalent resistance in the emitter of Q_{23A}, R_{eq2}, reflected to its base,

$$R_{i3} = r_\pi(Q_{23A}) + \beta R_{eq2} \qquad (15.28)$$

Resistance R_{eq2} is the sum of the diode resistances $r_d(Q_{18})$ and $r_d(Q_{19})$, and the parallel combination of the output resistance of Q_{13A} and the input resistance of Q_{14} designated as R_{eq3},

$$R_{eq2} = r_d(Q_{18}) + r_d(Q_{19}) + [r_o(Q_{13A}) \parallel R_{eq3}]$$

Resistance R_{eq3}, on the other hand, equals the sum of $r_\pi(Q_{14})$ and the load resistance reflected to its base,

$$R_{eq3} = r_\pi(Q_{14}) + \beta R_L$$

For $I_C(Q_{14}) = 2$ mA and $\beta = 250$, $r_\pi(Q_{14}) = 3.25$ kΩ and

$$R_{eq3} = 3.25 \text{ k}\Omega + 250(2 \text{ k}\Omega)$$
$$= 503 \text{ k}\Omega$$

Transistors Q_{23A} and Q_{13A} operate at a current of 180 μA. The diode resistances are small when compared to other series resistances and are neglected. For this current level, $r_o(Q_{13A}) = 278$ kΩ and $r_\pi(Q_{23A}) = 7.2$ kΩ, and transistors Q_{13} and Q_{23A} have a $\beta = 50$ and $V_A = 50$ V. Resistances R_{eq2} and R_{i3} evaluate to

$$R_{eq2} = 278 \text{ k}\Omega \parallel 503 \text{ k}\Omega$$
$$= 179 \text{ k}\Omega$$

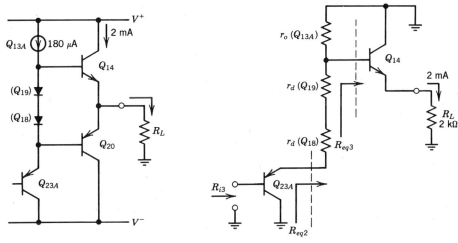

(a) Simplified output stage

(b) Small-signal equivalent circuit for calculation of R_{i3}

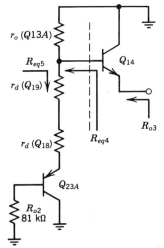

(c) Small-signal equivalent circuit for calculation of R_{o3}

Figure 15.9 Output stage.

and

$$R_{i3} = r_\pi(Q_{23A}) + \beta R_{eq2}$$
$$= 7.2 \text{ k}\Omega + 50(179 \text{ k}\Omega)$$
$$= 9.0 \text{ M}\Omega$$

We will now calculate the small-signal output resistance of the third stage using Fig. 15.9c. The resistance seen looking into the emitter of Q_{14} is the sum of $r_\pi(Q_{14})$

Circuit Description of the 741

and the equivalent resistance seen at its base (R_{eq4}), both reflected back to the emitter. To reflect a resistance in the base back to the emitter, we divide it by β. Thus,

$$R_{o3} = \frac{r_\pi(Q_{14}) + R_{eq4}}{\beta} \quad (15.29)$$

Resistance R_{eq4} is the parallel combination of the output resistance of Q_{13A} and R_{eq5},

$$R_{eq4} = r_o(Q_{13A}) \parallel R_{eq5}$$

where

$$R_{eq5} = r_d(Q_{19}) + r_d(Q_{18}) + \frac{r_\pi(Q_{23A}) + 81 \text{ k}\Omega}{\beta}$$

Note that the resistance in the base of Q_{23A} is the output resistance R_{o2} of the second stage and, thus, is 81 kΩ. For $r_\pi(Q_{23A}) = 7.2$ kΩ, $\beta = 50$ for the *pnp*, and $r_d(Q_{18}) = r_d(Q_{19}) = 144$ Ω,

$$R_{eq5} = 0.288 \text{ k}\Omega + \frac{7.2 \text{ k}\Omega + 81 \text{ k}\Omega}{50}$$
$$= 2.03 \text{ k}\Omega$$

Since $r_o(Q_{13A}) = 278$ kΩ and $R_{eq5} = 2.03$ kΩ, then

$$R_{eq4} = 278 \text{ k}\Omega \parallel 2.03 \text{ k}\Omega = 2.02 \text{ k}\Omega$$

and, using (15.19), we obtain

$$R_{o3} = \frac{3.25 \text{ k}\Omega + 2.02 \text{ k}\Omega}{250}$$
$$= 21 \text{ }\Omega$$

To determine R_{out}, the output resistance of the op amp when Q_{14} is conducting, we must add R_{o3} to the series limiting resistance R_6, which is 27 Ω. Thus,

$$R_{out} = R_{o3} + R_{CL} \quad (15.30)$$
$$= 48 \text{ }\Omega$$

The value of the output resistance is a function of the operating point of the devices involved and, of course, frequency. Theoretically, R_{out} is equal to the same value when Q_{20} is conducting and Q_{14} is off. Practically, this is not true because the output stage is not perfectly symmetric. The difference in R_{out}, however, is masked by other factors.

15.4.6 Voltage Gain, and Input and Output Resistances

The complete small-signal equivalent circuit for the 741 op amp, including the values calculated in Sections 15.4.3 through 15.4.5, is shown in Fig. 15.6. The overall voltage gain with $R_L = 2$ kΩ is

$$\begin{aligned}A = A_v &= G_{m1}(R_{o1} \| R_{i2})G_{m2}(R_{o2} \| R_{i3}) \\ &= (0.183 \text{ mS})(6.8 \text{ M}\Omega \| 5.7 \text{ M}\Omega)(6.8 \text{ mS})(81 \text{ k}\Omega \| 9.0 \text{ M}\Omega) \\ &= (567)(546) = 310{,}000\end{aligned}$$

Both stages contribute about the same amount of voltage gain. These gains are beta-dependent because the various input and output resistances are beta-dependent. Thus the overall voltage gain varies with temperature and process variations. The assumed values for the current gain of the transistors, especially for the input stage where the current levels are low, are optimistic and, hence, the calculated voltage gain is somewhat higher than in the actual device.

The nominal values of the input and output resistance of the 741 are

$$R_{in} = 1.5 \text{ M}\Omega \tag{15.33}$$

and

$$R_{out} = 47 \text{ }\Omega \tag{15.34}$$

The calculated values of the op amp's gain, bandwidth, input resistance, and output resistance are not precise. The variations in the active device parameters alone would prevent an exact analysis. At the circuit level, the variations in these parameters can be minimized using a number of circuit design techniques.

15.4.7 Frequency Response

The frequency response of the 741 is dominated by the 30 pF compensation capacitor connected between the base of Q_{16} and the collector of Q_{17}, as shown in Fig. 15.10a. The -3 dB frequency for the small-signal, open-loop gain of the amplifier can be estimated by considering the effect of this capacitor. The equivalent circuit used to estimate this frequency is shown in Fig. 15.10b.

Resistance R_{ic} and equivalent input capacitance C_m in Fig. 15.10b form a low-pass filter and are used to compute the -3 dB frequency,

$$f_{-3 \text{ dB}} = \frac{1}{2\pi R_{ic} C_m} \tag{15.35}$$

where

$$R_{ic} = R_{o1} \| R_{i2}$$

358 Circuit Description of the 741

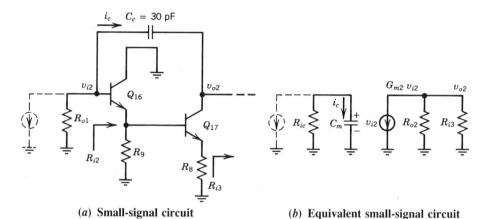

(a) Small-signal circuit   (b) Equivalent small-signal circuit

Figure 15.10 Compensation circuit.

Note that R_{o1} is the small-signal output resistance of the differential amplifier stage while R_{i2} is the small-signal input resistance of the common-emitter stage. From the data given in Sections 15.4.3 and 15.4.4,

$$R_{ic} = 6.8 \text{ M}\Omega \parallel 5.7 \text{ M}\Omega$$
$$= 3.1 \text{ M}\Omega$$

The reflected (Miller) capacitor C_m in Fig. 15.10b accounts for the current i_c through C_c in Fig. 15.10a. This current is given by

$$i_c = (v_{i2} - v_{o2}) s C_c = v_{i2} s C_m$$

Solving for C_m, we get

$$C_m = \left(1 - \frac{v_{o2}}{v_{i2}}\right) C_c$$

or

$$C_m = (1 - A_{v2})C_c$$

where A_{v2} is the voltage gain of the common-emitter stage given by

$$A_{v2} = -G_{m2}(R_{o2} \parallel R_{i3})$$

R_{o2} is the small-signal output resistance of the CE stage, and R_{i3} is the small-signal input resistance of the CC stage. Transconductance G_{m2} and resistances R_{o2} and R_{i3} were found to be 6.8 mS, 81 kΩ and 8 MΩ, respectively. Thus,

$$A_{v2} = -6.8 \text{ mS} (81 \text{ k}\Omega \parallel 9 \text{ M}\Omega)$$
$$= -546$$

and

$$C_m = (1 + 546)(30 \text{ pF})$$
$$= 16.4 \text{ nF}$$

The equivalent input capacitance is extremely large because of the high voltage gain of the common-emitter stage. The -3 dB frequency is then given by

$$f_{-3 \text{ dB}} = \frac{1}{2\pi(3.1 \text{ M}\Omega)(16.4 \text{ nF})}$$
$$= 3.1 \text{ Hz}$$

For a -3 dB frequency of 3.1 Hz and a small-signal voltage gain of 310,000, the gain-bandwidth product GB_f (in hertz) of the op amp is

$$GB_f = (3.1 \text{ Hz})(310,000) = 961 \text{ kHz} \tag{15.36}$$

Measured values of $f_{-3 \text{ dB}}$, A_v, and GB_f are closer to 5 Hz, 160,000, and 800 kHz. One reason for the discrepancy is because the beta of Q_{16} is much lower than the assumed value of 250 due to the lower value of its collector current.

We now turn our attention to the slew rate limit of the amplifier. Compensation capacitor C_c, connected across the Darlington pair of Q_{16} and Q_{17}, causes this stage to act as an integrator as shown in Fig. 15.11. Input current I_x is the output current from the first stage. The maximum current available to charge C_c to its appropriate level

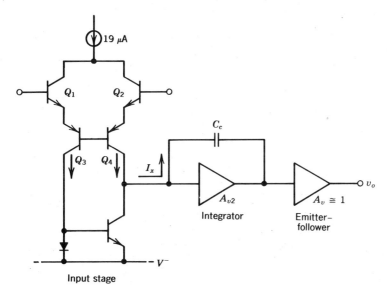

Figure 15.11 Simplified schematic of the 741 used for slew-rate calculations.

occurs when the input stage is overdriven (nonlinear operation), causing the Q_1–Q_3 pair to be cut off. Thus,

$$I_x = 19 \text{ μA}$$

With the second stage acting as an integrator, the maximum magnitude of the rate of change of the output voltage of the op amp, defined as its slew rate S_R, equals

$$S_R = -\left.\frac{dv_o}{dt}\right|_{max} = \frac{I_x}{C_c} = 0.63 \text{ V/μsec}$$

This theoretically calculated value is close to the value specified by the 741 manufacturers.

15.5 COMPARISON OF THE 741 AND THE 101

The schematic of the 101A shown in Fig. 15.12 shares many similarities with the one for the 741. The objectives in the design of both devices were to eliminate the major problem areas of the 709. Both designs solved these problems, made additional refinements, and today are the two leading general-purpose IC operational amplifiers.

The design philosophy of each is almost identical. Each amplifier has three stages; the differential, common-emitter, and emitter-follower amplifiers. The simplified, functional circuit of the 741 shown in Fig. 15.2 applies as well to the 101A. Both designs are historically important because they represent the next level of technology after the 709. They are also important conceptually for they still are used as the basis for an entire range of general-purpose devices, even those developed recently.

Similar circuits shared by the two include an input composite *pnp* differential pair, active loads in the first and second stages, a second-stage Darlington pair, an integrating capacitor (C_c) around the second stage, a class *AB* output circuit, and current biasing (although implemented differently).

The differences between the 741 and the 101A are in the frequency compensation, offset null, short-circuit protection, and current-biasing areas. The compensation capacitor C_c for the 101A is added externally and establishes a -3 dB frequency near 5 Hz. The user, however, has the option with the 101A of either increasing or decreasing C_c (to 3 pF) to suit particular application needs. The small-signal, open-loop frequency response for the 101A for C_c values of 30 and 3 pF is shown in Fig. 13.10.

A second difference between the 741 and the 101A is in the offset-null method and circuitry. The 741 unbalances the emitter currents of Q_5 and Q_6 (see Fig. 15.1) to force V_{OS} to zero. The 101A in Fig. 15.12, on the other hand, unbalances the collector currents of the same transistors. The external circuits used in each case are shown in Fig. 13.4.

Other differences between these devices are associated with the implementation of the short-circuit protection and current-biasing schemes. To the user of the op amp, these differences are transparent.

The 101, introduced in 1967, was followed by an improved version, the 101A, in

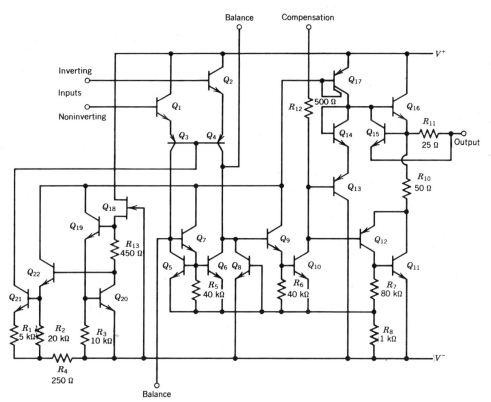

Figure 15.12 101A schematic.

1968. The changes were in the biasing circuit for the input stage. These changes provided a better control of the input current characteristics over the operating temperature range.

15.6 SUMMARY

The general-purpose, monolithic operational amplifier consists of three basic stages. The three basic, integrated-circuit amplifier circuits are like their discrete component counterparts in that their small-signal behaviors are similar. The differential amplifier circuit provides voltage gain, a high input resistance, level shifting, and a single-ended output signal. The common-emitter circuit provides voltage gain, high input resistance, and controls the bandwidth. The emitter-follower provides current gain, a low output resistance, and can sink or source current.

Unlike their discrete counterparts, the monolithic versions of these circuits are implemented differently. The implementations of the integrated-circuit versions take advantage of the positive features of IC technology and utilize current-source biasing, matched transistors and resistors, multiple-emitter and multiple-collector transistor structures, and inexpensive active loads.

Appendix A

Data Sheets: 741, 101A, 108A, OP-07, 111, 140, 555

Number
741
101A
108A
OP-07
111
140
555

Function
Operational amplifier
Operational amplifier
Operational amplifier
Operational amplifier
Comparator
Regulator
Timer

364 Appendix A

A Schlumberger Company

μA741
Operational Amplifier

Linear Products

Description
The μA741 is a high performance Monolithic Operational Amplifier constructed using the Fairchild Planar epitaxial process. It is intended for a wide range of analog applications. High common mode voltage range and absence of latch-up tendencies make the μA741 ideal for use as a voltage follower. The high gain and wide range of operating voltage provides superior performance in integrator, summing amplifier, and general feedback applications.

- **NO FREQUENCY COMPENSATION REQUIRED**
- **SHORT-CIRCUIT PROTECTION**
- **OFFSET VOLTAGE NULL CAPABILITY**
- **LARGE COMMON MODE AND DIFFERENTIAL VOLTAGE RANGES**
- **LOW POWER CONSUMPTION**
- **NO LATCH-UP**

Connection Diagram
8-Pin Metal Package

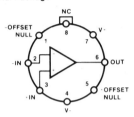

(Top View)

Pin 4 connected to case

Absolute Maximum Ratings
Supply Voltage
μA741A, μA741, μA741E ± 22 V
μA741C ± 18 V
Internal Power Dissipation
(Note 1)
Metal Package 500 MW
DIP 310 mW
Flatpak 570 mW
Differential Input Voltage ± 30 V
Input Voltage (Note 2) ± 15 V
Storage Temperature Range
Metal Package and Flatpak −65°C to +150°C
DIP −55°C to +125°C

Operating Temperature Range
Military (μA741A, μA741) −55°C to +125°C
Commercial (μA741E, μA741C) 0°C to +70°C
Pin Temperature (Soldering 60 s)
Metal Package, Flatpak, and
Ceramic DIP 300°C
Molded DIP (10 s) 260°C
Output Short Circuit Duration
(Note 3) Indefinite

Equivalent Circuit

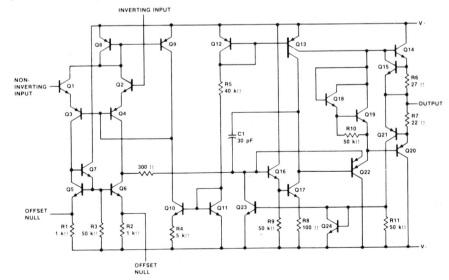

FAIRCHILD LINEAR INTEGRATED CIRCUITS • µA741

µA741A

ELECTRICAL CHARACTERISTICS ($V_S = \pm15V$, $T_A = 25°C$ unless otherwise specified)

PARAMETERS (see definitions)		CONDITIONS	MIN	TYP	MAX	UNITS
Input Offset Voltage		$R_S \leq 50\Omega$		0.8	3.0	mV
Average Input Offset Voltage Drift					15	µV/°C
Input Offset Current				3.0	30	nA
Average Input Offset Current Drift					0.5	nA/°C
Input Bias Current				30	80	nA
Power Supply Rejection Ratio		$V_S = +10, -20$; $V_S = +20, -10V$, $R_S = 50\Omega$		15	50	µV/V
Output Short Circuit Current			10	25	35	mA
Power Dissipation		$V_S = \pm20V$		80	150	mW
Input Impedance		$V_S = \pm20V$	1.0	6.0		MΩ
Large Signal Voltage Gain		$V_S = \pm20V$, $R_L = 2k\Omega$, $V_{OUT} = \pm15V$	50			V/mV
Transient Response	Rise Time			0.25	0.8	µs
(Unity Gain)	Overshoot			6.0	20	%
Bandwidth (Note 4)			.437	1.5		MHz
Slew Rate (Unity Gain)		$V_{IN} = \pm10V$	0.3	0.7		V/µs
The following specifications apply for $-55°C \leq T_A \leq +125°C$						
Input Offset Voltage					4.0	mV
Input Offset Current					70	nA
Input Bias Current					210	nA
Common Mode Rejection Ratio		$V_S = \pm20V$, $V_{IN} = \pm15V$, $R_S = 50\Omega$	80	95		dB
Adjustment For Input Offset Voltage		$V_S = \pm20V$	10			mV
Output Short Circuit Current			10		40	mA
Power Dissipation		$V_S = \pm20V$, $-55°C$			165	mW
		$+125°C$			135	mW
Input Impedance		$V_S = \pm20V$	0.5			MΩ
Output Voltage Swing		$V_S = \pm20V$, $R_L = 10k\Omega$	±16			V
		$R_L = 2k\Omega$	±15			V
Large Signal Voltage Gain		$V_S = \pm20V$, $R_L = 2k\Omega$, $V_{OUT} = \pm15V$	32			V/mV
		$V_S = \pm5V$, $R_L = 2k\Omega$, $V_{OUT} = \pm2V$	10			V/mV

NOTES
1. Rating applies to ambient temperatures up to 70°C. Above 70°C ambient derate linearly at 6.3mW/°C for the metal can, 8.3mW/°C for the DIP and 7.1mW/°C for the Flatpak.
2. For supply voltages less than ±15V, the absolute maximum input voltage is equal to the supply voltage.
3. Short circuit may be to ground or either supply. Rating applies to +125°C case temperature or 75°C ambient temperature.
4. Calculated value from: $BW(MHz) = \dfrac{0.35}{\text{Rise Time } (\mu s)}$

Operational Amplifiers/Buffers

LM101A/LM201A/LM301A Operational Amplifiers

General Description

The LM101A series are general purpose operational amplifiers which feature improved performance over industry standards like the LM709. Advanced processing techniques make possible an order of magnitude reduction in input currents, and a redesign of the biasing circuitry reduces the temperature drift of input current. Improved specifications include:

- Offset voltage 3 mV maximum over temperature (LM101A/LM201A)
- Input current 100 nA maximum over temperature (LM101A/LM201A)
- Offset current 20 nA maximum over temperature (LM101A/LM201A)
- Guaranteed drift characteristics
- Offsets guaranteed over entire common mode and supply voltage ranges
- Slew rate of 10V/μs as a summing amplifier

This amplifier offers many features which make its application nearly foolproof: overload protection on the input and output, no latch-up when the common mode range is exceeded, freedom from oscillations and compensation with a single 30 pF capacitor. It has advantages over internally compensated amplifiers in that the frequency compensation can be tailored to the particular application. For example, in low frequency circuits it can be overcompensated for increased stability margin. Or the compensation can be optimized to give more than a factor of ten improvement in high frequency performance for most applications.

In addition, the device provides better accuracy and lower noise in high impedance circuitry. The low input currents also make it particularly well suited for long interval integrators or timers, sample and hold circuits and low frequency waveform generators. Further, replacing circuits where matched transistor pairs buffer the inputs of conventional IC op amps, it can give lower offset voltage and drift at a lower cost.

The LM101A is guaranteed over a temperature range of $-55°C$ to $+125°C$, the LM201A from $-25°C$ to $+85°C$, and the LM301A from $0°C$ to $70°C$.

Schematic** and Connection Diagrams (Top Views)

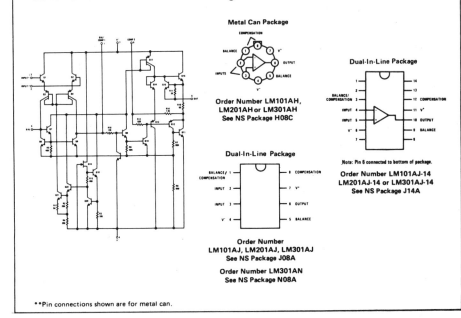

**Pin connections shown are for metal can.

Absolute Maximum Ratings

	LM101A/LM201A	LM301A
Supply Voltage	±22V	±18V
Power Dissipation (Note 1)	500 mW	500 mW
Differential Input Voltage	±30V	±30V
Input Voltage (Note 2)	±15V	±15V
Output Short Circuit Duration (Note 3)	Indefinite	Indefinite
Operating Temperature Range	−55°C to +125°C (LM101A) −25°C to +85°C (LM201A)	0°C to +70°C
Storage Temperature Range	−65°C to +150°C	−65°C to +150°C
Lead Temperature (Soldering, 10 seconds)	300°C	300°C

Electrical Characteristics (Note 4)

PARAMETER	CONDITIONS	LM101A/LM201A MIN	LM101A/LM201A TYP	LM101A/LM201A MAX	LM301A MIN	LM301A TYP	LM301A MAX	UNITS
Input Offset Voltage LM101A, LM201A, LM301A	$T_A = 25°C$, $R_S \leq 50\ k\Omega$		0.7	2.0		2.0	7.5	mV
Input Offset Current	$T_A = 25°C$		1.5	10		3.0	50	nA
Input Bias Current	$T_A = 25°C$		30	75		70	250	nA
Input Resistance	$T_A = 25°C$	1.5	4.0		0.5	2.0		MΩ
Supply Current	$T_A = 25°C$, $V_S = ±20V$		1.8	3.0				mA
	$V_S = ±15V$					1.8	3.0	mA
Large Signal Voltage Gain	$T_A = 25°C$, $V_S = ±15V$, $V_{OUT} = ±10V$, $R_L \geq 2\ k\Omega$	50	160		25	160		V/mV
Input Offset Voltage	$R_S \leq 50\ k\Omega$			3.0			10	mV
	$R_S \leq 10\ k\Omega$							mV
Average Temperature Coefficient of Input Offset Voltage	$R_S \leq 50\ k\Omega$		3.0	15		6.0	30	μV/°C
	$R_S \leq 10\ k\Omega$							μV/°C
Input Offset Current				20			70	nA
	$T_A = T_{MAX}$							nA
	$T_A = T_{MIN}$							nA
Average Temperature Coefficient of Input Offset Current	$25°C \leq T_A \leq T_{MAX}$		0.01	0.1		0.01	0.3	nA/°C
	$T_{MIN} \leq T_A \leq 25°C$		0.02	0.2		0.02	0.6	nA/°C
Input Bias Current				0.1			0.3	μA
Supply Current	$T_A = T_{MAX}$, $V_S = ±20V$		1.2	2.5				mA
Large Signal Voltage Gain	$V_S = ±15V$, $V_{OUT} = ±10V$, $R_L \geq 2k$	25			15			V/mV
Output Voltage Swing	$V_S = ±15V$							
	$R_L = 10\ k\Omega$	±12	±14		±12	±14		V
	$R_L = 2\ k\Omega$	±10	±13		±10	±13		V
Input Voltage Range	$V_S = ±20V$	±15						V
	$V_S = ±15V$		+15, −13		±12	+15, −13		V
Common-Mode Rejection Ratio	$R_S \leq 50\ k\Omega$	80	96		70	90		dB
	$R_S \leq 10\ k\Omega$							dB
Supply Voltage Rejection Ratio	$R_S \leq 50\ k\Omega$	80	96		70	96		dB
	$R_S \leq 10\ k\Omega$							dB

Note 1: The maximum junction temperature of the LM101A is 150°C, and that of the LM201A/LM301A is 100°C. For operating at elevated temperatures, devices in the TO-5 package must be derated based on a thermal resistance of 150°C/W, junction to ambient, or 45°C/W, junction to case. The thermal resistance of the dual-in-line package is 187°C/W, junction to ambient.

Note 2: For supply voltages less than ±15V, the absolute maximum input voltage is equal to the supply voltage.

Note 3: Continuous short circuit is allowed for case temperatures to 125°C and ambient temperatures to 75°C for LM101A/LM201A, and 70°C and 55°C respectively for LM301A.

Note 4: Unless otherwise specified, these specifications apply for C1 = 30 pF, ±5V $\leq V_S \leq$ ±20V and −55°C $\leq T_A \leq$ +125°C (LM101A), ±5V $\leq V_S \leq$ ±20V and −25°C $\leq T_A \leq$ +85°C (LM201A), ±5V $\leq V_S \leq$ ±15V and 0°C $\leq T_A \leq$ +70°C (LM301A).

National Semiconductor

Operational Amplifiers/Buffers

LM108A/LM208A/LM308A, LM308A-1, LM308A-2 Operational Amplifiers

General Description

The LM108/LM108A series are precision operational amplifiers having specifications about a factor of ten better than FET amplifiers over their operating temperature range. In addition to low input currents, these devices have extremely low offset voltage, making it possible to eliminate offset adjustments, in most cases, and obtain performance approaching chopper stabilized amplifiers.

The devices operate with supply voltages from ±2V to ±18V and have sufficient supply rejection to use unregulated supplies. Although the circuit is interchangeable with and uses the same compensation as the LM101A, an alternate compensation scheme can be used to make it particularly insensitive to power supply noise and to make supply bypass capacitors unnecessary. Outstanding characteristics include:

- Offset voltage guaranteed less than 0.5 mV
- Maximum input bias current of 3.0 nA over temperature
- Offset current less than 400 pA over temperature
- Supply current of only 300 µA, even in saturation
- Guaranteed 5 µV/°C drift.
- Guaranteed 1 µV/°C for LM308A-1

The low current error of the LM108A series makes possible many designs that are not practical with conventional amplifiers. In fact, it operates from 10 MΩ source resistances, introducing less error than devices like the 709 with 10 kΩ sources. Integrators with drifts less than 500 µV/sec and analog time delays in excess of one hour can be made using capacitors no larger than 1 µF.

The LM208A is identical to the LM108A, except that the LM208A has its performance guaranteed over a −25°C to 85°C temperature range, instead of −55°C to 125°C. The LM308A devices have slightly-relaxed specifications and performance guaranteed over a 0°C to 70°C temperature range.

Compensation Circuits

Standard Compensation Circuit

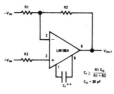

Alternate* Frequency Compensation

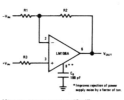

Feedforward Compensation

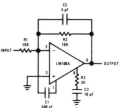

Typical Applications

Sample and Hold

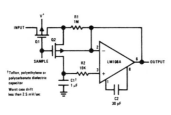

High Speed Amplifier with Low Drift and Low Input Current

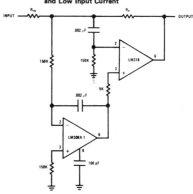

LM108A/LM208A
Absolute Maximum Ratings

Supply Voltage	±20V
Power Dissipation (Note 1)	500 mW
Differential Input Current (Note 2)	±10 mA
Input Voltage (Note 3)	±15V
Output Short-Circuit Duration	Indefinite
Operating Temperature Range LM108A	−55°C to 125°C
LM208A	−25°C to 85°C
Storage Temperature Range	−65°C to 150°C
Lead Temperature (Soldering, 10 sec)	300°C

Electrical Characteristics (Note 4)

PARAMETER	CONDITIONS	MIN	TYP	MAX	UNITS
Input Offset Voltage	$T_A = 25°C$		0.3	0.5	mV
Input Offset Current	$T_A = 25°C$		0.05	0.2	nA
Input Bias Current	$T_A = 25°C$		0.8	2.0	nA
Input Resistance	$T_A = 25°C$	30	70		MΩ
Supply Current	$T_A = 25°C$		0.3	0.6	mA
Large Signal Voltage Gain	$T_A = 25°C, V_S = ±15V$ $V_{OUT} = ±10V, R_L \geq 10\ k\Omega$	80	300		V/mV
Input Offset Voltage				1.0	mV
Average Temperature Coefficient of Input Offset Voltage			1.0	5.0	µV/°C
Input Offset Current				0.4	nA
Average Temperature Coefficient of Input Offset Current			0.5	2.5	pA/°C
Input Bias Current				3.0	nA
Supply Current	$T_A = +125°C$		0.15	0.4	mA
Large Signal Voltage Gain	$V_S = ±15V, V_{OUT} = ±10V$ $R_L \geq 10\ k\Omega$	40			V/mV
Output Voltage Swing	$V_S = ±15V, R_L = 10\ k\Omega$	±13	±14		V
Input Voltage Range	$V_S = ±15V$	±13.5			V
Common Mode Rejection Ratio		96	110		dB
Supply Voltage Rejection Ratio		96	110		dB

Note 1: The maximum junction temperature of the LM108A is 150°C, while that of the LM208A is 100°C. For operating at elevated temperatures, devices in the TO-5 package must be derated based on a thermal resistance of 150°C/W, junction to ambient, or 45°C/W, junction to case. The thermal resistance of the dual-in-line package is 100°C/W, junction to ambient.

Note 2: The inputs are shunted with back-to-back diodes for overvoltage protection. Therefore, excessive current will flow if a differential input voltage in excess of 1V is applied between the inputs unless some limiting resistance is used.

Note 3: For supply voltages less than ±15V, the absolute maximum input voltage is equal to the supply voltage.

Note 4: These specifications apply for $±5V \leq V_S \leq ±20V$ and $−55°C \leq T_A \leq 125°C$, unless otherwise specified. With the LM208A, however, all temperature specifications are limited to $−25°C \leq T_A \leq 85°C$.

370 Appendix A

OP-07
ULTRA-LOW OFFSET VOLTAGE OPERATIONAL AMPLIFIER

FEATURES
- Ultra-Low V_{OS} .... $10\mu V$
- Ultra-Low V_{OS} Drift .... $0.2\mu V/°C$
- Ultra-Stable vs Time .... $0.2\mu V$/Month
- Ultra-Low Noise .... $0.35\mu V_{p-p}$
- No External Components Required
- Replaces Chopper Amps at Lower Cost
- Single-Chip Monolithic Construction
- Large Input Voltage Range .... ±14.0V
- Wide Supply Voltage Range .... ±3V to ±18V
- Fits 725, 108A/308A, 741, AD510 Sockets

GENERAL DESCRIPTION

The OP-07 Series represents a breakthrough in monolithic operational amplifier performance — V_{OS} of $10\mu V$, TCV_{OS} of $0.2\mu V/°C$ and long-term stability of $0.2\mu V$/month are achieved by a low-noise, chopper-less bipolar input transistor amplifier circuit. Complete elimination of external components for offset nulling, frequency compensation and device protection permits extreme miniaturization and optimization of system Mean-Time-Between-Failure Rates in high-performance aerospace/defense and industrial applications. Excellent device interchangeability provides reduced system assembly time and eliminates field recalibrations.

True differential inputs with wide input voltage range and outstanding common mode rejection provide maximum flexibility and performance in high-noise environments and non-inverting applications. Low bias currents and extremely-high input impedances are maintained over the entire temperature range.

The OP-07 provides unparalleled performance for low noise, high-accuracy amplification of very low-level signals in transducer applications. Devices are available in chip form for use in hybrid circuitry. The OP-07 is a direct replacement for 725, 108A/308A*, and OP-05 amplifiers; 741-types may be directly replaced by removing the 741's nulling potentiometer.

*TO-99 package only. For Matched Dual see OP-207.

PIN CONNECTIONS & ORDERING INFORMATION

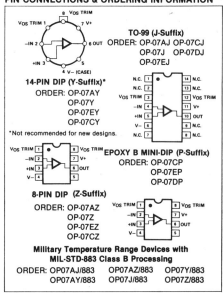

TO-99 (J-Suffix)
ORDER: OP-07AJ OP-07CJ
OP-07J OP-07DJ
OP-07EJ

14-PIN DIP (Y-Suffix)*
ORDER: OP-07AY
OP-07Y
OP-07EY
OP-07CY
*Not recommended for new designs.

EPOXY B MINI-DIP (P-Suffix)
ORDER: OP-07CP
OP-07EP
OP-07DP

8-PIN DIP (Z-Suffix)
ORDER: OP-07AZ
OP-07Z
OP-07EZ
OP-07CZ

Military Temperature Range Devices with MIL-STD-883 Class B Processing
ORDER: OP07AJ/883   OP07AZ/883   OP07Y/883
       OP07AY/883   OP07J/883   OP07Z/883

SIMPLIFIED SCHEMATIC

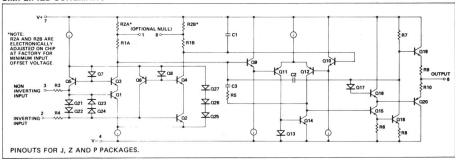

PINOUTS FOR J, Z AND P PACKAGES.

The premium performance of this product is achieved through an advanced processing technology. All Precision Monolithics products are guaranteed to meet or exceed published specifications.   ©1981 PMI   TECHNICAL SPECIFICATIONS SEPTEMBER 1981

Data Sheet OP-07

OP-07 ULTRA-LOW OFFSET VOLTAGE OPERATIONAL AMPLIFIER

ELECTRICAL CHARACTERISTICS at $V_S = \pm 15V$, $-55°C \leq T_A \leq +125°C$, unless otherwise noted.

PARAMETER	SYMBOL	CONDITIONS	OP-07A MIN	OP-07A TYP	OP-07A MAX	OP-07 MIN	OP-07 TYP	OP-07 MAX	UNITS
Input Offset Voltage	V_{OS}	(Note 1)	—	25	60	—	60	200	μV
Average Input Offset Voltage Drift Without External Trim	TCV_{OS}	(Note 3)	—	0.2	0.6	—	0.3	1.3	μV/°C
With External Trim	TCV_{OSn}	$R_p = 20k\Omega$	—	0.2	0.6	—	0.3	1.3	
Input Offset Current	I_{OS}		—	0.8	4.0	—	1.2	5.6	nA
Average Input Offset Current Drift	TCI_{OS}	(Note 3)	—	5	25	—	8	50	pA/°C
Input Bias Current	I_B		—	±1.0	±4.0	—	±2.0	±6.0	nA
Average Input Bias Current Drift	TCI_B		—	8	25	—	13	50	pA/°C
Input Voltage Range	IVR		±13.0	±13.5	—	±13.0	±13.5	—	V
Common Mode Rejection Ratio	CMRR	$V_{CM} = \pm 13V$	106	123	—	106	123	—	dB
Power Supply Rejection Ratio	PSRR	$V_S = \pm 3V$ to $\pm 18V$	94	106	—	94	106	—	dB
Large Signal Voltage Gain	A_{VO}	$R_L \geq 2k\Omega$, $V_O = \pm 10V$	200	400	—	150	400	—	V/mV
Maximum Output Voltage Swing	V_{OM}	$R_L \geq 2k\Omega$	±12.0	±12.6	—	±12.0	±12.6	—	V

TYPICAL OFFSET VOLTAGE TEST CIRCUIT

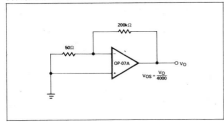

TYPICAL LOW FREQUENCY NOISE TEST CIRCUIT

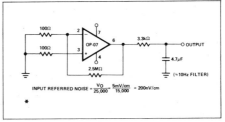

OPTIONAL OFFSET NULLING CIRCUIT

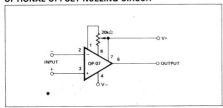

BURN-IN CIRCUIT

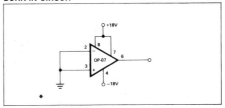

National Semiconductor

Voltage Comparators

LM111/LM211 Voltage Comparator[†]

General Description

The LM111 and LM211 are voltage comparators that have input currents nearly a thousand times lower than devices like the LM106 or LM710. They are also designed to operate over a wider range of supply voltages: from standard ±15V op amp supplies down to the single 5V supply used for IC logic. Their output is compatible with RTL, DTL and TTL as well as MOS circuits. Further, they can drive lamps or relays, switching voltages up to 50V at currents as high as 50 mA. Outstanding characteristics include:

- Operates from single 5V supply
- Input current: 150 nA max. over temperature
- Offset current: 20 nA max. over temperature

[†]See application hints LM311

- Differential input voltage range: ±30V
- Power consumption: 135 mW at ±15V

Both the inputs and the outputs of the LM111 or the LM211 can be isolated from system ground, and the output can drive loads referred to ground, the positive supply or the negative supply. Offset balancing and strobe capability are provided and outputs can be wire OR'ed. Although slower than the LM106 and LM710 (200 ns response time vs 40 ns) the devices are also much less prone to spurious oscillations. The LM111 has the same pin configuration as the LM106 and LM710.

The LM211 is identical to the LM111, except that its performance is specified over a $-25°C$ to $85°C$ temperature range instead of $-55°C$ to $125°C$.

Auxiliary Circuits**

**Note: Pin connections shown on schematic diagram and typical applications are for TO-5 package.

Offset Balancing

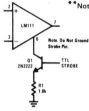
Strobing

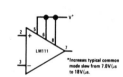
Increasing Input Stage Current*

*Increases typical common mode slew from 7.0V/µs to 18V/µs.

Typical Applications**

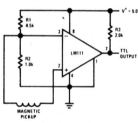
Detector for Magnetic Transducer

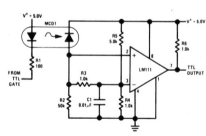
Digital Transmission Isolator

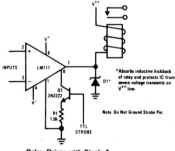
Relay Driver with Strobe*

*Absorbs inductive kickback of relay and protects IC from severe voltage transients on V^{++} line.

Note. Do Not Ground Strobe Pin.

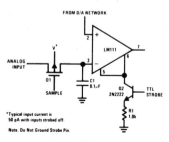
Strobing off Both Input* and Output Stages

*Typical input current is 50 pA with inputs strobed off.

Note. Do Not Ground Strobe Pin.

Absolute Maximum Ratings

Total Supply Voltage (V_{84})	36V
Output to Negative Supply Voltage (V_{74})	50V
Ground to Negative Supply Voltage (V_{14})	30V
Differential Input Voltage	±30V
Input Voltage (Note 1)	±15V
Power Dissipation (Note 2)	500 mW
Output Short Circuit Duration	10 sec
Operating Temperature Range LM111	−55°C to 125°C
LM211	−25°C to 85°C
Storage Temperature Range	−65°C to 150°C
Lead Temperature (soldering, 10 sec)	300°C
Voltage at Strobe Pin	$V^+ - 5V$

Electrical Characteristics (Note 3)

PARAMETER	CONDITIONS	MIN	TYP	MAX	UNITS
Input Offset Voltage (Note 4)	$T_A = 25°C$, $R_S \leq 50k$		0.7	3.0	mV
Input Offset Current (Note 4)	$T_A = 25°C$		4.0	10	nA
Input Bias Current	$T_A = 25°C$		60	100	nA
Voltage Gain	$T_A = 25°C$	40	200		V/mV
Response Time (Note 5)	$T_A = 25°C$		200		ns
Saturation Voltage	$V_{IN} \leq -5$ mV, $I_{OUT} = 50$ mA $T_A = 25°C$		0.75	1.5	V
Strobe ON Current (Note 6)	$T_A = 25°C$		3.0		mA
Output Leakage Current	$V_{IN} \geq 5$ mV, $V_{OUT} = 35V$ $T_A = 25°C$, $I_{STROBE} = 3$ mA		0.2	10	nA
Input Offset Voltage (Note 4)	$R_S \leq 50k$			4.0	mV
Input Offset Current (Note 4)				20	nA
Input Bias Current				150	nA
Input Voltage Range	$V^+ = 15V$, $V^- = -15V$, Pin 7 Pull-Up May Go To 5V	−14.5	13.8,−14.7	13.0	V
Saturation Voltage	$V^+ \geq 4.5V$, $V^- = 0$ $V_{IN} \leq -6$ mV, $I_{SINK} \leq 8$ mA		0.23	0.4	V
Output Leakage Current	$V_{IN} \geq 5$ mV, $V_{OUT} = 35V$		0.1	0.5	µA
Positive Supply Current	$T_A = 25°C$		5.1	6.0	mA
Negative Supply Current	$T_A = 25°C$		4.1	5.0	mA

Note 1: This rating applies for ±15V supplies. The positive input voltage limit is 30V above the negative supply. The negative input voltage limit is equal to the negative supply voltage or 30V below the positive supply, whichever is less.

Note 2: The maximum junction temperature of the LM111 is 150°C, while that of the LM211 is 110°C. For operating at elevated temperatures, devices in the TO-5 package must be derated based on a thermal resistance of 150°C/W, junction to ambient, or 45°C/W, junction to case. The thermal resistance of the dual-in-line package is 100°C/W, junction to ambient.

Note 3: These specifications apply for $V_S = \pm 15V$ and Ground pin at ground, and $-55°C \leq T_A \leq +125°C$, unless otherwise stated. With the LM211, however, all temperature specifications are limited to $-25°C \leq T_A \leq +85°C$. The offset voltage, offset current and bias current specifications apply for any supply voltage from a single 5V supply up to ±15V supplies.

Note 4: The offset voltages and offset currents given are the maximum values required to drive the output within a volt of either supply with a 1 mA load. Thus, these parameters define an error band and take into account the worst-case effects of voltage gain and input impedance.

Note 5: The response time specified (see definitions) is for a 100 mV input step with 5 mV overdrive.

Note 6: Do not short the strobe pin to ground; it should be current driven at 3 to 5 mA.

National Semiconductor

Voltage Regulators

LM140L/LM340L Series 3-Terminal Positive Regulators

General Description

The LM140L series of three terminal positive regulators is available with several fixed output voltages making them useful in a wide range of applications. The LM140LA is an improved version of the LM78LXX series with a tighter output voltage tolerance (specified over the full military temperature range), higher ripple rejection, better regulation and lower quiescent current. The LM140LA regulators have ±2% V_{OUT} specification, 0.04%/V line regulation, and 0.01%/mA load regulation. When used as a zener diode/resistor combination replacement, the LM140LA usually results in an effective output impedance improvement of two orders of magnitude, and lower quiescent current. These regulators can provide local on card regulation, eliminating the distribution problems associated with single point regulation. The voltages available allow the LM140LA to be used in logic systems, instrumentation, Hi-Fi, and other solid state electronic equipment. Although designed primarily as fixed voltage regulators, these devices can be used with external components to obtain adjustable voltages and currents.

The LM140LA/LM340LA are available in the low profile metal three lead TO-39 (H) and the LM340LA are also available in the plastic TO-92 (Z). With adequate heat sinking the regulator can deliver 100 mA output current. Current limiting is included to limit the peak output current to a safe value. Safe area protection for the output transistor is provided to limit internal power dissipation. If internal power dissipation becomes too high for the heat sinking provided, the thermal shutdown circuit takes over, preventing the IC from overheating.

For applications requiring other voltages, see LM117 Data Sheet.

Features

- Line regulation of 0.04%/V
- Load regulation of 0.01%/mA
- Output voltage tolerances of ±2% at $T_J = 25°C$ and ±4% over the temperature range (LM140LA) ±3% over the temperature range (LM340LA)
- Output current of 100 mA
- Internal thermal overload protection
- Output transistor safe area protection
- Internal short circuit current limit
- Available in metal TO-39 low profile package (LM140LA/LM340LA) and plastic TO-92 (LM340LA)

Output Voltage Options

LM140LA-5.0	5V	LM340LA-5.0	5V
LM140LA-12	12V	LM340LA-12	12V
LM140LA-15	15V	LM340LA-15	15V

Equivalent Circuit

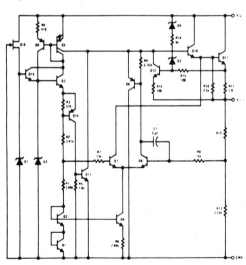

Connection Diagrams

TO-39 Metal Can Package (H)

Order Number:
LM140LAH-5.0   LM340LAH-5.0
LM140LAH-12   LM340LAH-12
LM140LAH-15   LM340LAH-15
See Package H03A

TO-92 Plastic Package (Z)

Order Number:
LM340LAZ-5.0
LM340LAZ-12
LM340LAZ-15
See Package Z03A

Absolute Maximum Ratings

Input Voltage
    5.0V, 12V, 15V Output Voltage Options   35V
Internal Power Dissipation (Note 1) Internally Limited
Operating Temperature Range
    LM140LA $-55°C$ to $+125°C$
    LM340LA $0°C$ to $70°C$
Maximum Junction Temperature $+150°C$
Storage Temperature Range
    Metal Can (H package) $-65°C$ to $+150°C$
    Molded TO-92 $-55°$ to $+150°C$
Lead Temperature (Soldering, 10 seconds) $+300°C$

Electrical Characteristics (Note 2)

Test conditions unless otherwise specified
$T_A = -55°C$ to $+125°C$ (LM140LA)
$T_A = 0°C$ to $+70°C$ (LM340LA)
$I_O = 40$ mA
$C_{IN} = 0.33 \mu F$, $C_O = 0.01 \mu F$

OUTPUT VOLTAGE OPTION			5.0V			12V			15V			
INPUT VOLTAGE (unless otherwise noted)			10V			19V			23V			UNITS
PARAMETER		CONDITIONS	MIN	TYP	MAX	MIN	TYP	MAX	MIN	TYP	MAX	
V_O	Output Voltage	$T_j = 25°C$	4.9	5	5.1	11.75	12	12.25	14.7	15	15.3	
	Output Voltage Over Temp. (Note 4)	LM140LA LM240LA, $I_O = 1-100$ mA, $I_O = 1-40$ mA and $V_{IN} = (\ )V$	4.8 (7.2-20)		5.2	11.5 (14.5-27)		12.5	14.4 (17.6-30)		15.6	V
		LM340LA, $I_O = 1-100$ mA or $I_O = 1-40$ mA and $V_{IN} = (\ )V$	4.85 (7-20)		5.15	11.65 (14.3-27)		12.35	14.55 (17.5-30)		15.45	
ΔV_O	Line Regulation	$I_O = 40$ mA, $T_j = 25°C$, $V_{IN} = (\ )V$	(7-25)	18	30	(14.2-30)	30	65	(17.3-30)	37	70	mV
		$I_O = 100$ mA, $V_{IN} = (\ )V$	(7.5-25)	18	30	(14.5-30)	30	65	(17.5-30)	37	70	
	Load Regulation	$T_j = 25°C$, $I_O = 1-40$ mA		5	20		10	40		12	50	mV
		$I_O = 1-100$ mA		20	40		30	80		35	100	
	Long Term Stability			12			24			30		mV 1000 hrs
I_Q	Quiescent Current	$T_j = 25°C$		3	4.5		3	4.5		3.1	4.5	mA
		$T_j = 125°C$			4.2			4.2			4.2	
ΔI_Q	Quiescent Current Change	$T_j = 25°C$, ΔLoad $I_O = 1-40$mA			0.1			0.1			0.1	mA
		ΔLine			0.5			0.5			0.5	
		$V_{IN} = (\ )V$	(7.5-25)			(14.3-30)			(17.5-30)			
V_N	Output Noise Voltage	$T_j = 25°C$ (Note 3), $f = 10$ Hz $- 10$ kHz		40			80			90		μV
$\Delta V_{IN} / \Delta V_{OUT}$	Ripple Rejection	$f = 120$ Hz, $V_{IN} = (\ )V$	55 (7.5-18)	62		47 (14.5-25)	54		45 (17.5-28.5)	52		dB
	Input Voltage Required to Maintain Line Regulation	$T_j = 25°C$, $I_O = 40$ mA	7			14.2			17.3			V

Note 1: Thermal resistance of the Metal Can Package (H) without a heat sink is 40°C/W junction to case and 140°C/W junction to ambient. Thermal resistance of the TO-92 package is 180°C/W junction to ambient with 0.4 inch leads from PC board and 160°C/W junction to ambient with 0.125 inch lead length to a PC board.

Note 2: The maximum steady state usable output current and input voltage are very dependent on the heat sinking and/or lead length of the package. The data above represent pulse test conditions with junction temperatures as indicated at the initiation of tests.

Note 3: It is recommended that a minimum load capacitor of $0.01 \mu F$ be used to limit the high frequency noise bandwidth.

Note 4: The temperature coefficient of V_{OUT} is typically within $0.01\% V_O/°C$.

LM555/LM555C Timer

Industrial Blocks

General Description

The LM555 is a highly stable device for generating accurate time delays or oscillation. Additional terminals are provided for triggering or resetting if desired. In the time delay mode of operation, the time is precisely controlled by one external resistor and capacitor. For astable operation as an oscillator, the free running frequency and duty cycle are accurately controlled with two external resistors and one capacitor. The circuit may be triggered and reset on falling waveforms, and the output circuit can source or sink up to 200 mA or drive TTL circuits.

- Adjustable duty cycle
- Output can source or sink 200 mA
- Output and supply TTL compatible
- Temperature stability better than 0.005% per °C
- Normally on and normally off output

Applications

- Precision timing
- Pulse generation
- Sequential timing
- Time delay generation
- Pulse width modulation
- Pulse position modulation
- Linear ramp generator

Features

- Direct replacement for SE555/NE555
- Timing from microseconds through hours
- Operates in both astable and monostable modes

Schematic Diagram

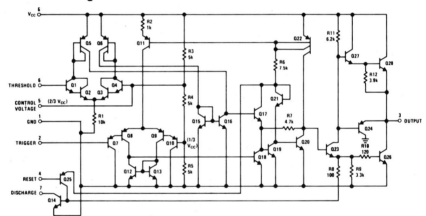

Connection Diagrams

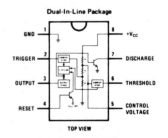

Metal Can Package — Order Number LM555H, LM555CH
See NS Package H08C

Dual-In-Line Package — Order Number LM555CN
See NS Package N08B
Order Number LM555J or LM555CJ
See NS Package J08A

Absolute Maximum Ratings

Supply Voltage	+18V
Power Dissipation (Note 1)	600 mW
Operating Temperature Ranges	
LM555C	0°C to +70°C
LM555	−55°C to +125°C
Storage Temperature Range	−65°C to +150°C
Lead Temperature (Soldering, 10 seconds)	300°C

Electrical Characteristics (T_A = 25°C, V_{CC} = +5V to +15V, unless otherwise specified)

PARAMETER	CONDITIONS	LM555 MIN	LM555 TYP	LM555 MAX	LM555C MIN	LM555C TYP	LM555C MAX	UNITS
Supply Voltage		4.5		18	4.5		16	V
Supply Current	V_{CC} = 5V, R_L = ∞		3	5		3	6	mA
	V_{CC} = 15V, R_L = ∞ (Low State) (Note 2)		10	12		10	15	mA
Timing Error, Monostable								
Initial Accuracy			0.5			1		%
Drift with Temperature	R_A, R_B = 1k to 100 k, C = 0.1μF, (Note 3)		30			50		ppm/°C
Accuracy over Temperature			1.5			1.5		%
Drift with Supply			0.05			0.1		%/V
Timing Error, Astable								
Initial Accuracy			1.5			2.25		%
Drift with Temperature			90			150		ppm/°C
Accuracy over Temperature			2.5			3.0		%
Drift with Supply			0.15			0.30		%/V
Threshold Voltage			0.667			0.667		x V_{CC}
Trigger Voltage	V_{CC} = 15V	4.8	5	5.2		5		V
	V_{CC} = 5V	1.45	1.67	1.9		1.67		V
Trigger Current			0.01	0.5		0.5	0.9	μA
Reset Voltage		0.4	0.5	1	0.4	0.5	1	V
Reset Current			0.1	0.4		0.1	0.4	mA
Threshold Current	(Note 4)		0.1	0.25		0.1	0.25	μA
Control Voltage Level	V_{CC} = 15V	9.6	10	10.4	9	10	11	V
	V_{CC} = 5V	2.9	3.33	3.8	2.6	3.33	4	V
Pin 7 Leakage Output High			1	100		1	100	nA
Pin 7 Sat (Note 5)								
Output Low	V_{CC} = 15V, I_7 = 15 mA		150			180		mV
Output Low	V_{CC} = 4.5V, I_7 = 4.5 mA		70	100		80	200	mV
Output Voltage Drop (Low)	V_{CC} = 15V							
	I_{SINK} = 10 mA		0.1	0.15		0.1	0.25	V
	I_{SINK} = 50 mA		0.4	0.5		0.4	0.75	V
	I_{SINK} = 100 mA		2	2.2		2	2.5	V
	I_{SINK} = 200 mA		2.5			2.5		V
	V_{CC} = 5V							
	I_{SINK} = 8 mA		0.1	0.25				V
	I_{SINK} = 5 mA					0.25	0.35	V
Output Voltage Drop (High)	I_{SOURCE} = 200 mA, V_{CC} = 15V		12.5			12.5		V
	I_{SOURCE} = 100 mA, V_{CC} = 15V	13	13.3		12.75	13.3		V
	V_{CC} = 5V	3	3.3		2.75	3.3		V
Rise Time of Output			100			100		ns
Fall Time of Output			100			100		ns

Note 1: For operating at elevated temperatures the device must be derated based on a +150°C maximum junction temperature and a thermal resistance of +45°C/W junction to case for TO-5 and +150°C/W junction to ambient for both packages.
Note 2: Supply current when output high typically 1 mA less at V_{CC} = 5V.
Note 3: Tested at V_{CC} = 5V and V_{CC} = 15V.
Note 4: This will determine the maximum value of R_A + R_B for 15V operation. The maximum total (R_A + R_B) is 20 MΩ.
Note 5: No protection against excessive pin 7 current is necessary providing the package dissipation rating will not be exceeded.

Appendix B

Butterworth, Chebyshev, and Bessel Charts

CASCADED FILTER CHARTS

Butterworth

	First Stage		Second Stage		Third Stage		Fourth Stage	
Order	$1/Q$	f_c factor	$1/Q$	f_c factor	$1/Q$	f_c factor	$1/Q$	f_c factor
2	1.414	1						
3	1.00	1	1.00	1				
4	1.848	1	0.765	1				
5	1.00	1	1.618	1	0.618	1		
6	1.932	1	1.414	1	0.518	1		
7	1.00	1	1.802	1	1.247	1	0.445	1
8	1.962	1	1.663	1	1.111	1	0.390	1

CASCADED FILTER CHARTS (Continued)

Chebyshev, 0.5 dB Ripple

Order	First Stage 1/Q	First Stage f_c factor	Second Stage 1/Q	Second Stage f_c factor	Third Stage 1/Q	Third Stage f_c factor	Fourth Stage 1/Q	Fourth Stage f_c factor
3	1	0.626	0.586	1.069				
4	1.418	0.597	0.340	1.031				
5	1	0.362	0.849	0.690	0.220	1.018		
6	1.463	0.396	0.552	0.768	0.154	1.011		
7	1	0.256	0.916	0.504	0.388	0.823	0.113	1.008
8	1.478	0.296	0.621	0.599	0.288	0.861	0.087	1.006

Chebyshev, 1 dB Ripple

Order	First Stage 1/Q	First Stage f_c factor	Second Stage 1/Q	Second Stage f_c factor	Third Stage 1/Q	Third Stage f_c factor	Fourth Stage 1/Q	Fourth Stage f_c factor
3	1	0.494	0.496	0.997				
4	1.275	0.529	0.281	0.993				
5	1	0.289	0.715	0.655	0.180	0.994		
6	1.314	0.353	0.455	0.747	0.125	0.995		
7	1	0.205	0.771	0.480	0.317	0.803	0.092	0.996
8	1.328	0.265	0.511	0.584	0.234	0.851	0.702	0.997

Chebyshev, 2 dB Ripple

Order	First Stage 1/Q	First Stage f_c factor	Second Stage 1/Q	Second Stage f_c factor	Third Stage 1/Q	Third Stage f_c factor	Fourth Stage 1/Q	Fourth Stage f_c factor
3	1	0.369	0.392	0.941				
4	1.076	0.471	0.218	0.964				
5	1	0.218	0.563	0.627	0.138	0.976		
6	1.109	0.316	0.352	0.730	0.096	0.983		
7	1	0.155	0.607	0.461	0.243	0.797	0.070	0.987
8	1.206	0.238	0.395	0.572	0.179	0.842	0.054	0.990

Chebyshev, 3 dB Ripple

Order	First Stage 1/Q	First Stage f_c factor	Second Stage 1/Q	Second Stage f_c factor	Third Stage 1/Q	Third Stage f_c factor	Fourth Stage 1/Q	Fourth Stage f_c factor
3	1	0.299	0.326	0.916				
4	0.929	0.443	0.179	0.950				
5	1	0.178	0.468	0.614	0.113	0.967		
6	0.958	0.298	0.289	0.722	0.078	0.977		
7	1	0.126	0.504	0.452	0.199	0.792	0.057	0.983
8	0.967	0.224	0.325	0.566	0.147	0.839	0.044	0.987

CASCADED FILTER CHARTS (Continued)

Bessel

	First Stage		Second Stage		Third Stage		Fourth Stage	
Order	$1/Q$	f_c factor	$1/Q$	f_c factor	$1/Q$	f_c factor	$1/Q$	f_c factor
3	1	2.322	1.447	2.483				
4	1.916	2.067	1.241	1.624				
5	1	3.647	1.775	2.874	1.091	2.711		
6	1.959	2.872	1.636	3.867	0.977	3.722		
7	1	4.972	1.878	3.562	1.513	5.004	0.888	4.709
8	1.976	3.701	1.787	4.389	1.407	0.637	0.816	5.680

Note: On all odd-order filters the first stage is a first-order filter stage. All other stages are second-order filters.

Low-pass filters:
$f_c = f_{-3\,dB} \, (f_c \text{ factor})$

High-pass filters:
$f_c = f_{-3\,dB} / (f_c \text{ factor})$

where

$f_c = f_c$ used in calculations for stage

$f_{-3\,dB} = f_{-3\,dB}$ desired.

Appendix C

Answers to Selected Problems

CHAPTER 3

3.1 9, 200 Ω, 1.8 kΩ
3.2 -50 μA, -150 μA, -200 μA
3.3 (a) -6, 15.56 dB
(b) 1.2%
3.7 10 kΩ, 50 kΩ, 1 nF
3.9 1 kΩ, 14 kΩ, 10 kΩ
3.15 2, 2.5 kΩ, 200 rad/sec, 6 V (at the output of the op amp)
3.19 $R_1 = 10$ kΩ, $R_2 = 50$ kΩ, $R_3 = R_4 = 10$ kΩ, $C_1 = 0.2$ μF, $C_2 = 10$ μF
3.21 1 μA, 1 μA, 99 μA, 98 μA, 99 μA, -99 μA
3.22 2 mA, 2 mA, 2 mA, 4 mA, 1.8 mA, 0.2 mA, -0.2 mA
3.23 10 V, 9.9 V, 10.45 V, 9.45 V, -10 V
3.24 $-R_1(R_3 + R_4)/R_2R_4$
3.25 $-R_2/(2R_2 + R_3)$
3.26 $-(R_2R_4 + R_3R_4 + R_2R_3)/R_1R_3$
3.27 46.9 Hz, 1.739 Hz, 76.2 dB, 44.8 dB

381

CHAPTER 4

4.2 5 mA, 2.5 mA, 1.25 mA, 0.625 mA, 9.375 mA, -300 V

4.5 $V_{s1}\left(\dfrac{R_2}{R_1 + R_2}\right)\left(\dfrac{R_3 + R_4}{R_4}\right) - V_{s2}\dfrac{R_4}{R_3}$

4.6 $R_1 = R_2 = R_3 = R_4/3 = 10\ \text{k}\Omega$

4.10 ± 0.2 mA

4.11 1 kΩ, 2 kΩ, 0.01 μF

4.19 -2 V, -1 V, 0.1 mA

4.22 $V_o = -V_{s1} - V_{s2} + V_{s3} + V_{s4}$

4.25 $-1/R^2C^2s^2$

4.26 $v_o = -(v_{s1} + v_{s2} + v_{s3})/R_1C_1s$

4.27 $V_o = (V_{s2} - V_{s1})/R_1C_1s$

4.29 $A = 1$, $B = -R_5/R_4$, $C = R_5/R_3$, $D = R_5/R_2$

4.30 $V_o = -V_T \ln(V_s/RI_s)$

4.32 $V_o = -I_s R_4 (V_{s1}/V_{s2}) e^{R_3/R_2}$

CHAPTER 5

5.1 (a) $V_o = -IR + V_R$
(b) 100 kΩ

5.2 0.45 V

5.4 $-100\ \mu$A, $-50\ \mu$A, -0.15 mA, -7 V

5.11 $V_{LL} = \dfrac{V_s R_2 + V_o(\min) R_1}{R_1 + R_2}$, $V_{UL} = \dfrac{V_s R_2 + V_o(\max) R_1}{R_1 + R_2}$

5.13 $V_o \cong 400 V_R\ \Delta R/R$

5.16 57.1 mA, 0.5714 mA, 6.286 V

5.20 $V^+ = 9$ V, $V^- = -17$ V

5.22 $v_1 = 0.1[\cos 30° + \cos(4\pi 10^3 t + 30°)]$

CHAPTER 6

6.1 $\omega RC = 1$

6.2 $R_4 = 4R_3$

6.6 $v_3(t) = 1.414 \sin(2\pi 10^4 t - 45°)$, $v_{o2}(t) = -2\cos(2\pi 10^4 t)$

6.9 66.48 kHz, 10 V, 73%

6.17 12 kΩ, 0.1 μF
6.18 $R_2/R_1 = L_2/L_1$, $\omega_o = 1/\sqrt{(L_1 + L_2)C_1}$
6.19 $R_2/R_1 = C_1/C_2$, $\omega_o = \sqrt{(C_1 + C_2)/L_1C_1C_2}$
6.20 $R = 500$ Ω, $R_1 = 50$ kΩ, $R_2 = 1.45$ MΩ, $C = 0.13$ μF
6.22 $C = 0.1$ μF, $R = 18.2$ kΩ, D_1, D_2:4.4 V zeners, $R_1 = R_2 = 10$ kΩ, $R_3 = 250$ Ω

CHAPTER 7

7.2 -4 V, -1
7.3 8 V, -0.5
7.5 No change
7.7 $v_o = 10 - 0.213t$
7.10 Diodes should be reversed.
7.11 $R_1 = R_2 = 10$ kΩ, $R_3 = 5$ kΩ, $R = 10$ kΩ, $C = 0.1$ μF
7.12 $R_4 = R_3 = 2R_1 = 4R_2$, $k = 2$
7.13 $C = 0.1$ μF, $R = 1$ kΩ, $R_1 = 14.4$ kΩ, $C_1 = 1$ μF
7.15 $R_2 = 2R_1$
7.16 $R_3 = 5R_1 = 12.5R_2$
7.19 $R_2 = 0.2R_3 = 0.25R_1$

CHAPTER 8

8.1 $R_1 = 16$ kΩ, $R_2 = 64$ kΩ, $C = 311$ pF
8.2 $T(s) = -\dfrac{R_2}{R_1} \dfrac{s}{s + (R_2/L)}$
8.4 $R_1 = 20$ kΩ, $R_2 = 40$ kΩ, $C = 40$ nF
8.5 $T(s) = -\dfrac{R_2}{R_1} \dfrac{R_1/L}{s + (R_1/L)}$
8.7 $T(s) = (1 + K) \dfrac{s}{s + (1/R_1C_1)}$
8.8 $T(s) = -\dfrac{1}{R_1C_2} \dfrac{s}{s^2 + s\left(\dfrac{1}{R_1C_1} + \dfrac{1}{R_2C_2}\right) + \dfrac{1}{R_1R_2C_1C_2}}$
8.9 $R_1C_1 = R_2C_2$, $R_1 = 796$ Ω, $R_2 = 15.9$ kΩ, $C_1 = 0.2$ μF, $C_2 = 0.01$ μF
8.11 6.59 kHz

8.12 38.42 kHz

8.13 (a) $R_1 = R_2 = R_3 = 10$ kΩ, $C_{\text{load}} = 10$ nF (b) $R_1 = R_2 = R_3 = R_4 = 10$ kΩ, $C = 0.01$ μF

8.14 (a) R^2/sL (b) R^2/R_L (c) $R^2/(R_L + 1/sC)$

8.17 $T(s) = \dfrac{Ks^2}{s^2 + s\dfrac{R}{L} + \dfrac{1}{LC}}$

8.18 $T(s) = \dfrac{Ks}{RC\left(s^2 + s\dfrac{1}{RC} + \dfrac{1}{LC}\right)}$

8.19 $T(s) = \dfrac{Ks^2}{s^2 + s\dfrac{1}{RC} + \dfrac{1}{LC}}$

CHAPTER 9

9.1 Y_2 (2 possible configurations)

9.2 $R_1 = 1$ kΩ, $C_2 = 0.1$ μF, $R_3 = 166.7$ Ω, $R_4 = 10$ kΩ, $C_5 = 2.38$ μF

9.3 $C_1 = 0.1$ μF, $R_2 = 1876$ Ω, $C_3 = 0.1$ μF, $C_4 = 1$ μF, $R_s = 13.5$ kΩ

9.9 Low-pass, $A = -\dfrac{R_{4A} + R_{4B}}{R_{2A} + R_{2B}}$, $\omega_o = \dfrac{1}{\sqrt{R_{4A}R_{4B}C_{4A}C_{4B}}}$, $Q = \dfrac{\sqrt{\dfrac{R_{4A}R_{4B}C_{4B}}{C_{4A}}}}{R_{4A} + R_{4B}}$

9.12 $A_o = \dfrac{R_2}{R_1}$, $\omega_o = \dfrac{1}{C_1\sqrt{R_3R_5}}$, $Q = \dfrac{R_2}{\sqrt{R_3R_5}}$

9.14 $T(s) = \dfrac{R_4}{R_3 + R_4} \cdot \dfrac{s^2 - s\dfrac{2}{R_2C} + \dfrac{1}{R_1R_2C^2}}{s^2 + s\dfrac{2}{R_2C} + \dfrac{1}{R_1R_2C^2}}$

9.15 $T(s) = \dfrac{R_4}{R_3 + R_4} \cdot \dfrac{s^2 + \dfrac{1}{R_1R_2C_1C_2}}{s^2 + s\dfrac{C_1 + C_2}{R_2C_1C_2} + \dfrac{1}{R_1R_2C_1C_2}}$

9.16 $T(s) = \dfrac{K}{(1-K)RC} \cdot \dfrac{s}{s^2 + s\dfrac{3}{(1-K)RC} + \dfrac{1}{(1-K)R^2C^2}}$

9.18 $f_o = 60.3$ Hz, $A_o = 50.3$, $Q = 50.3$

9.19 Band-reject response

9.20 $A_o = 1$, $\omega_o = 17.5$ krad/sec, $Q = 15$

9.21 $\dfrac{V_1}{V_s} = \dfrac{R_3 s}{R_2 R_4 C_2 \left(s^2 + s\dfrac{1}{R_1 C_1} + \dfrac{R_3}{R_1 R_2 R_4 C_1 C_2}\right)}$

CHAPTER 10

10.3 $I_L = V_c/R$

10.5 $I_L = V_c/R$

10.6 $R_1 = 20$ kΩ, $R_2 = 113.3$ kΩ, $R_3 = R_4 = 10$ kΩ, $R_{CL} = 2.5$ Ω, $V_B = 269$ mV

10.7 387 Ω

10.8 $-10.3 \leq V_o \leq -5.2$ V

10.10 3 Ω

10.11 6.064453125 V

10.12 4.2578125 mA

10.15 1100000010

10.16 $40 \leq f_L \leq 49.9$ kHz, $50.1 \leq f_U \leq 60$ kHz

10.18 $\dfrac{V_o}{V_s} = -\dfrac{R_1 C_1}{V_c}\left(s + \dfrac{V_c}{R_1 C_1}\right)$

10.19 $C = 1$ μF, $R_A = 115.942$ kΩ, $R_B = 14.493$ kΩ

CHAPTER 11

11.1 1010.1, 99

11.2 -1009.1, 99

11.3 9900

11.4 9999

11.5 (a) $-\dfrac{AR(R_1 - R_2)}{(R_1 + R)(R_2 + R)}$ (b) $R_1 > R_2$

11.6 $\dfrac{A}{1 + AF}$, $\dfrac{-R_2 R_3 + R_1 R_4}{(R_1 + R_2)(R_3 + R_4)}$

11.7 $1/(1 + AF)$

11.8 $-\dfrac{A(R_3 R_4 + R_2 R_3 + R_2 R_4)}{(1 + A)R_1 R_3 + R_2 R_3 + R_1 R_4 + R_2 R_4 + R_3 R_4}$,

$-\dfrac{R_3 R_4 + R_2 R_3 + R_2 R_4}{R_1 R_3}$, $\dfrac{AR_1 R_3}{R_1 R_3 + R_2 R_3 + R_1 R_4 + R_2 R_4 + R_3 R_4}$

11.9 9.901 kHz

11.10 9.901 kHz

11.11 $\omega_1 \left[1 + \dfrac{AR_1R_2}{R_1R_2 + R_1R_3 + R_2R_3} \right]$

11.13 $R_2 - R_1 < \dfrac{(R_1 + R)(R_2 + R)}{AR}$

11.14 $\omega_1 \dfrac{AR(R_1 - R_2) + (R_1 + R)(R_2 + R)}{(R_1 + R)(R_2 + R)}$

11.15 $\omega_1 \left(1 + \dfrac{AR_1}{R_1 + R_2} \right)$

11.16 $\omega_1 \left(1 + \dfrac{AR_1R_3}{R_1R_3 + R_2R_3 + R_1R_4 + R_2R_4 + R_3R_4} \right)$

CHAPTER 12

12.1 (a) 9999 (b) 2 kΩ, 18 kΩ (c) 19,998 MΩ

12.2 (a) 9090.91 (b) 2 kΩ, 20 kΩ (c) 0.2 Ω

12.3 (a) $-\dfrac{AR_i}{R + R_i}$ (b) $T(s) = -\dfrac{ARR_i}{(1 + A)R_i + R}$ (c) $R << R_i$

12.5 (a) 1 Ω (b) 10.9 rad/sec

12.6 0.0011 Ω

12.7 $\sqrt{R_iG'/R_o}$

12.8 1 kΩ, 99 kΩ

12.9 $\sqrt{R_i|G'_i|R_o}$

12.10 1 kΩ, 100 kΩ

12.12 $AF < \dfrac{(\omega_1 + \omega_2 + \omega_3)(\omega_1\omega_2 + \omega_1\omega_3 + \omega_2\omega_3)}{\omega_1\omega_2\omega_3} - 1$

12.13 $\omega_1 < 60$ rad/sec

12.14 $C_L \leq 10$ pF

CHAPTER 13

13.2 ± 0.625 V

13.7 -0.985 V

13.8 -1.5 mV

13.10 10.01 or 9.99 V, 10 V
13.11 The output will be distorted.
13.13 2.7 V
13.14 1515 hr
13.16 -5 V, -3.25 V, -7 V, -5.25 V
13.17 -5 mV, 0.18 mA, 0.10 mA, 0.08 mA

References

The following key identifies the type of information to be found in each reference:

(1) Operational amplifiers and applications (general)
(2) Operational amplifier active filters
(3) Integrated circuit operational amplifiers

Al-Nasser, Farouk
Tables Speed Design of Low-Pass Filters,
EDN, March 1971   (2)

Al-Nasser, Farouk
Tables Shorten Design Time for Active Filters,
Electronics, 1972   (2)

Bannon, E.
Operational Amplifiers: Theory and Servicing,
Reston Publishing, 1975   (1)

Brandstedt, Bjorn
Tailor the Response of Your Active Filters,
EDN, March 1973   (2)

References

Budak, Aram
Passive and Active Network Analysis and Synthesis,
Houghton Mifflin, 1974   (2)

Burr-Brown Research Corporation
Handbook of Operational Amplifier Applications, 1963   (1)
Handbook of Operational Amplifier Active RC Networks, 1966   (2)

Clayton, G. B.
Operational Amplifiers, Second Edition,
Newnes-Butterworths, 1979   (1)

Coughlin, Robert, and Driscoll, Frederick
Operational Amplifiers and Linear Integrated Circuits, Second Edition,
Prentice-Hall, 1982   (1)

Cowdell, Robert
Bypass and Feedthrough Filters; Graphic Technique,
Electronics Design, August 1975   (2)

Daryanani, G.
Principles of Active Network Synthesis and Design,
John Wiley, 1976   (2)

Doyle, Norman
Swift, Sure Design of Active Bandpass Filters,
EDN, January 1970   (2)

Faulkenberry, Luces
Introduction to Operational Amplifiers with Linear IC Applications,
John Wiley, 1982   (1)

Geffee, Philip
Designers Guide to Active Bandpass Filters,
EDN, June 1974   (2)

Graeme, Jerald, Tobey, Gene, and Huelsman, Lawrence
Operational Amplifiers; Design and Applications,
McGraw-Hill, 1971   (1)

Graeme, Jerald
Applications of Operational Amplifiers,
McGraw-Hill, 1973   (1)

Gray, Paul and Meyer, Robert
Analysis and Design of Analog Integrated Circuits, Second Edition,
John Wiley, 1977   (3)

Herpy, Miklos
Analog Integrated Circuits,
John Wiley, 1980   (3)

Huelsman, Lawrence
Theory and Design of Active RC Circuits,
McGraw-Hill, 1968   (2)

Huelsman, Lawrence
Active Filters; Lumped, Distributed, Integrated, Digital and Parametric,
McGraw-Hill, 1970   (2)

Irvine, Robert
Operational Amplifier Characteristics and Applications,
Prentice-Hall, 1981   (1)

Johnson, David, and Jayakumar, V.
Operational Amplifier Circuits; Design and Applications,
Prentice-Hall, 1982   (1)

Johnson, D., Johnson, R., and Moore, H.
A Handbook of Active Filters,
Prentice-Hall, 1980   (2)

Jung, Walter
IC Op Amp Cookbook,
Howard W. Sams, 1974   (1)

Kincaid, Russell, and Shirley, Frederick
Active Bandpass Filter Is Made Easy with Computer Program,
Electronics, May 1974   (2)

Leach, Donald, and Chan, Shu-Park
A Generalized Method of Active RC Network Synthesis,
IEEE Transactions on Circuit Theory, November 1971   (2)

Meyer, Robert
Integrated Circuit Operational Amplifiers,
IEEE Press, 1978   (3)

Patterson, Marvin
Designer's Guide to Active Filters,
EDN, March 1974   (2)

Roberge, James
Operational Amplifiers; Theory and Practice,
John Wiley, 1975   (1)

Russell, Howard
Single Amplifier Active Filters Give Stable Q,
EDN/EEE, January 1972   (2)

Sallen, R. P., and Key, E. L.
A Practical Method of Designing RC Active Filters,
Institute of Radio Engineers; Transactions on Circuit Theory,
March 1955   (2)

Schaeffer, Lee
Op Amp Active Filters-Simple to Design Once You Know the Game,
EDN, April 1976   (2)

Sedra, A., and Brackett, P.
Fiter Theory and Design; Active and Passive,
Matrix Publishers, 1978   (2)

Solomon, J. E.
The Monolithic Op Amp; A Tutorial Study,
IEEE Journal of Solid State Circuits, December 1974   (3)

Thomas, Lee
The Biquad; Some Practical Design Considerations,
IEEE Transactions on Circuit Theory, May 1971   (2)

Wait, John, Huelsman, Lawrence, and Korn, Granino
Introduction to Operational Amplifier Theory and Applications,
McGraw-Hill, 1975   (1)

Widlar, Robert
Design Techniques for Monolithic Operational Amplifiers,
IEEE Journal of Solid State Circuits, August 1969   (3)

Zicko, Howard
Optimize Second Order Active Filters,
Electronic Design, 1972   (2)

Index

Ac-to-dc convertor, 59
Adder, 46, 69
Amplifier:
  absolute value, 57, 70
  ac, 33
  antilog, 61
  averaging, 69
  bandpass, 33
  bridge, 69, 93
  characteristics, 25
  common-emitter, 16, 352
  dc, 26
  differential, 15, 350
  emitter-follower, 18, 354
  instrumentation, 36
  inverting, 29, 35, 246, 251, 263, 269
  logarithmic, 60, 70
  noninverting, 27, 33, 48, 244, 260, 266
  operational, 3
  power, 211
  single supply, 33
  summing, 46, 70
Analog computation, 66, 70
Analog-to-digital convertor, 226
Analog divider, 65, 70
Analog multiplier, 62
Analog switch, 140

Bandwidth, 243, 248
Bypass capacitors, 316

Capacitance multiplier, 70
Capacitive load, 324
Charge-to-voltage convertor, 78
Closed-loop:
  bandwidth, 248
  gain, 10, 243
  impedances, 259
Common-emitter circuit, 16, 352
Common-mode rejection ratio, 286

393

Comparator, 80, 322
Compensation:
    capacitor, 253, 358
    feedforward, 255
    one-pole, 253, 358
    two-pole, 254
Conditional stability, 272
Convertors, 75
Cube circuit, 65
Current constraint, 20
Current noise, 293
Current source, 212
Current-to-voltage convertor, 76, 92

Data sheet parameters, 277
DC analysis of 741, 340
DC restorer, 144
Demodulator, 230
Design examples, 311
Detector, 81, 92
Device protection, 323
Difference amplifier, 49, 350
Differencing integrator, 70
Differential input voltage, 282
Differentiator, 55
Digital-to-analog convertor, 93, 221
Distortion, 291
Divider, 65, 70
Double integrator, 69

Electronic switch, 140
Emitter follower circuit, 18, 354

Feedback, 8, 97
Filter:
    active, 155, 176, 325
    all-pass, 159, 205
    bandpass, 158, 169, 181, 194, 322
    band-reject, 194, 205
    Bessel, 154
    Butterworth, 154
    characteristics, 149
    Chebyshev, 154
    components, 153
    frequencies, 152
    gyrator, 161
    high-pass, 157, 167, 179, 187, 194, 332
    IGMF, 177, 328
    IGSF, 182
    low-pass, 156, 171, 187, 201, 329
    state variable, 190, 205
    transfer function, 150
    tunable, 238

Frequency compensation schemes, 253
Frequency response of 741, 357
Function generator, 113
Function synthesizer, 131

Gain-bandwidth product, 17, 249, 264, 359
Gilbert multiplier cell, 63
Grounding problems, 316
Gyrator, 161

Hybrid pi model, 346
Hysteresis, 88

Ideal operational amplifier, 19
IGMF filter, 177, 328
IGSF filter, 182
Input bias current, 16, 282, 342
Input constraints, 19
Input impedance, 262, 265
Input offset current, 283, 342
Input offset voltage, 16, 280, 342
Input resistance, 16, 259, 357
Input voltage, 282
Integrator, 51, 320
Interfacing to loads, 85

KRC filter, 198, 205

Large signal frequency response, 291
Level detector, 82
Limitations of op amp, 19, 241
Limiter, 123, 143
LM101A operational amplifier, 12, 278, 360
LM108A operational amplifier, 12
Load considerations, 270
Loop gain, 9
Lower cutoff frequency, 25

Mathematical operator, 46
Modulator, 228
Monostable multivibrator, 140
uA709 operational amplifier, 12
uA741 operational amplifier, 12, 337, 360
Multiplexer, 138
Multiplier, 62

NAB preamplifier, 41
Negative feedback, 8
Noise, 293

Offset current, 283, 342
Offset null circuit, 281
Offset voltage, 16, 280, 342

Index

Open-loop gain, 8
Operating temperature range, 291
Operational amplifier:
    block diagram, 13
    characteristics, 4
    definition, 3
    ideal, 19
    specifications, 5, 278
Oscillator:
    characteristics, 99
    Colpitts, 118
    Hartley, 118
    phase shift, 119
    quadrature, 103
    rectangular, 106, 119
    sawtooth, 115
    sinusoidal, 99
    triangular, 113
    twin-T, 119
    voltage controlled, 115
    Wien-bridge, 99
Output current, 289
Output impedance, 268
Output resistance, 18, 266, 357
Output voltage range, 289
Output voltage swing, 289

Packages, 6
Peak detector, 136, 144
Phase detector, 90
Phase locked loop, 227
Phase margin, 254
Piecewise linear circuit, 131
Positive feedback, 97
Power dissipation, 290
Power supply rejection ratio, 290
Precision rectifier, 58
Protection schemes, 324

Ramp generator, 312
Rectifier circuit, 58
RIAA preamplifier, 172

Sample-and-hold circuit, 134, 144
Sample-hold-and subtract circuit, 144

Schmitt trigger, 87, 93
Secondary parameters, 277
Sensitivity functions, 203
Sensitivity to gain, 247
Simulated inductor, 161, 165
Slew rate, 18, 291, 360
Square circuit, 65
Square root circuit, 65
Stability considerations, 272
Step generator, 144
Summer, 46, 69, 316
Summing integrator, 70
Supply bypassing, 316
Supply current, 290
Supply voltage range, 290
Supply voltage rejection ratio, 290

Temperature, 8, 291
Timer, 232
Transfer characteristic, 6, 124
Transfer function, 150
Transient response, 292

Unity gain frequency, 17, 249, 359

V-to-I convertor, 93, 212
Voltage amplifier, 25
Voltage constraint, 19
Voltage controlled current source, 212, 237
Voltage controlled voltage source, 199, 210
Voltage follower, 28
Voltage gain, 15, 243, 302, 357
Voltage noise, 293
Voltage reference, 217
Voltage regulator, 216, 238

Wien-bridge oscillator, 99
Window detector, 83, 92

Zener diode limiter, 126
Zero-crossing detector, 81

Operational Amplifiers: Analysis, Design and Engineering Applications.

About the Authors

Chuck Wojslaw (pronounced Woy-slaw) received his BSEE and MSEE degrees from the University of California, Berkeley and the University of Santa Clara. Prior to joining the faculty of San Jose City College, he worked in industry for 16 years as a design and supervising engineer. He is the author of two books and numerous articles, and has been a lecturer at the University of Santa Clara and San Jose State University.

Evan Moustakas a native of Greece, is Professor of Electrical Engineering at San Jose State University where he has been teaching since 1961. He received his BSEE and MSEE degrees from Oregon State University and his PhD from the University of Santa Clara. He is the coauthor of two books and several papers in the fields of operational amplifiers, electronics, and active filter design.